George Owen Squier

George Owen Squier

U.S. Army Major General, Inventor, Aviation Pioneer, Founder of Muzak

PAUL W. CLARK *and*
LAURENCE A. LYONS

McFarland & Company, Inc., Publishers
Jefferson, North Carolina

LIBRARY OF CONGRESS CATALOGUING-IN-PUBLICATION DATA

Clark, Paul W., 1935–
  George Owen Squier : U.S. Army major general, inventor, aviation pioneer, founder of Muzak / Paul W. Clark and Laurence A. Lyons.
      p.   cm.
  Includes bibliographical references and index.

  ISBN 978-0-7864-7635-0 (softcover : acid free paper) ∞
  ISBN 978-1-4766-1557-8 (ebook)

  1. Squier, George Owen, 1865–1934. 2. Military engineers—United States—Biography. 3. Aeronautical engineers—United States—Biography. 4. Generals—United States—Biography. 5. United States. Army—Officers—Biography. 6. Inventors—United States—Biography. I. Lyons, Laurence A., 1939– II. Title.
UG128.S75C55 2014
355.0092—dc23
[B]   2014010534

BRITISH LIBRARY CATALOGUING DATA ARE AVAILABLE

© 2014 Paul W. Clark and Laurence A. Lyons. All rights reserved

No part of this book may be reproduced or transmitted in any form or by any means, electronic or mechanical, including photocopying or recording, or by any information storage and retrieval system, without permission in writing from the publisher.

On the cover: G.O. Squier at the radio laboratory, 1922 (U.S. Army Communications-Electronics Command Historical Office)

Printed in the United States of America

McFarland & Company, Inc., Publishers
  Box 611, Jefferson, North Carolina 28640
      www.mcfarlandpub.com

To the three most loving women in my life:
Aunruen (Anna), my loyal wife and supportive companion
born into a remarkable Thai family; Tamosan Kikuchi,
my former abbess and a renowned modern Japanese saint;
and Wilhelmina Smith, my protective maternal
grandmother whose love truly warmed my life.
P. W. C.

To Eugenia Anne Lyons, my dear wife,
best friend and greatest supporter.
L. A. L.

Table of Contents

Acknowledgments ix
Abbreviations and Acronyms xi
Preface 1
Introduction 3

1. Early Life and West Point — 9
2. Soldier-Scientist — 20
3. An Electrical Laboratory — 34
4. Soldier-Entrepreneur — 45
5. The Philippine Cables — 60
6. Founding of the Signal School — 71
7. Origins of Army Aviation — 81
8. Radio Over Telephone Lines — 100
9. Science and Syndicate — 112
10. Secret Missions — 123
11. The Biggest Thing of the War — 138
12. New Weapons — 161
13. Science Joins the Army — 171
14. Voice-Commanded Squadrons — 186
15. Retirement Years and Legacy — 205

Appendix: Technical Information 217
Chapter Notes 225
Bibliography 261
Index 273

Acknowledgments

Paul W. Clark wishes to express his appreciation to Colonel Alfred F. Hurley, Colonel Roland E. Thomas, and Professor Melvin Kranzberg for making his attendance at Case Western Reserve University possible. He owes Colonel Hurley a debt of gratitude for his special caring about subordinates and for providing opportunities to grow professionally and personally. Paul also appreciates Dean of the Faculty Brigadier General William T. Woodyard's liberal research policies, which generously permitted time off from other duties to pursue this work. He is pleased to acknowledge a grant from the Office of the Chief of Air Force History which relieved him of much of the financial burden of numerous and far-ranging research trips. The support given this project at Air Staff level by Major General Robert Ginsburgh, Brigadier General Brian Gunderson, and Dr. Thomas Belden is also deeply appreciated.

There are so many colleagues and friends who have helped Paul in his course of research that there is inadequate space to mention them here. Yet some have helped so unselfishly that he wishes to acknowledge publicly their contributions: Mrs. Betty Fogler, the Air Force Academy Library; Mr. James Walker, the National Archives; Mr. George Raynor Thompson, former Chief Historian of the Signal Corps; Dr. Paul Scheips, Office of the Chief of Military History; Colonel George Pappas, Army Historical Research Collection at Carlisle Barracks, Pennsylvania; the lovely Mrs. Loren Babcock, General Squier's "niece"; Mrs. Joseph O. Mauborgne, Jr., wife of former Chief Signal Officer of the Army; and Mrs. Joyce Price, Paul's indefatigable typist.

Paul is profoundly grateful to his teachers, especially to Professor Edwin T. Layton, for their dedication to teaching, patience with their students, and love of learning. Finally, he is indebted to his family for their good-natured patience and understanding of the demands which this study has imposed on them.

Both of the authors wish to thank Duane and Jan Chiswell of the Dryden

Historical Society, Dryden, Michigan, for many photographs and memorabilia of General Squier's early and later life. Catherine Barry-Orth of the Waterford Genealogical Society, Waterford, Michigan, provided valuable information about Squier's family background and, in particular, called the authors' attention to the autobiography he wrote while at West Point. We are also greatly indebted to Floyd Hertweck, Chrissie Reilly and Susan Thompson (Command Historian) of the U.S. Army Communications-Electronics Command (CECOM) for an abundance of photographs, articles and documents relating to General Squier's career. Mr. Hertweck authors a blog on Squier for an Army publication and is truly expert on all facets of the general's achievements. Finally, we are very grateful to Jeff Lynn, who helped us compile the index.

Abbreviations and Acronyms

AGO	Office of the Adjutant General
AIEE	American Institute of Electrical Engineers
ASS	Army Signal School
AT&T	American Telephone and Telegraph Company
BAAS	British Association for the Advancement of Science
BAP	Bureau of Aircraft Production
BOF	Board of Ordnance and Fortification
CECOM	U.S. Army Communications-Electronics Command
C/Ord	Chief of Ordnance
C/S	Chief of Staff
CSO	Chief Signal Officer
C/WCD	Chief of the War College Division
MIT	Massachusetts Institute of Technology
NACA	National Advisory Committee for Aeronautics
NAS/NRC	National Academy of Sciences/National Research Council
NBS	National Bureau of Standards
OCSO	Office of the Chief Signal Officer
OIC	Officer in Charge
SecNav	Secretary of the Navy
SecState	Secretary of State
SecWar	Secretary of War
SO	Special Order
WCD	War College Division

Preface

Major General George Owen Squier was a remarkable man. During the 1920s and '30s, he was one of the most famous men in America and abroad, as a scientist, soldier, military strategist, electrical communications expert and inventor, aeronautical pioneer, diplomat, and philanthropist. He had risen, like a Horatio Alger hero, from humble beginnings in the countryside of Michigan to the position of Chief Signal Officer of the United States Army, the first Chief Signal Officer to attain the rank of Major General while in that office.[1] Despite his many achievements, no biography of George Squier has heretofore been published. We hope this book will bring that exceptional man the attention he deserves.

George Squier lived in two parallel worlds. He had, by any measure, a very successful military career. But he was also a scientist and inventor, with many publications and inventions to his name. He managed to successfully fulfill the demands of both, though many times his pursuits were in conflict. In order to properly tell his story we need to provide some of the details of his scientific work in order to underline the importance of his contributions, which we benefit from even today. To avoid interrupting the narrative we have relegated the more technical information to an appendix.

As a public man and high-ranking officer, Squier was a key contributor to aviation, military aviation in particular. In the years after World War II service bias caused U.S. Air Force historians to minimize his role in the development of aviation, but it was substantial. He also led the effort to equip American forces with modern communications, the first belligerent in World War I to do so. The most enduring monuments to his career-long crusade for scientific and engineering research within the Army are the laboratories at Fort Monmouth and Langley Field (although the former was recommended for closure in 2005, with many of its functions transferred to Aberdeen Proving Ground, Maryland). Although he is not well known today compared to his contemporaries Alexander Graham Bell and the Wright Brothers, these men knew him well and respected his intellect and originality. Yet his inventions in communications technology are fundamental to today's telephone system and were the technical basis for the company he founded, Muzak.

The most definitive and only study of Squier's life is the doctoral dissertation written by one of the authors, Paul W. Clark, in 1974.[2] The dissertation's extensive bibliography and footnotes reflect the thorough research in many archives and multiple sources consulted. Though unpublished at the time, it is the most cited source for all subsequent scholarship on the General. This book is based on that document, but supplements it with personal information about him that was omitted at the time. In addition, this current work provides new scholarship on his scientific achievements. Like the dissertation, the book is organized both chronologically and by subject matter. The narrative flows from the beginning of his life to his death, but when a particular subject is introduced—Squier's visits to the Western Front, to cite one example—the narrative follows that subject to its conclusion, even though it overlaps in time with his attempts to convince the British to use his novel approach to submarine cable communications.

Major General George O. Squier, Chief Signal Officer of the Army (courtesy CECOM Historical Office).

Introduction

George Owen Squier contributed significantly to the development of two great technological revolutions of his time—radio and airplanes—and their application to warfare. He was recognized for these and other achievements, not only in America, but also throughout the world. The Franklin Institute awarded him its prestigious Franklin Medal, and he was elected to membership in the National Academy of Sciences. He was knighted by the British, appointed a Commander of the Cross by Italy, selected a Commander of the Legion of Honor by France, and awarded the Distinguished Service Medal by the United States. He made it possible for the Americans to equip their Army and Air Service with modern radios and personnel trained in their use for the 1919 campaign, should it have come about.[1] Through his technical vision and leadership, Americans accomplished in 18 months what the British and French (or any other belligerent) were unable to do in four years.[2] He led the effort to equip the United States and its allies with American-made airplanes and engines, an effort which started slowly, but, at the time of the Armistice, was rapidly coming to fruition. Through unconventional methods he secured for the fledgling American aviation program the largest single congressional appropriation in history up to that time.

Background and Military Career

The following is a brief overview of Squier's education, technical achievements and military career up to the time when he was appointed to be Chief Signal Officer of the U.S. Army. He was born on March 21, 1865, in Dryden, Michigan. He won a competitive appointment to the United States Military Academy in 1883. An excellent student, he graduated seventh in his class and was assigned to a coast artillery unit at Fort McHenry in Baltimore, Maryland. While there he enrolled in a doctoral program at Johns Hopkins University.

He was awarded a Ph.D. in electrical engineering in 1893, the first U.S. Army officer to hold a doctorate.[3] He was then assigned to Artillery School at Fort Monroe, Virginia, where he played a key role in the establishment of the Army's first electrical engineering laboratory. He was also one of a group of young officers who established the *Journal of the United States Artillery* in 1892. From 1895 to 1898 Squier supervised a program of research on artillery problems at Fort Monroe. Collaborating with the civilian scientist Albert Crehore, whom he had met at Hopkins, he jointly invented a new ballistic instrument, the polarizing photochronograph, to measure projectile velocity within the barrel and in the vicinity of the muzzle of a cannon. He and Crehore also patented a new method of range finding for coastal artillery guns. (At that time, Army officers/researchers were allowed to take out patents in their own names.) In addition, he and Crehore conducted research in radio and improved submarine cable telegraph communications using alternating current.

In 1898, he transferred to the Signal Corps and continued his research despite the outbreak of war with Spain. Chief Signal Officer General Adolphus W. Greely strongly supported the work being done by Squier and Crehore and, in 1899, Greely sent Squier to London to study radio with the legendary Guglielmo Marconi as his mentor. Squier was recalled a year later to command a relief expedition to the Philippines after the Army's cable-laying ship U.S.S. *Hooker* had run aground. Over the course of a year he supervised the laying of 30 cables by the cable ship U.S.S. *Burnside* connecting the major islands of the Philippines, then under the jurisdiction of the United States as a result of the Spanish American War. In 1903, Squier, now a Captain, became Chief Signal Officer of the Department of California. After two years in San Francisco Greely placed Squier in charge of establishing the new Army Signal School at Fort Leavenworth, Kansas. In this assignment he also encouraged the study of Army aeronautics, a Signal Corps responsibility since the late 19th century. Squier closely followed the progress being made by the Wright Brothers in solving the problem of heavier-than-air flight. During the period from 1905 to 1909, Squier spent much of his time developing aviation for the Army, including efforts to persuade Congress to appropriate funds to purchase aircraft and perform aviation-related projects. Squier helped prepare the specifications for the first American military aircraft, supervised the purchase of the aircraft from the Wright brothers, and personally set an endurance record on a ten-minute flight with Orville Wright in 1908.

Squier returned to communications-related research in 1909, establishing a new research laboratory at the Bureau of Standards in Washington, D.C. The focus was not on radio communications but the improvement of landline communications. It was Squier's idea, in fact, to apply the techniques of radio communications to a wired circuit. Radio communications use carrier multiplexing, which separates communication channels by assigning them to different frequencies, thereby increasing the amount of information that can pass from transmitter to receiver without interference. Squier saw that the high frequencies generated by the recently developed Alexanderson alternator could also be used as carrier frequencies on wired circuits.[4] Carrier multiplexing applied to wired circuits (patented by him in 1910) was the invention that led to the great expansion of telephone service throughout the country, and, indeed, the world. In the patent he stipulated that the invention was for the free use of the public, a stipulation that subsequently and inadvertently resulted in an unfair stain on his reputation.

In 1912, somewhat to his and everyone's surprise, Squier was appointed American military attaché in London. His appointment was unexpected because the position of military attaché, particularly in major capitals, was usually reserved for officers with a social pedigree. It appears that General Leonard Wood, Army Chief of Staff, personally intervened to send an officer with the background and education needed to understand the complexities of modern warfare. American ambassador Walter Hines Page had specifically requested the assignment of such an individual on the eve of an anticipated European war.

Squier's technical ability and his acceptance by the British General Staff more than fulfilled Wood's and Page's expectations. He was asked to testify before Parliament in its investigation of the Marconi scandal (accusations that the Marconi Company had bribed government officials), and he worked closely with the leading British scientists to conduct experiments on radio phenomena. He reported back to the U.S. Army on developments in military aviation, with specific reference to its role in the Balkan wars. He also found time to invent new techniques for radio and submarine cable telegraphy.

With the outbreak of war, the importance of America as a potential ally prompted the British government to extend special privileges to the American military attaché. Lord Kitchener arranged for him to make a secret trip to the Western Front when no other attaché was allowed to do so. Squier was allowed to go anywhere, ask any questions and interview anyone he wished. In a highly unusual move, Kitchener allowed Squier to keep a personal diary of his travels in France, including detailed information about supplies, artillery, aviation, transportation, staff organization, combat, cavalry, morale and, of great interest to him, the application of science to war. He organized and summarized this information in reports to Ambassador Page and the War Department in Washington. He also wrote a special report on the British radio service, an organization focused on intercepting enemy communications rather than on providing the British army with modern wireless technology. As a career soldier, Squier had observed the European war more closely than virtually any other American officer. He was well aware of the high casualties and poor coordination caused by inadequate communications.[5] He would eventually work to rectify this inadequacy for the American military.

In May 1916, Squier returned to Washington, D.C., to take charge of the Army's troubled aviation program. Aviation was treated as an unwanted stepchild at that time, particularly in America. Dissension within the ranks had led to the dismissal of the chief of the aviation section of the Signal Corps, and a reprimand of the Chief Signal Officer. In the press's opinion, Squier was the ideal man to take charge. His scientific reputation, experience in the European war, and early involvement in aviation made him the natural choice for the position. He assumed his duties with characteristic energy and set about to remedy the situation by restoring morale within the section and by establishing close ties with aviation organizations inside and outside the government. He was so successful in this endeavor that, in February 1917, Squier was promoted from Lieutenant Colonel to Brigadier General and appointed Chief Signal Officer of the Army. He held that rank until 1923.

General Squier's Appearance and Character

This book contains several photographs of George Squier, but a few comments about his appearance and character are in order. He was shorter than many of his contemporaries, which may not be apparent from his photos. He did not at any time exhibit the famous "Napoleon Complex" but worked harmoniously with his colleagues, superiors and subordinates. People wanted to work with and for him because he was dedicated to accomplishing whatever task he or his organization needed to perform. He devoted equal energy and time to both scientific work and Army assignments. It's not known whether he "suffered fools" badly or easily. Most likely he ignored them if they were not in the way, and if they were, he managed to work around them. He was, admittedly, criticized for skirting the edge of insubordination. This criticism was, as the reader will see, entirely justified. From the beginning of his career to the end he would try any path that might get him what he wanted. He would take multiple approaches to obtain the desired result, whether it be permission to travel to present a scientific paper, or obtain a massive congressional appropriation. In doing so, he certainly could have damaged his Army career, but usually the results obtained stifled attempts at punishment, including two courts-martial. Much of what he accomplished came from taking a bold approach: he did work within the system, but he also made the system work for him.

In his youth, Squier was a notably handsome man who aged well, despite experiencing male-pattern baldness. Although a hat served to hide this affliction, there are enough photos of him in laboratories, in meetings, and in his retirement to indicate that he was not sensitive about his lack of hair. He remained fit and well groomed throughout his life.

Squier never married. Whereas, in today's environment that might lead to speculation about his possibly being gay, there is no evidence to support this. At the time being unmarried was not unusual, particularly for an individual dedicated to his profession. In Squier's case he had two professions and was exceptionally successful at both. It was (and still is) difficult for a career military officer with a family to manage the frequent moves and assignments necessary for advancement. For Squier, this difficulty would have been compounded by managing a simultaneous career as a scientist, inventor, and entrepreneur. He would have had little time for a family. As it was, Squier was devoted to his sister, Mary, and to her family. After she died, her adopted daughter Lavinia inherited the family memorabilia that is now housed in university and government archives, and which have been extensively used in researching this book.

1
Early Life and West Point

George Owen Squier was born on 21 March 1865 in Dryden, Michigan. Nathaniel Squier (1752–1832), his great-grandfather on his father's side, was born in Connecticut in 1752.[1] He fought in the Revolutionary War, was captured in the Danbury Raid in Danbury, Connecticut, and was taken to New York as a prisoner of the British in 1777. In 1800 he married Jemima Dilno, the daughter of Thomas Dilno of Cornwall, Vermont, a woman 28 years younger. Soon after their marriage they moved to London, Oxford Township, Ontario, Canada, where most of their seven children were born. In order of birth, the seven children were a girl, Sarah, five boys, Luman, Ethan, Thomas (who died at the age of 12), Hiram, George, and another girl, Tena.[2] Ethan Squier (1804–1892), George's grandfather, was the smallest in stature of the boys, but he was not the smallest intellectually and spiritually. Luman (1800–1877), though four years older than he, usually consulted him before making any important decision, financial or otherwise, throughout his life.[3]

Nathaniel's family was settled on a farm that was so poor he had to work every day to provide a meager living. Jemima worked hard, taking in washing from people in London. At 36 she contracted tuberculosis, or consumption, as it was then called.[4] Alarmed at her condition, Nathaniel moved from Canada to Vermont, where Jemima's parents lived, to see if a change of climate would cure her. In February 1811 they arrived at their destination, but Jemima's health only grew worse; just three weeks later she died. After the funeral Nathaniel, with his sons Luman, Ethan, and Hiram, returned to Canada.

In February of 1812 the rest of the family moved to a new farm in Canada. Nathaniel, thinking that his family of small children needed a mother, married a much younger woman, Sarah Messinger, born in Massachusetts. According to George, she did not treat the boys very well, often punishing them when their father was away at work.[5] Nathaniel and Sarah had 12 children, one of whom, Lodemia (also known as Lodama), died in 1917, at the age of 96.

In the War of 1812 Nathaniel entered the British army as a volunteer.

He was in the thick of the fight, at Lundy's Lane, and was present at several minor engagements during that campaign. The day before the Battle of Thames, he left the army for good: he had applied for leave, stating that his grain needed cutting and that his family was starving. The General in command refused to grant him permission. Despite this, he left camp and went home.[6]

In the meantime, reports of the natural wealth and advantages of Michigan had been circulated throughout the entire province. Nathaniel, anxious to secure a home in which to spend his old age, decided to claim some of the land the United States government was offering new settlers. He traveled alone to Michigan to select a suitable site. This turned out to be a 40-acre parcel located two miles east of Utica, Macomb County, and about 25 miles north of Detroit. Less than a month after his purchase he and his sons had cleared half an acre and built a mud-plastered log cabin.

The family stayed together until Luman, then Ethan, came of age, and according to custom, left home to be on their own. Nathaniel gave Ethan a yoke of oxen as compensation for the labor he had done in clearing the farm.

Ethan Squier, George Squier's grandfather (courtesy Dryden Historical Society).

Ethan did not have a pair of boots until he was 25 but, with his newly acquired cattle, he considered himself very rich. He immediately began to train them to plow his father's farm. After working for his father about six months, he purchased 20 acres of land located a mile from his father's property, for two hundred dollars and his prized cattle. The two hundred dollars was to be paid in six-month installments of $50 each. He would gladly have started to clear his own land but, as he did not have enough cash to cover his first installment, he earned money cutting wood for a neighbor. Sometime during that long winter Ethan first met his future wife Lovina Hundey. By applying himself, he managed to save enough to pay his first installment nearly a month before it was due. He immediately began to save money for the second installment and even received a five-dollar discount for paying it two months early. After this he sold the land for $150 to a man who agreed to pay the remaining two installments. Ethan rented a small house on the eastern outskirts of Utica, and after being wed to Lovinia, he invested his money in stock, which he kept at his father's place. He continued to work as a day laborer to save money. In 1832 a son was born to the couple, Almon Justice Squier (1832–1905), George's father.

Ethan wanted to have a farm of his own. He traveled north, stopping over at the little village of Lakeville, Macomb County, Michigan, when they came to the County of Lapeer. He had been told that this particular township had high, dry rolling land and fertile soil, thickly covered with beech and maple timber. In all, he purchased 160 acres—the south half of section No. 11 township of Dryden. The only consideration for the purchase was that he needed to live on it for five years and clear and sow five acres a year.

When Almon was about 18 years old, his father, wanting him to become something other than a farmer, managed to secure a position for him at a store owned by John M. Lamb, in whom Ethan had much confidence. Almon had all the education that his father could afford, and though it was nothing beyond reading, writing, spelling, and a little arithmetic, Ethan considered his son a well-educated youth. Almon did so well as a clerk that he soon sur-

Main Street, Dryden, Michigan, c. 1890 (courtesy Dryden Historical Society).

passed Lamb himself. Their business gradually increased, supplying the lumber camps farther north, and in less than two years Dryden became a thriving little village. Unfortunately, however, by the age of 20, Almon "formed habits which were destined to make him a ruined and degraded man ... habits that he formed then and that will bring him to a drunkard's grave."[7]

Lamb could drink and gamble nearly all night and the next day be ready and as shrewd as ever in his business operations. But "after closing the store at night [he] played cards either in the back office of the store with the blinds to the windows all fastened.... Almon hereafter at first used to only sit by and watch the games and never thought of tasting the liquor. But as time rolled on he began to play and finally to drink a glass at the opponents' expense when he would be successful enough to win."[8] Neither Lovinia nor Ethan knew anything of this for nearly eight months because Almon was then boarding at the store for the reason, he said, of being more convenient to his work. When Ethan was first told of this arrangement he ridiculed the idea. But as he always had the greatest confidence in Lamb he refused to believe anyone who claimed that Almon had taken to drink.[9] However, after a few more reports, Ethan ordered Almon to leave his job at once and sent him back to school. Almon was successful in his studies, particularly in mathematics, but he often stayed up until midnight, drinking and gambling at the old place with the same company as before. "The fact was that he had formed an appetite for strong drink and that appetite had such a strong hold on him that he could no more control it than he could control the winds."[10]

George Squier's mother, Emily Gardner, was the eldest of four children, born to a carpenter and his wife, both of whom hailed from Connecticut. Mrs. Gardner was not well and died shortly after her last child was born. Although Emily knew that Almon drank and gambled before she married him, he was very personable, a good dancer, and Ethan's son, facts which made all the girls in the neighborhood vie for his company. The two married at Almont, a little village six miles east of Dryden.

Emily Gardner Squier, George Squier's mother (courtesy Dryden Historical Society).

As a wedding gift, Ethan gave

his son 80 acres of his farm, located on the west side, and built the couple a comfortable little one-story house. He also furnished him with a team of oxen, plows, harrows and everything needed to begin farming. Ethan even furnished the young couple's home and stocked their cellar with a barrel of well-cured pork, a bin full of potatoes, fruit, pickles, maple sugar, and other essentials for housekeeping. Almon and Emily moved into the house in the latter part of December, with Almon determined to become a farmer in earnest. However, he spent his evenings at the barroom instead of at home. This greatly upset Emily, who was afraid to be alone after dark.

Almon failed as a farmer. While he spent up to a week at a time on benders, his untended crops were in poor condition. Ethan had to step in and harvest the corn for him to keep it from spoiling. After nearly a year's experience at farming Almon concluded that he was more inclined to be a storekeeper. Entering into a contractual agreement with a partner, Almon was to furnish half of the money to buy the stock while the other person was to furnish the other half, keep the books and be allowed the use of the building free of rent. Almon failed at this venture as well. Back to his old habits, he gambled and drank every night. Business slackened and credit became unavailable. Less than a year after they began, the partnership was dissolved, the goods sold at a great sacrifice, and the debts turned over to Detroit lawyers for collection. Almon's share, which was nearly three hundred dollars, had to be paid by Ethan. The next business he tried was saloon keeping. This seemed to suit him better than anything he had yet attempted as he had the liquor and the cards right at his disposal. As George Owen Squier admitted, "I hate to write it, but he became in a short time after going into the business a perfect drunken sot."[11]

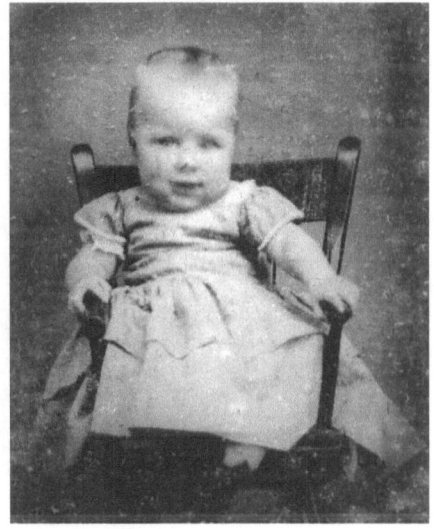

George Squier as an infant (courtesy CECOM Historical Office).

Almon and Emily's first child was a son named Herbert. Two years later a daughter, Sarah, was born. Both died early in childhood. Herbert became ill while playing on a Sunday afternoon and died on the following Friday. Less than two months later, Sarah picked up and ate a poisonous bug while playing and she, too, became sick, succumbing almost immediately. The March after Sarah's death George Owen Squier was born. The name George was selected by his mother; Owen came from a character in a novel that his father had just read. Two years later saw the arrival of his sister, Mary Adel, so named because of her resemblance to her aunt Mary Gardner.

Almon continued to farm, but "it was not an uncommon thing for him to draw a load of wheat to market and then go straight to Lapeer or Almont and drink and gamble till every cent was gone. Grandfather always had to help him harvest his crops after finishing his own, to keep them from lying on the ground and wasting."[12]

Emily's health began to fail, prompting Almon to take her to the State Medical Institution at Flint. While she was there, George's grandfather, Ethan, began to look after him, not trusting Almon with his care. Emily stayed at Flint several months, gradually wasting away from tuberculosis. Since she was not recovering at Flint, Almon moved her back home. She died from pneumonia shortly afterward, in 1872, when George was seven years old.[13] She was buried on the family lot, about a mile and a half north of Dryden village.

Now a widower, Almon decided to run the farm by himself. George and Mary moved in with Ethan. George went to school regularly and was a good student, particularly in mathematics. The school was considered very large for a village the size of Dryden and was staffed with a principal and two female assistants. George ascribed his diligence as

George Squier with his sister, Mary (courtesy Dryden Historical Society).

a student to his grandfather: "I have always considered what a blessing it was to me that I came under the care and management of such a grandfather as Ethan Squier was to me. He was so very prompt himself in all his dealings that when I came under his care he applied all his principles and practices to me."[14]

George specifically mentions a Mr. Messer, the principal, as "the best teacher I ever had and to him I owe my first idea of an education at any cost, and of an upright, temperate, Christian character. He remained in Dryden for three consecutive years and I don't believe there ever was a man who worked harder and with better success than he did.... [He ran] a school that was to produce the best teachers in the County of Lapeer and as good as in any county in the whole state ... the school year from September 1878 to June 1879 was perhaps the most decisive and profitable year I ever spent. I began to know my abilities and capabilities and began to stretch out in my imagination into some ideal that I wished to become like."[15]

In June 1879, when he was 14 years old, George graduated from the equivalent of today's eighth or ninth grade. He was advised by his uncle and his grandfather not to go to school anymore; they insisted that he had enough education and that it was high time he chose an occupation. George disliked the idea of leaving school, but he was pressed so hard that he finally began to believe he was as well educated as most.

Jacob C. Lamb (for whom Almon had worked many years earlier) and his son, Edwin, were just starting a business at Imlay City, about eight miles north of Dryden. They talked with George's uncle Jim and his grandfather to see if the youngster would like to be an apprentice in their mercantile business. Ethan and Jim advised him to take advantage of this opportunity; there would be a good prospect for him to work his way up and possibly become a partner in the firm. They argued that, as his father was leading a reckless drunken life and paying no attention to his children, he would soon have to take care not only of himself, but of Mary also. Giving in to the weight of argument, George decided to go. He stayed with Lamb for two years, advancing from chore boy in the back grocery department to a position behind the showcases and in the silk and satin department. Lamb believed that George was born to be a merchant; he often told him that he did the best of any boy with whom he had ever worked. The first year that he was in the store he was content, for he had made up his mind (he thought) to become a merchant. But sometime around September 1880 he began to get restless. He thought over his prospects and projected that if he stayed with Lamb he would clerk until he was offered a partnership in the business. He would be a wealthy merchant and then "retire a gray headed man with nothing as a result of my having lived but a few houses, stores and the like."[16] He resolved then and there to quit and get an education at any price. The strong will George displayed at the age of fifteen prefigured the firm sense of purpose he would demonstrate over and over again whenever he set out to accomplish something.

He entered Lapeer High School the following fall, working on a farm for his room and board in the home of the Honorable Joshua Mauwaring where his sister, Mary, lived.[17] That winter's announcement of competitive examinations for applicants to West Point appeared in the local newspaper. He left the mercantile trade in order to become a "man of letters."[18] Enrollment in West Point offered a golden opportunity for a poor country boy to realize his ambitions. Without informing anyone (not even his beloved sister), he slipped away and took the examinations with 30 competitors for the coveted appointment. Soon afterward Mary received the following message[19]:

> Dear Sister,
> I was in Port Huron last week to write the competitive examination for West Point. Thirty boys wrote, and will you believe it, *I was the lucky one*. Break the news to dear old Grandpa—
> George

Soon everyone in the village heard the news and rejoiced for George, but they wondered how he would fare on his chosen course with such an impoverished background and so little consistent schooling. They had, however, underestimated his determination.

Unable even to defray the expenses of preparing for enrollment at West Point, George withdrew from Lapeer High School and took a teaching position at the Terry School, a three-mile walk from his home. His sister's devotion to him showed itself in her daily preparation of breakfast at four a.m. for his early morning departure for school. She also made articles of bedding and other necessities required as part of his initial outfit upon arrival at the Point.[20] When the day arrived for his departure many friends and townspeople bade him farewell from the platform of the train station.

West Point

Founded in 1802 during Thomas Jefferson's presidency, West Point trained men under the direction of the Corps of Engineers. Before the Civil War, the Point established a brilliant record as an engineering school. Graduates of West Point undertook the principal scientific work of the government on railroad surveys, coast and geodetic surveys, and river and harbor improvements. After the War Between the States, scientific functions in the military began to decline.[21] Transfer of the weather service to the Department of Agriculture and creation of the Geological Survey constituted a loss of important scientific services from the Army. Although West Point escaped the general decline infecting the rest of Army science, it ceased to be exclusively an engineering school. The school stagnated.[22]

West Point's curriculum in post–Civil War years was an anachronism. Descriptive geometry, for example, was taught from Albert E. Church's work until the 1930s. In the 60 years since 1843 the Corps of Cadets had only doubled. Despite new weapons and equipment introduced during the Civil War, West Point received old, condemned equipment and weapons. The entire faculty came exclusively from the alumni, who kept to the old ways. Promising

George Squier as a West Point cadet (courtesy Dryden Historical Society).

students were returned to the academy after one tour of duty with the active Army. There they could remain for the next 30 or more years without venturing again outside the college walls. Most cadets studied under the pre-war professors of West Point and used the original class notes and texts. Regardless of General William Tecumseh Sherman's apparent desire to improve the officer corps of the Army, he was satisfied to "preserve the Academy in its pre–Civil War purity."[23]

Thus, one strains to imagine West Point being a cauldron of change. Yet Squier's ambitions and love of physics were fired in that environment. The influence of the Academy's leading professor, Peter S. Michie, may be glimpsed in Squier's subsequent research and military interests.[24] Michie taught at West Point from 1867 until his death in 1901. Although his texts on analytical mechanics, astronomy, and wave-motion physics replaced all others to become the core of instruction, wave-motion physics intrigued Squier the most. He often took his sections into a lecture hall for special demonstrations in the wave phenomena of light and sound.[25] The noted military historian R. Ernest Dupuy implied that Squier's consuming interest in wave motion physics stemmed from Michie's teaching.[26]

Although Michie served for so long as an academic man, he had also earned wide respect as a soldier in the Civil War. A Point graduate in 1863, he returned four years later as a brevet brigadier general. His military study at the academy centered on coast defense. His work on *The Life and Letters of Emory Upton* revealed an additional interest in Army reform.[27] Michie's book long remained a standard on the life of one of the Army's greatest reform thinkers. Wave-theory physics, coast defense and reform all characterized Squier's Army career, too. He disciplined his mind and body in the service of achievement. In his West Point class of 64 cadets he was ranked first on the order of merit in discipline, and seventh in general merit.[28] He had come a long way from his grandfather's farm in Michigan.

True to his religious heritage, George allotted his time in a most methodical manner. In later years he enjoyed telling of losing a half hour in his freshman year and not regaining it until his senior year.[29] He recorded only one occasion when he dated girls[30]; he kept to his goal of becoming a man of letters. But intellectual attainments were only part of his scheme for self-improvement. Like many of his age and background he possessed a burning passion for assurance of personal salvation. Typically, his concern extended to others. He regularly attended Wednesday evening prayer meetings, encouraging friends to accompany him. He gave personal testimony of the saving grace of Jesus Christ and dedicated his working hours to the justification of faith in God.

Just before graduation he requested a pass to spend the summer touring Europe with companions.[31] From mid–June until the end of September 1887, he and his friends rode bicycles about England, Scotland, France, Switzerland and Germany. With the stirring addresses of Generals Sherman and Sheridan still fresh in mind,[32] the grandeur of European art and science inspired him to higher intellectual attainments in order to be of greater service to the Army. Even before entering elementary school George had been fascinated by physical phenomena. When asked just prior to

George Squier as a young lieutenant at Fort McHenry (courtesy CECOM Historical Office).

graduation from West Point to select a branch of the Army, he chose one of the most technical: artillery. An artillery officer required special schooling in electricity, ballistics, physics, metallurgy, and chemistry. Midway through his European travels he wrote a friend from Richtersweil, Switzerland, that he expected to be assigned to a regiment at San Francisco, California.[33] Few things, though, are certain in the service. While touring Europe, Squier received his assignment to a battery in the Third Artillery, located at Fort McHenry, in Maryland. He told his sister how "very much pleased" he was with the appointment.[34] A month later he decided how he would use the leisure time so characteristically available to service men on 19th-century garrison duty. He confided to his sister about the "fine chance at Johns Hopkins University if I choose."[35]

2
Soldier-Scientist

Founded a little over a decade before Squier began his studies there, The Johns Hopkins University was one of America's major centers for graduate instruction in physics and electrical engineering. Graduate instruction, creative scholarship, and original research marked Hopkins as a unique American institution. President Daniel Coit Gilman attracted to the science faculty such luminaries as Henry Rowland in physics, Ira Remsen in chemistry, and Simon Newcomb in astronomy. These men were still vigorously productive when Squier attended. He was in strong company. Other young men attending in the same time period were Frederick J. Turner, Josiah Royce, John Dewey, Henry C. Adams, Joseph S. Ames, Walter Reed, Simon Flexner, and Woodrow Wilson. Squier could have hoped for no better assignment to further his ambitions.

The roots of Squier's desire to attend The Johns Hopkins University are obscure. Perhaps he hoped to qualify himself for transfer to a technical service, Ordnance, Engineers, or Signals, since everyone else in his West Point class was assigned to Cavalry, Infantry, or Artillery. Perhaps, too, the inspiration of Michie's classes motivated him to seek more knowledge of physics. Whatever his reasons for becoming a soldier-scientist, he took an unprecedented course of action in the military by trying for a Ph.D. It was highly unusual, even for civilians. Of 51 men whose names appear in both categories of soldiers and engineers in the *Dictionary of American Biography*, only two were electrical engineers. There was none identified as a mechanical engineer. The remainder appeared as general or civil engineers, a reflection of the rather broad engineering responsibilities required of soldiers and the rather undifferentiated character of engineering practice during much of the 19th century. As the engineering profession became more professionalized and specialized toward the end of the century, the number of soldier/engineers in the *Dictionary* decreased. Although there were about as many notable soldier/engineers living in the first quarter of the 20th century, there were only two specialized soldier/engineers. One was Eugene Griffen.

Griffen was graduated from the Academy in 1875 and subsequently joined the Corps of Engineers. After working on a surveying party in Colorado, New Mexico, Arizona, and Texas, he returned to West Point as an assistant professor of military and civil engineering. In 1887, as assistant to the engineer commissioner of Washington, D.C., Griffen published the results of his investigation of telephones, telegraphs, arc lights, incandescent lights, and underground electric wires. In his report he proposed the use of electricity to power streetcars. As a result of this report he received an offer to assume the post of general manager of Thomson-Houston Electric Company, a leading electrical traction firm. He accepted the offer and resigned from the Army in 1889. Three years later he was elected vice-president of General Electric Company, formed from Edison and Thomson-Houston interests. In the succeeding year Griffen became president of the Thomson-Houston International Electric Company.[1] One of only two specialized soldier/engineers, Griffen had no formal education in electrical engineering and left the Army early in his career for more favorable business opportunities.

Squier was the other engineer/soldier listed in the *Dictionary*. His decision to attend Hopkins as a doctoral student was unprecedented. His choice of specialized engineering education outside an Army school was equally unusual. Reasons for his decision, however, remain speculative.

Fort McHenry and The Johns Hopkins University

With enthusiasm and verve Squier reported to his first military assignment. As soon as he familiarized himself with his responsibilities at the post he enrolled as a part-time graduate student in the university. He resolved schedule conflicts by offering to stand post duties on weekends in exchange for the necessary free time during the week to attend classes.[2] Young officers were much in demand in Baltimorean Society on the weekends, so the arrangement was quite acceptable to his fellow lieutenants.

During his first year at Fort McHenry George combined military duties with academic studies. His studies began with an electrical course taught by Dr. Louis Duncan, chairman of the Department of Electricity. Besides his assignment to one of the Fort's batteries, Squier, like many young military officers who have followed him, carried a varied and heavy load of extra duties. He sat on courts-martial panels, boards of survey,[3] and burial details. He served as inspector of units, rifle-range officer, and court-martial defense counsel. In time he was even asked to proctor entrance examinations at The Johns Hopkins University.[4]

When school closed for the summer excusal from routine duties ended, too, and he rejoined his battery for summer maneuvers. Squier nearly failed to survive the summer of 1890. His battery had been transferred to Old Point Comfort, Virginia, where they were engaged in artillery practice. Squier was in Captain Thurston's battery when it was assigned a 4.25-inch Rodman rifle, which had been sent to that post during the Civil War, for the day's practice firing. One of the first of that class manufactured by the government, it was mounted on a wooden platform on the beach. Fired for several years, everyone considered it safe, but that morning the breech split in two during firing and spewed fragments of gun metal in all directions. A private under Squier's command sustained injuries and Squier himself barely escaped death from chunks of metal flying past both sides of his body in his position just to the rear of the rifle.[5] His duties for the remainder of the summer were considerably less dangerous and far more glamorous.

Colonel E. C. Brush, commanding officer of the 1st Light Artillery, Ohio National Guard, requested Squier by name to serve on detail as inspector and instructor in military science and tactics during the month of August.[6] George always made a particularly good impression outside his immediate unit. His discreet manner, cheerful attitude, and obvious competence placed him in demand for speaking, advising and teaching engagements. Upon finishing his duties with the Ohio National and at their encampment near McConnellisville, Colonel Brush wrote the Secretary of War that Squier "carried away with him the hearty good will of all and the Army is to be congratulated upon having so accomplished an officer."[7]

A year after he arrived, Squier applied for permission to attend Johns Hopkins on a full-time basis to pursue a Ph.D. in electrical engineering. Aware of the length of time needed for the academic work required to earn a doctorate, and of how often the Army reassigned their officers, he knew it would be difficult to complete all the requirements while attending school on a part-time basis. However, his professors respected his talent and they encouraged his full-time attendance. Ira Remsen especially encouraged him to participate fully in the university community.[8] Unknown to Squier, Professor Duncan wrote to the Secretary of War, asking for Squier's assignment to Hopkins.[9] The Adjutant General of the Army denied the request, citing paragraph 37 of the Army Regulations, which stipulated service of three years with an active line unit before such assignments could be granted.[10]

Squier assured the Adjutant General that his "object in wishing to continue his studies is not with any intention of resigning from the service, but because I hope the course will fit me for greater usefulness to the service in the future."[11] The Major General Commanding, J. M. Schofield, replied that he might reapply after three years of service.

In March 1890, Squier resubmitted his request for a detail of two years, to commence in October of the same year. This time he strengthened his case on two points. He called the Adjutant General's attention to provisions made by the Navy for the detail of an ensign to Johns Hopkins. In a letter recommending Squier's fulltime attendance, Professor Ira Remsen, one of the founding fathers of the American Chemical Society, noted how naval officers had attended Hopkins since 1883 "studying branches kindred to ... [their] profession."[12] Remsen advised the Adjutant General that Squier's action met "with the hearty approval of the authorities of the University." General Schofield now approved, but Secretary of War A.M. Keever disapproved, recommending instead that Squier's name be placed on a list of officers recommended for student detail to colleges.[13] The Secretary of War further directed that the class situation at Fort Monroe Artillery School be investigated with a view to assigning Squier there if he showed as much talent as his professors suggested.[14] The school's commanding officer said there would be room for him, but Squier said he preferred Hopkins to Monroe at that time.[15]

George Squier, new Johns Hopkins Ph.D. in electrical engineering (courtesy CECOM Historical Office).

As he did when he was 15, Squier did not give up in the face of discouragement. He tried a new tack. Finding the Secretary unwilling to detail him, especially in preference to officers who had been waiting longer for such a desirable assignment, he requested a relief from duties for a period of eight months, to commence about 5 October 1890.[16] His request was approved by the Secretary of War, at the discretion of the Post Commander of Fort McHenry.[17] The Post Commander permitted Squier to attend school during the week in exchange for his performing the duties of other officers on the weekends. Squier's ability to maintain this rigorous schedule is indicative of to his stamina and intellectual endurance. The Army's supportive attitude certainly fostered Squier's ambitions.

During his years in full-time attendance at Hopkins, Squier learned from men who would remain close to him throughout their careers. He studied thermodynamics, magnetism, and electricity under Henry Rowland, and optics under Professor Kimball. He took theory of functions under Professor Craig, mechanics under Joseph Ames, and physical optics under Henry Rowland. Professor Duncan was chairman of the Electrical Engineering Department, and Professor Rowland directed the Physical Laboratory. Squier described his early elation over experimenting to a reporter many years later:

> I was always tinkering around with spools and twine and a piece or two of wire if I could find any. If I could turn a crank and see something that I made with my own hands move, twist or vibrate or show the slightest animation, I felt I had invented something and was happy.[18]

Working for one of America's greatest physicists in one of the finest American laboratories was an exhilarating experience. He attributed his later inventing activity to the excitement he found in the classrooms of The Johns Hopkins University.

By spring 1891, Squier's work at Hopkins had gained serious attention. President Gilman appointed him a Fellow of Johns Hopkins for the succeeding academic year. An emolument of five hundred dollars accompanied the appointment.[19] Dr. Duncan endorsed Squier's election to membership in the American Institute of Electrical Engineers.[20] His election to a learned society was the first of many such elections (over 30 by his life's end). They led to the establishment of important bridges of communication between professional science and the Army. His success within the university community established the first tentative link between his career in the Army and institutional civilian science. Regarded by now as one of Hopkins' prize pupils, Squier came to the attention of other teaching institutions.

The University of South Dakota at Vermillion asked the Adjutant General if Squier would desire a detail to their institution.[21] The Military Academy also began eyeing him for assignment. He even received orders assigning him to the Academy at West Point in August.[22] Squier's attachment to Johns Hopkins was far too strong by this time to be disrupted, so he obtained a release.

In early June 1891, Squier applied to the Adjutant General for an assignment to the university itself or for continuation of authority for partial relief from his post and battery duties. The latter proposal was approved for another eight-and-a-half months, effective 1 October 1891.[23] When the school closed for the summer, Squier's excusal from routine duties ended, too, so he rejoined his battery for summer maneuvers.

The summer having ended, Squier soon settled back into the routine of duties at Fort McHenry and his studies at the university. It was not long, however, before West Point authorities began to press again for his assignment. In January, and again in March 1892, Tasker Bliss, aide-de-camp to the Major General Commanding, acted as broker between Squier and West Point officials.[24] Just when it appeared that Squier had succeeded in convincing the

Army that everyone's best interests would be served by allowing him to remain with his battery at Fort McHenry, his Battery Commander, Captain Thurston, informed him that he headed the list of those officers who should be nominated, under the ordinary rules of selection, for attendance at the Fort Monroe Artillery School, as all others had attended. He offered to rearrange assignments so Squier could remain at Fort McHenry and continue attending Hopkins. Thurston's offer of help was significant in view of an anticipated transfer of the 3d Artillery to Atlanta in the fall.[25] Frequent attempts to reassign Squier to new locations or move him closer to field duties elicited expressions of genuine pique in his personal notebook. Squier had already begun to consider seriously the impact of America's burgeoning electrical industry on the conduct of war and he was determined to help provide the needed expertise to the Army.

Squier believed that the War Department failed to appreciate the value and nature of scientific endeavors and the preparation required to pursue them. He wrote that the "War Department cannot make an electrical engineer by a general order."[26] Older officers and peers voiced resentment, feeling that he was seeking special privileges for the study of subjects that were civilian in character. Their attitudes moved him to acknowledge that "objection will be made by some to such details [school assignments]—that the younger officers want to shirk company duties—that they are anxious to shine in society—and what not: but such reasons seem to me on too low a plane to merit serious consideration."[27]

In June 1892, Squier submitted his thesis entitled *Electro-Chemical Effects Due to Magnetization* to the faculty. In the preface, he acknowledged his indebtedness to the Honorable Redfield Proctor, United States Senator from Vermont and Secretary of War, for relieving him from routine military duties in order to complete his course of study at Hopkins.

Squier's thesis research followed the lead taken by Rowland in the study of electro-chemical effects of magnetism. After he submitted his thesis for approval, it was passed by his examining committee and he was certified for graduation the following summer.[28]

With Squier's thesis finished, it must have appeared to his contemporaries that he would finally return to the life of a young artillery officer and pick up his full share of fortress duties. When, instead, he applied for permission to attend the International Electrical Congress which was to be held in conjunction with the Columbian Exposition in Chicago, even normally patient Captain Thurston was moved to express his aggravation.[29] Not only did Squier ask for time to attend his graduation ceremonies, to visit the Congress, and to tour the electrical displays at the Exposition when almost certainly his regiment would be off on disagreeable maneuvers, he also asked for a further extension of his assignment to Fort McHenry in order to continue his research. Approval of his request would, of course, have meant extended relief from routine duties. Thurston observed with evident displeasure that Squier "has been practically detached from his battery and duties most of the time for the past two years to the detriment of other officers." He added that should Squier's requests be granted he will "not only avoid the discomforts of a Southern Station with his regiment but the course at the Artillery School which he has never taken."[30] Although Thurston disapproved, the regimental commander interceded with an approval.[31] The Major General Commanding of the Army concurred in allowing Squier to remain at McHenry until March 1893, when he would consider a re-application for detail to the Exposition. In his statement of approval he remarked on the pleasure it gave him in "encouraging young officers, who as in the present case, devote their time to hard work for the purpose of fitting themselves still further for the military service of the Government."[32]

Captain Thurston's ire was not so easily mollified. When the Post Adjutant, Lieutenant Colonel LeR. Irwin, departed in August, Squier was immediately assigned duties which, in addition to the adjutancy, included Post Treasurer, Acting Signal Officer, and Superintendent of the Post School.[33] Squier willingly performed these tasks until October, when the doors of Hopkins opened for the new year. With noticeable impatience, he requested to be relieved from his post and battery duties, except on Saturdays and Sundays, as "authorized by the War Department."[34] The controversy extended over a period of several months, ending toward the end of December 1892, when he was finally relieved of his duties and allowed to resume his full-time research in electrical engineering in the Physical Laboratory.[35]

Founding the Artillery Journal

The rapid development of electrical industries and the impending construction of the Cataract Company's Niagara Fall's dynamo led a young group of reform-minded artillery officers to consider the impact of electrical engineering upon warfare, especially upon seacoast fortifications. Among them were Lieutenants Charles D. Parkhurst, Henry C. Davis, Jr., John W. Ruckmann, and Squier.

Described as a "remarkably gifted man," Ruckmann possessed a natural bent for mathematics and physics. After seven years of routine artillery service following graduation from West Point, he entered the Artillery School and Fort Monroe. He remained for seven years. Dividing his time between artillery and the Inspector General's office, he eventually commanded the Coast

Defenses of Manila Bay. When World War I broke out he received an appointment as Major General of the National Army. By then he had acquired the reputation as the finest technical artilleryman in the service.[36] Davis, an 1883 West Point classmate of Ruckmann, also displayed a penchant for mathematics. While assigned to the Agricultural and Mechanical College of Mississippi as an instructor of Military Science and Tactics, he also served as a professor of mathematics at the invitation of the college president. An honor graduate of Fort Monroe Artillery School in 1892, he was one of three men selected to pursue special study. With the exception of a brief absence during the war, Davis remained at the Artillery School until 1905. Following a series of command assignments in various artillery districts and forts, Davis was detailed as commander of Manila's coastal defenses. After one year in that post he resigned because of ill health.[37] Parkhurst was the oldest of the group of lieutenants, having graduated from West Point in 1872. Engaged in Indian campaigns as a cavalryman for the first decade of his career, he joined the Artillery Corps in 1884. An occasional contributor of articles on pistols, horses, and sabres to the recently established *Cavalry Journal*, he acquired an interest in the state of organization and training in the Artillery Corps after his assignment to the Artillery School.[38] Little more is known about Parkhurst in this period aside from his continuing interest in publishing on artillery matters. During the Spanish-American War he was seriously wounded in the battle of San Juan. For his gallantry in action he received the Silver Star medal. He retired in 1909 as a colonel.[39]

As a member of this group Squier pointed out that electricity threatened "to revolutionize our whole heavy artillery organization."[40] The introduction of the telephone and writing telegraph were envious benefits. Less obvious was protection of the power plant by its distant removal from the firing range of enemy guns. He advocated the use of alternating current power because of its greater efficiency in long-distance transmission. Referring to the Lauffen-Frankfort line in Europe,

George Squier as a captain (courtesy CECOM Historical Office).

he reported that 300 horsepower had been transmitted 112 miles at voltages between 16,000 volts and 30,000 volts with an efficiency of 74 percent. The availability of such protected power would make possible the use of enormous engines of war weighing many tons and costing thousands of dollars. Searchlights and electrically operated range-finder systems could augment the operation of such guns. The presence and capability of artillery pieces maneuvered, aimed, and fired electrically posed challenging implications in military doctrine. The young officers realized their desire to explore questions of doctrine, maneuver practice, adherence to general orders, electrical studies, devices, and machines applied to coastal defenses in a professional and cooperative spirit by establishing the *Journal of the United States Artillery*, with the support of Colonel Royal T. Frank, School Commandant.[41]

The establishment of the *Journal* provided a medium for the young officers to discuss the application of science to problems in artillery. It also served to raise the morale and quality of the artillery service which was then regarded contemptuously as "a joke."[42] The founding of the artillery journal may be viewed as part of the institutionalization of science throughout the Navy and the Army, as well as the professionalization of the armed services. Notable steps in these processes were the founding of the United States Naval Institute in 1873, later famous for its *Proceedings*; The United States Cavalry Association and the *Cavalry Journal* in 1888; and the Association of Military Surgeons, which published *The Military Surgeon* beginning in 1891.[43] These journals represented a realization of a program put forward by one of the Army's most innovative commanding generals, William Tecumseh Sherman. The journals were both innovative in military institutions and imitative of European military practice. The Military Service Institution, for example, was patterned on the Royal United Service Institution of Great Britain.[44] German-speaking officers communicated in the *Organ der Militär-wissenschaftlichen Verein in Wien*, founded in 1870. General military information was reported in the *Militär-Wochenblatt*, published in Berlin since 1815, and the *Militärische Blätter*. The U.S. Army *Cavalry* and *Artillery Journals* departed significantly from publications of general military interest by appealing to specific elements within the Army. The British army followed this American trend in 1906 with a publication entitled *Cavalry Journal, Horsed and Mechanized*, which was later absorbed by the *Royal Armoured Corps Journal*. Issues raised in the *Artillery Journal* were intended to stimulate discussion on developments within the Artillery Corps, especially the impact of electricity on artillery operations.

One issue discussed in the new artillery journal was the role of an electrical engineer in an artillery unit. The employment of engineers aboard naval vessels provided practical illustrations for Squier on how electrical engineers should be detailed in the artillery. Just as one officer on naval vessels has charge of all electrical apparatus, he suggested, so an electrical engineer, directly responsible to the senior artillery officer in command, should supervise "each group of guns ashore, with its electric lighting system, search light system, range finding system, generators and motors."[45] Therefore, he urged, the War Department should "speedily educate a limited number of officers as electrical engineers at our best institutions of learning."[46] The use of modern electro-mechanical machines of war pointed the way toward fewer men possessing more skills.

The Columbian Exposition

Squier's efforts to enhance his own professional standing and to publicize developments in electrical engineering within the military were illustrated by his re-application for detail to the International Electrical Congress in April 1893. It was scheduled to be held in the fall in conjunction with the Columbian Exposition at Chicago.[47] The Army delayed answering his request. By now, however, Squier knew influential men who could plead his case; he did not hesitate to call on them for assistance. The men to whom he appealed were professional scientists with bureaucratic ties, rather than politicians. Thomas C. Mendenhall, a friend from Hopkins, was such a man.

Mendenhall, Superintendent of the United States Coast and Geodetic Survey and Chairman of the Committee on Congress Program, wrote to the Secretary of War requesting Squier's attendance. Captain Thurston's earlier apprehensions were being realized. The battery had been assigned to Fort McPherson, near Atlanta, for summer maneuvers. And now it appeared Squier would be in Chicago with scientists instead of following his military profession. For six years he had missed field duties to follow scientific studies. Mendenhall told the Secretary he knew that the Army would have observers at Chicago and he thought Squier's presence would be important since his "accurate knowledge of electricity ... would ensure a standing in the Congress of which the department might be well proud."[48] Squier wanted to leave for Chicago in mid-summer, after his graduation ceremonies, in time for the meetings which began 15 August. The Secretary was uncertain if Squier should be relieved so soon and asked Mendenhall for a suitable reporting date to place in Squier's orders. One can only imagine the extent of Captain Thurston's ire when, in late May, Mendenhall wrote "at once if possible."[49] So almost at once Squier went off to Chicago as the Army's observer where, in his own enthusiastic words, "Never so many men distinguished in the domain of electrical science had been assembled at one time for mutual conference, and benefit."[50] An account of his visit appeared on the first page of the next issue of the *Artillery Journal*.

Dr. Hermann von Helmholtz was the natural honorary president of the Congress and considered by many to be its father. The Congress divided itself into three sections: pure theory, headed by Professor Rowland; theory and practice, chaired by Professor Charles R. Cross of the Massachusetts Institute of Technology; and pure practice, headed by Professor Edwin J. Houston of Philadelphia.[51] The Chamber of Deputies, composed of official delegates of participating governments, approved adoption of definitions and values of fundamental units of resistance, current, electromotive force, magnetic units, units of self-induction, light, energy, and other units. From this vitally important conference came international agreement on the use of the electrical unit terms of ampere, volt, coulomb, farad, joule, watt, and henry.[52] Squier developed lifelong public and private relationships with the British delegates, notably W. H. Preece, W. E. Ayrton, S. P. Thompson, and Alexander Siemens. Sylvanus P. Thompson's paper on "Ocean Telephony" attracted perhaps the most attention. In it he predicted the imminent use of submarine cables to convey speech.

Squier considered that the most important and valuable discussion dealt with polyphase motors and transmission of power. Central to this discussion was Charles Steinmetz's elucidation of the law of hysteresis. Steinmetz was a German émigré who became one of General Electric's most brilliant engineers. It thus became possible to predict accurately the electrical losses in dynamos and electrical machinery due to the magnetic properties of iron and steel. He thought the Artillery Corps could benefit greatly through an intelligent application of this scientific knowledge to practical gunnery. His faith in the value of scientific intercourse for military and industrial development was evident in his comment on how "the essential oil from the abstruse memoir of to-day, lubricates the commercial machine of to-morrow."[53] Despite American advances and prestige in the electrical industry, he advised others to look to Europe for examples in applying electricity to the problems of coastal defense, particularly to the work of Messrs. Sautter, Harle & Company of Paris. Their catalogue described how motors and other devices of their design were constructed for the propulsion of ships, submarine navigation, loading hoists, command of the helm, the aiming of search-lights, and the maneuvering of armored turrets.[54] Squier promoted a plan for applying electricity to artillery problems in a way most suitable to the interests of proper coastal defense as viewed by the professional artilleryman.

Squier suggested starting this important work at the Artillery School, which he regarded as the proper "fountainhead of artillery information, instruction and experimentation."[55] Furthermore, education of electro-technical personnel should be decisively carried forward by enrolling a limited number of officers in the principal electrical schools of America for at least a year's study.

He recommended an apprenticeship in workshops abroad after their graduation, such as in those of Sautter, Harle & Company. His plan envisioned a great electrical laboratory at Fortress Monroe, a vision he articulated more explicitly in several months time.

A Proposal for an Army Electrical Engineering Laboratory

Soon after returning to his battery, which was still on summer maneuvers at Fort McPherson, Georgia, he applied for assignment to another battery at Fort McHenry so he could continue his experiments at Hopkins. As justification for this continued privilege he expressed the belief that "this class of specialists [electrical engineers] more than any other is what the Artillery needs at this time."[56] As to why he should be selected, he noted that his "research in electro-chemistry during the academic year 1891-92 has attracted some attention among physicists in Europe as well as in this country."[57] Major General Oliver O. Howard, Commanding Officer of the Department of the East and former West Point Superintendent, disapproved and wanted to know "why he can not take the course at Fort Monroe and continue his researches there. Duty with an officer's battery, except for good reasons to the contrary, is considered of primary importance."[58] Colonel LaRhett L. Livingston, Commanding Officer of the 3d Artillery, revealed his compassionate support of Squier by suggesting he re-apply and bring to the General's attention that the Fort Monroe school would not open until fall of 1894, a year away.[59] Following Livingston's advice, Squier resubmitted his request with a recommendation that an electrical research laboratory be established at Fort Monroe. He asserted that "the Artillery will shortly need expert Electrical Engineers among her officers to develop and care for the numerous electrical plants which will surely be installed in our seacoast forts."[60] Colonel Livingston appended a strong endorsement, praising Squier's devotion to duty. As if answering dissent in advance he stated "it cannot justly be said he did not do his full share, except in the matter of drills. His desire now should not be considered personal, but rather the ambition and zeal of an Artillery officer for his arm of service."[61] This time the request obtained the approval of the Major General Commanding and the Secretary of War.[62] So, for the seventh fall season in his seven-year career in the Army, Squier returned to the halls and laboratories of The Johns Hopkins University.

Research at Hopkins

Squier turned his attention to the effects of magnetizing currents on steel used in the new heavy guns being supplied to Artillery units. From his friend

Lieutenant Charles Parkhurst he obtained a breech-loading rifle for his experiments. These experiments were natural for a student of Rowland who had made his reputation in developing laws of magnetism. Inasmuch as they utilized the law of hysteresis presented by Charles Steinmetz at Chicago, Squier's experiments also showed the influence of the Electrical Congress upon his research. They were probably of marginal value to the Army, although tests on magnetic properties of iron were becoming as important as tests for tensile strength and elastic limit. Squier suggested magnetic tests on gun steel since the new coastal guns, maneuvered by electrical machinery, were subjected to magnetizing electric currents.[63]

Squier's method of experimentation was a modification of one of Rowland's techniques.[64] Comparing his experimental results on gun steel to published data on wrought iron and average steel castings, Squier concluded that the steel found in the new guns possessed superior magnetic properties.[65] He noted the improvement in gun steel produced by forging rather than simple casting and predicted the exclusive use of forging for marine dynamos and gun-training motors.[66]

In the course of his research on magnetization of gun steel several other ordnance-related subjects presented themselves for investigation. In a notebook of research possibilities he pondered the improvement of armor piercing projectiles by powerfully magnetizing them during manufacture, thereby producing a fibrous arrangement of molecules directed along the length of the projectile. To test the idea he magnetized the edge of a razor with the expectation that it would better preserve its cutting edge, a practice adopted some 60 years later by several razor blade manufacturers.[67] His results are unrecorded. In spring 1894, he became interested in what magnetic effects were produced in heavy cannon fired in a north-south orientation, noting they should be magnetized.[68]

His interests, however, began to drift away from metallurgical questions and toward more conventional artillery researches. By mid–April Squier had devised a method of electrical firing of cannon. Using his method he figured that he could conduct closed iron cylinder firing experiments. Further investigation along these lines was postponed because of his departure for a European tour with Professor Craig, who had been his mathematics instructor at Hopkins.

A Visit to Helmholtz in Berlin

Squier's application for a summer leave to visit Europe in 1894 met with considerable hostility, extending even to the enlisted ranks. His superiors had been keeping a detailed memorandum on the amount of service he had performed in the battery and how much time he had spent at the university. When his application left the Department of the East on its way to the Adjutant General of the Army, someone attached a copy of the memorandum, but apparently the wrong one. For on the one that went forward a clerk had penned the caustic observation that "Lt. Squier is one of those scientific officers who apparently has no relish for line duties." This slur against the character of an officer produced outrage in the Adjutant General, who promptly brought it to the attention of the Major General Commanding. He indignantly directed that the clerk be reprimanded and all subordinates be reminded of proper military courtesy.[69] Squier's request won immediate approval.[70]

Squier was untiring in his professional pursuits, even on leave. Always he engaged in making plans to advance his scientific stature, to promote scientific research in the Army, or develop personal business interests. His trip to Europe was intended to fulfill the first two mentioned purposes. In the company of Professor Craig he developed many new professional contacts among European scientists, among them the renowned dean of electrical scientists, Professor Hermann von Helmholtz. On Squier's behalf Professor Craig arranged a year's study program in the famous doyen's laboratory in Berlin.[71] When Squier applied to the Major General Commanding for an extension of nine months to his three months' summer leave of absence one can well imagine how the air might have turned blue at Army headquarters. Tasker Bliss, aide-de-camp to the Major General Commanding, endorsed Squier's letter with the recommendation:

> Letters received at this headquarters from Lieutenant Squier during the prolonged indulgence extended to him at the Johns Hopkins University stated that after the completion of his course at that place he very much wished to take the course at the Artillery School, and for that reason he was, on several occasions, excused from other details. He has now been detailed for the course at the Artillery School, and the Major General Commanding does not think that he should be longer excused from this and other details which his brother officers are obliged to take.[72]

Permission was denied.

Squier accepted the decision without appeal and traveled in Russia until July, when he made his way to Berlin.[73] In somewhat surprised tone the post adjutant at Fort McHenry asked the Adjutant General if Squier was still attached to his personnel. The answer was no; Squier would be transferred to Fort Monroe upon his return from Europe at the end of August.

3

An Electrical Laboratory

Squier's assignment to Fort Monroe allowed him to conduct research with minimal conflict from competing military duties. From 1895 to 1898, he superintended a vigorous program of research on artillery problems in the Electrical Laboratory at Fort Monroe. In artillery he investigated interior and exterior ballistics, developed electrically operated range and position finders, photographically examined gun-recoil motion, and studied the use of searchlights in coastal artillery emplacements. Squier received generous support from the commanding officer of the Artillery School, Colonel Royal T. Frank, and the Board of Ordnance and Fortification, established in 1885 as the first funding agency of research activities for the War Department. After his artillery course he was appointed Chief Instructor in the Department of Electricity and Mines in the Artillery School.

The plans laid out by Ruckman, Davis, Parkhurst and Squier for applying electrical engineering to artillery problems and for improving the quality of the artillery service were advancing. A journal was regularly published and an electrical laboratory was made available to Squier and his colleagues for their researches. These projects gave promise of what new technologies and scientific institutions could do for the Army.

Artillery Research

In one of their frequent journal discussions Lieutenant Davis suggested three divisions to an adequate definition of artillery practice: ballistic firing, target firing, and tactical firing.[1] Squier commended Davis's division of the problem and suggested that such an approach held the key to advancing the rational development of a new weapon. He argued:

> Before beginning battery target practice with any gun we should know all we can learn about its performance from the most careful and intelligent ballistic firing by

a competent board of artillery experts, and then when this information is at hand and in its most available form we are ready to undertake target firing by battery; which should be conducted as nearly under service conditions as possible. A servant is not the man to discover some new fact about jump,[2] nor is the gun the proper place to work out allowances for deviating causes. All this is the work of ballistic firing and should be done prior to any practice by enlisted men.... This is the secret of small arms success.[3]

Reflecting on the previous six seasons of artillery practice he observed that the "plain truth is, we rushed into target firing before we knew anything about ballistic firing. We attempted to teach gunners what we knew nothing about ourselves."[4] Squier thus reminded his colleagues of the proper relationship between research, development, and combat employment of new weapons.[5] His views on this subject provide worthwhile insights into the value he placed upon scientific research within the Army and how he conceived its role in assisting a combat arm in fulfilling its mission. Squier devoted his research to providing American artillerymen with the necessary ballistic data so that target and tactical firing might be improved.

In his artillery research Squier relied less upon his military colleagues than upon civilian associates. Dating from Squier's assignment to Fort Monroe, indeed, there existed no research relationship among Ruckmann, Davis, Parkhurst or Squier. Squier's principal civilian associate was Albert Cushing Crehore. Squier and Crehore, a relative of the famous surgeon, Dr. Harvey Cushing, became fast friends while still at Johns Hopkins. Upon graduation Crehore accepted an assistant professorship in physics at Dartmouth University. Before departing Hopkins, they made a gentlemen's agreement to let the other know if "anything of unusual interest in physics should come our way...."[6] Squier's inventive work of the next five or six years cannot be understood apart from Crehore's contributions and collaboration.

A New Gun Chronograph: The Polarizing Photochronograph and Exterior Ballistics

While Squier was travelling in Europe with Dr. Craig in the summer of 1894, Crehore presented a paper before a meeting of the American Institute of Electrical Engineers. He reported on light polarization experiments he conducted with two prisms, separated by a liquid-filled vessel. He described how the angle of the plane of polarization could be rotated by applying an electromagnetic field around the vessel.[7] After his return from Europe, Squier contacted Crehore to propose using his experimental principle as the basis of a ballistics device to measure the velocity of ordnance projectiles.[8] Squier recognized in Crehore's experiments the possibility of constructing a radically different chronograph or sensitive detector of small time intervals in which no "ponderable matter" had to be moved. The proposed device would provide a new freedom from mechanical inertia. This is important because, in the measurement of high velocities by optical means, i.e., with a camera, the speed by which the shutter can be opened or closed must be much faster than the speed of the object being measured. This, in turn, means that the shutter mechanism must be of extremely low mass. This problem was dealt with much later, in the 20th century. One answer was the development of very high-speed cameras, using electronic methods to rapidly open and close a shutter-like mechanism. Another approach was, in essence, to leave the shutter open and turn the light source on and off very rapidly, a difficult problem whose solution led to stroboscopic photography. Both of these approaches were pioneered by Harold Edgerton and used to photograph the progress of nuclear detonations. He also produced many remarkable images familiar to readers of popular magazines and other phenomena, like the famous photographs of droplets of milk splashing on a flat surface.[9]

Ballistic investigations attracted the attention of many physicists following the first practical experiments conducted in the 18th century by the British scientists Benjamin Robins, Count Rumford, and Charles Hutton. Although ballistic theory was formulated as early as the work of Galileo and Newton, the artillerist needed practical information about the effects of bore length, bore surface, muzzle shape, composition and amount of explosive charge, shape of projectile, and angle of firing upon projectile velocity and range. The range was simple to determine, but not the velocity. Two classes of experiments led to the determination of an instant velocity and an average velocity during an interval of time.

In the first class of experiments, fragments of a bullet were fired from a gun into a ballistic pendulum, which consisted of a large weight suspended by a rope. Applying the principle of conservation of momentum, the bullet's velocity at the instant of impact was derived from the distance it caused the pendulum to move.[10] By placing the pendulum at fixed distances from the muzzle of the gun, a graph of the changes in velocity in relation to the distance from the gun could be plotted. The second class of ballistic experiments employed electro-mechanically controlled timing devices, called chronographs.[11] Accurate determination of the time interval was essential. Chronographs recorded the precise beginning and end of an interval of time. The intervals were recorded on smoked glass by a scriber. Some scribers responded to the regular movements of a tuning fork. By knowing the frequency of vibration, i.e., the number of cycles the fork vibrated in one second, the experimenter could compute the length of time separating two events. He simply counted the number of cycles between the recorded events and divided by the fork's frequency of vibration, a constant number. In 1853, Helmholtz substituted the scriber with a sparking device to make small punctures on a moving paper record. In each case knowledge of the rate of movement of the record was used to compute the time interval. This principle was adopted years later, during the First World War by Alfred Loomis in constructing the Aberdeen Chronograph, which served as the Army standard even during the Second World War. With the exception of Helmholtz's sparking chronograph, which did not win wide acceptance, all chronographs before 1894 employed stylus recorders of time and event. Such stylus recorders were inaccurate because of the sluggish response resulting from the movement of a physical stylus possessing mass.

A device which substituted the physical stylus with a massless recording system would eliminate inaccuracies in measurement. If such a massless measuring system could be constructed, accurate measurement of extremely small time intervals would be significantly improved. In expectation of achieving such a goal, Crehore immediately agreed to Squier's proposal. (The technical details of their new measuring instrument, which they named a polarizing photochronograph, are presented in the Appendix.)

Working in his spare time at Dartmouth, Crehore began building part of their experimental apparatus, particularly the camera. In the meantime, Squier prepared the optical bench and instruments associated with the prism. On the 19th of December, Crehore shipped the completed components to Fort Monroe. He and Squier feverishly worked on the project during Christmas recess at the Artillery School. Colonel Royal T. Frank,[12] Commandant of the Artillery School, welcomed Crehore's interest in artillery problems and placed all the facilities of the school, including a special proving ground, at their disposal.[13] The winter of 1894-95 was one of the worst on record for that part of Virginia. Both Squier and Crehore suffered with the intense cold to which they were frequently exposed in tending to their field rifle, situated near the laboratories.

The Polarizing Photochronograph and Measurement of Muzzle Velocities

Squier and Crehore, with their new device, began studying the variation in projectile velocities near the muzzle of a gun. Artillerists had long suspected that projectile velocity increased *after* leaving the muzzle, but according to Squier and Crehore's literature search, the supposition was experimentally unverified. So, using an experimental setup similar to their previous ballistic measurements, the investigation proceeded in a new direction. The results they obtained confirmed their suspicion about muzzle velocities.[14]

Careful evaluation of their results showed projectile velocities did indeed increase for a space of about five feet beyond the muzzle.[15] These findings prompted Squier and Crehore to urge a revision in the method of calculating average velocities of projectiles.[16] With earlier chronographs, velocities had to be determined 100 to 200 feet from the piece and reduced back to the muzzle by means of artillery tables, based on derived formulas. The smallest measurable time interval required 100 feet of projectile travel. Squier and Crehore maintained that the customary formulas and tables required revision since they assumed maximum velocity occurred at the gun. They argued, because of their results, that all average velocities and muzzle velocities of high-powered guns were incorrectly estimated. Aside from consequences in firing accuracy, gun powder purchasing criteria counted heavily on when maximum velocity was achieved. If incorrectly determined, superior powders might be eliminated from consideration. Hence, firing accuracy and combat effectiveness depended upon accurate knowledge of ballistic phenomena which could only be obtained through scientific research.

News of their achievements soon reached The Johns Hopkins University. Squier received an invitation from Joseph Ames, acting president of the university, to present an account of their chronograph work before the Scientific Association of the University.[17] Squier hurriedly wrote Crehore about the invitation, adding he had better go as "Rowland has taken a great interest in the subject and wants me to come."[18]

Early the next month Crehore commenced patent application proceedings. He claimed credit for the invention based on his work with prisms, but assured Squier of one-half interest in the patent through codicils.[19] Patents on the polarizing chronograph were quickly secured in the United States, Canada, Great Britain, France, Belgium, Italy, Germany, Australia, Hungary, Switzerland, Spain, Russia, Norway, Sweden, Brazil, and Australasia.[20]

In May, Squier applied to the Board of Ordnance and Fortification for an appropriation of $2,225 to cover the remaining costs of developing the photochronograph. The expenditure was approved without question.[21] Subsequently, the Board adopted a strange policy. They approved granting patent protection to Squier and Crehore contingent upon reimbursement of the expenses incurred in the January trials (shells, powder, etc.), a total of $161.73. They also required a signed agreement that, if the invention were successful, the government could manufacture it without payment of royalty.[22] The Board also agreed to give money for further trials, so Squier and Crehore began preparing for experiments during August.

Squier and Crehore planned to measure projectile velocities within the bore, hitherto unobtainable without mutilation of the gun. But the pressure for time free from school and military duties once again confronted Squier. Colonel Frank accommodatingly relieved him of all duties in early July so that he could prepare for the August trials.[23]

Polarizing Photochronograph and Measurement of Interior Velocities

Preparations for the August trials intensified toward the end of July. The problem of measuring velocities within the bore of a gun without mutilating it had never been successfully overcome. Some earlier chronographs could measure interior velocities at the expense of piercing holes along the bore and inserting electrical circuits. These circuits were interrupted as the projectile emerged from the gun.[24] The literature on this subject was meager and incomplete. That which was available cited derived data and failed to specify what experiment or set of experiments provided the original data. Furthermore, few formulas were presented in describing the experiments. The formulas presented were derivative and expressed relations between travel and velocity or pressure, but not between travel and time, which were the only two parameters directly observable.[25] Squier and Crehore intended measuring the critical time and travel parameters.

Squier and Crehore employed a variation of a French Marine experimental design in which a cylindrical rod was attached to the projectile. Copper bands were placed around the rod at measured intervals. The rod, in turn, maintained contact with a single-ring electrode attached to a wooden collar secured in the bore at the front of the gun. The gun itself served as a second, circuit-completing electrode because the projectile maintained continuous electrical contact with the bore, a fact determined experimentally by Crehore and Squier.[26] As the shell emerged, contact was alternately made and broken with the copper sleeves. These events were photographically recorded by the chronograph. Squier and Crehore's experiment produced a series of discrete points on a distance-versus-time graph, rather than a continuous recording. So the copper bands were relocated, still at their previously fixed intervals, to new positions along the rod in order to provide many more data points and thereby approximate a continuous curve of projectile travel through the bore. The experiment provided direct observation of time-and-distance parameters. From such parameters it was possible to plot a curve of distance traveled by the projectile through the bore as a function of time. These important parameters were obtained for the first time without mutilating the bore of the gun.

The Board of Ordnance and Fortification allotted $2,225 for the perfection of the photochronograph, then only a laboratory contrivance. The August trials served the dual purpose of investigating interior ballistics and completing a device suitable for commercial production. Even as the trials proceeded, commercial-quality instruments were being readied by the celebrated optician and instrument-maker, John A. Brashear of Allegheny, Pennsylvania. The order to build the special instrument used in measuring angles of light traced on the photographic plate was placed with the firm of Warner and Swasey, Cleveland, Ohio, industrialists.[27] (Figure A-1 in the Appendix is a photograph of the manufactured device.)

Squier and Crehore briefly turned their attention to the experimental study of alternating current.[28] Since the photochronograph could record a varying current, it also offered splendid opportunities for studying the influence of capacitive and inductive reactance upon the magnitudes and phases of voltages and currents in electrical circuits. These were relations well developed in complex variable theory, but difficult to demonstrate experimentally, especially with high-frequency currents. A record of that period of time immediately following the connection or "make" of an electrical circuit was of special interest, too. In this very brief time, transient currents flowed. The total flow of current was not regular but followed a decaying exponential curve until the regular pattern predominated (usually a sine or constant-value wave). The transient value then became negligible in magnitude and effect. Because the photochronograph was ideally suited to recording phenomena during short time intervals, it was usefully employed in this investigation. In the days before oscilloscopes that could display the wave forms of currents present in an electrical circuit, a device which could make visual records of signal wave forms was a valuable instrument in verifying theoretical findings. The press of ordinary duties, however, prevented them from pursuing further this line of investigation until a year later.[29]

Squier and Crehore claimed that their instrument opened a new range of experiments by making possible the "reliable measurement of minute intervals of time."[30] The Board of Ordnance and Fortification agreed, recommending that the polarizing photochronograph be adopted as "the type for instruments of this class, for the use of the United States Military Service."[31] Six months later the Chief of Ordnance recommended purchasing the photochronograph for the Army's ordnance test facility at Sandy Hook, in northern New Jersey. General Daniel W. Flagler told the Secretary of War that "it is reported that the instrument [is] sensitive and accurate and fitted for the solution of numerous ballistic questions that arise." Crehore accepted the contract and received payment of $4,250.[32]

Despite the ingenious design and accuracy of the polarizing photochronograph, it remained a laboratory instrument. The previously mentioned

Aberdeen (Loomis) chronograph was admirably suited to field use because of its portability and ease of manufacture. Hundreds of them were built in World War I and, as a result of electronic improvements added in World War II, it became standard U.S. Army and Navy equipment.[33]

An Alternating Current Horizontal Base Range Finder System

During the fall of 1896 Crehore and Squier began outlining requirements for an improved range finder to use in coastal fortifications. Stimulated by a crisis between Great Britain and the United States over the Venezuelan boundary dispute, national interest in coastal defenses was at a high level. The Board of Ordnance and Fortification had plentiful funds available for research on improving coastal defenses. Crehore and Squier's fascination with alternating current led to planning a horizontal base range finder based on a Wheatstone bridge with a movable iron core transformer, which provided electrical resistance and inductive reactance in each leg of the bridge.[34]

Professor Joseph J. Thomson accomplished the first mathematical analysis of a Wheatstone bridge with resistance and reactance.[35] Since his analysis involved the extremely time-consuming simultaneous solution of differential equations, Squier and Crehore provided a graphical solution. Satisfied with the utility of their graphical analysis of alternating currents in a Wheatstone bridge and sure that an entire range-finding system could be constructed on such a principle of operation, Crehore and Squier applied to the Board for financial support in perfecting the system.[36]

The Board approved the project, even with Squier's proviso that he and Crehore retain patent protection. Squier and Crehore agreed to permit manufacture of the range finder by the United States Government as long as it was constructed in a government shop or arsenal.[37]

A horizontal base range finding system of the type that Squier and Crehore proposed was built around two observation posts separated by a long base line. Maximum accuracy was obtained when the posts were situated a mile or more apart. Effective operation of this system clearly depended upon some reliable, rapid means of communication between posts. Locating and fixing upon a target worked in the following manner. When an observer spotted a target, a man in each station turned a sighting telescope on it. Each observer noted the azimuth of the ship from his end of the base line. Lines were then drawn on a platting table according to the reported azimuths. The lines originated at points representing the spotters stationed at each end of the base line. The platted lines intersected at the location of the enemy ship.

Knowing the position and range of the target in relation to coastal defense batteries, accurate firing data could then be relayed to the gunners.

The chief difficulty of a horizontal base system lay in obtaining parallelism between the sighting telescope at the distant end of the base line and a platting arm at the home end, which represented the line of sight of the distant telescope. Firing data were calculated and orders given at the home end. A small movement of the telescope at the distant end of the base line was generally difficult to "parallel" or duplicate on the platting table at the home end. By utilizing the balancing property of a Wheatstone bridge it was supposed that a small movement of the spotting telescope, fixed to the movable core of a transformer at the distant station could be paralleled at the home station by simply rebalancing the bridge. Hence, one of the legs of a bridge was located at the distant end of the base line and the other three were located at the home end. An artillery officer at the platting table could maintain the bridge in balance by moving the platting arm, until a null current was indicated on the galvanometer connected across the bridge legs. Parallelism was thus rapidly established.[38]

If a target vessel could travel more quickly than range and azimuth data could be calculated and passed to the gunners, coastal defenses possessed little practical value. Thus, the organization of the platting table and the quantity of mensural implements required to obtain accurate firing information also weighed heavily in selecting a suitable range-finding system. The chief novelty of the Squier-Crehore range finder was its substitution of shadow lines for mechanical arms previously used on the platting table.[39] By means of a lamp, condensing and focusing lenses, and straight edge, a shadow was cast upon the platting board. The intersection of two such shadow lines, each representing a sighting from one end of the base line, provided an instantaneous location of the target. Squier and Crehore's old fascination with massless measuring systems thus manifested itself again.[40]

The use of alternating current between base line stations offered one decided advantage over systems using direct current. For technical reasons, direct current systems required ready access to the wire connecting home and distant stations. This requirement, though, made their intercommunication vulnerable to enemy shell fire. Alternating current systems, on the other hand, could operate without access to any wires once the base line stations were connected. Thus, once the connecting wires were laid they could be heavily encased to make them invulnerable to enemy action. Such protection, afforded by an alternating current range-finding system, might make coastal defenses more reliable in battle. In accepting Squier's proposal, the Board allotted $3,300 for the installation of an experimental system at Fort Monroe.[41] Shortly thereafter Squier commenced work on the project with representatives from three private contractors.[42]

Searchlights for Coast Defense

Electricity brought a revolution in the use of artillery. It led to a marked change in the accuracy and speed of gun-laying. Searchlights and range and position finders served the same purpose of assisting artillerymen to locate and fire on their targets more accurately. Thus, research on searchlights became an important part of the program to improve coastal defenses. Squier acquired and installed at Fort Monroe one of the most powerful searchlights available, having an estimated intensity of 194,000,000 candle power. In a report to the Board he stressed the leading role France, England and Germany held in the development of "projectors" for coast defense and urged support for American research on their operation and maintenance.[43]

The tests that Squier suggested revealed his grasp of operational demands. Proposed tests included determining the most effective height above sea level for maximum light coverage; the extent to which the presence of a light betrays its position to the enemy when mounting height is unknown; feasibility of using an automatic, oscillating governor, such as is used on a Gatling gun, to cause the beam to continuously sweep a given area; practicability of mounting a searchlight on a flatcar and operating the unit at intervals from continually changing positions behind a line of defense positions. He appended to his report a large bibliography drawn from international sources on the military use of searchlights.[44]

There is little record of searchlight investigations by Squier during the succeeding year, 1897. But under the threat of war with Spain, in 1898, research was pushed forward by the Board of Ordnance and Fortification. The Board placed Squier in charge of searchlight experiments for the Army, supporting his work with an allotment and shipping another searchlight from Sandy Hook to Fort Monroe for his use.[45]

Radio Research

The Board went beyond supporting Squier's recommendations for a program of scientific research and development in the Artillery Corps by placing him in charge of the electrical plants at Fort Monroe and of all the experimental work conducted there for the Board.[46]

One of the projects in which the Board showed special interest was the feasibility of radio for military purposes. In early February the Board asked Squier to submit cost estimates for procuring apparatus to fully test the Marconi system. By 1898, when the Board asked Squier to investigate radio, Marconi had already established a station for transmitting between the shore and a lightship.[47] Squier himself had used a radio he built in 1897 to ring bells, light lamps, fire cannon, detonate mines, and start machinery by radio control.[48] Consequently, he was already familiar with radio and its operation.

Squier replied that he could build a working transmitter and receiver for the Board for $1,050. On the basis of his own previous radio experiments, he recommended that "the military and naval possibilities of this system should be thoroughly investigated."[49] Military duties connected with the Spanish-American War caused a temporary suspension of further radio research.

4
Soldier-Entrepreneur

Squier's personal research interests exposed another dimension of the conflict between military and scientific demands on Squier's time and loyalties. Working on official research in his laboratory at Fort Monroe, he invented devices which possessed commercial promise as well as military value. A high-speed telegraphic transmitter, which the inventors called the Synchronograph, was the most important of his inventions. Had he chosen to resign from the Army and exploit it commercially, no conflict of interest need have arisen. But he tried to balance the two interests and maintain his loyalty to the Army. To a situation where there were already frictions enough between his dual roles of scientist and soldier, Squier added the third role of inventor-entrepreneur.

This attempt was initially unobjectionable to those Army officers who recognized the need for obtaining scientific talent in studying military problems. If the Synchronograph had a place in the commercial market, the Army could also use its high-speed transmission capability.

The Synchronograph

While Crehore and Squier were using their polarizing photochronograph in ballistics research in 1896, it occurred to Crehore that the device might also be applied to telegraphy. By modifying the photochronograph, it could receive telegraph messages at a much faster rate than existing equipment. To confirm the possibility of using the photochronograph as a high-speed telegraphic receiver, they conducted a series of trials at Fort Monroe when Crehore could take leave from Cornell.

Squier confided their discovery to the Board of Ordnance and Fortification, asking them not to divulge it until he and Crehore could prepare a formal presentation.[1] He stated that the limits upon faster telegraphic and submarine cable speeds were imposed by the electrical characteristics of the line and the nature of the sending and receiving instruments employed. His instrument, as presently configured for government service as a gun chronograph, was capable of overcoming the limits and receiving high-speed messages in the laboratory. Squier enclosed copies of messages received by the photochronograph. The messages were received at the rate of 1,200 words per minute, which was about five times faster than reception with any equipment in use. He asserted a higher speed could be attained if they used a faster sending apparatus. Requiring a wider discussion of the principles involved in their proposed system, the two inventors approached established engineers.

They visited Louis Duncan, formerly a professor at Hopkins; Michael Pupin of Columbia University; and Nichols and Frederick Bedell of Cornell University.[2] They all concurred in the feasibility of the project and encouraged further development.[3] Soon afterward Pupin sent a high-frequency alternator of his own design to them.[4] (The Appendix presents a technical description of the new telegraphic transmitter, based on the Pupin alternator, which the inventors christened the Synchronograph, somewhat confusingly, because they were using their other invention the [polarizing] photochronograph, as a receiver.)

Squier predicted great commercial possibilities for the Synchronograph in an address before the annual meeting of the American Institute of Electrical Engineers. He emphasized the benefit of using alternating current power. Since the use of alternating current in the Synchronograph resulted in very low power losses, it should be possible to communicate simultaneously with a network of receiver stations from one transmitter station. Having successfully demonstrated this concept in his Electrical Laboratory at Fort Monroe, Squier concluded that the Synchronograph admirably suited the large news reporting services.[5] After carefully analyzing American message traffic volume and revenue statistics, he declared that consumers would be willing to pay proportionately more for telegraphic messages than for mail or telephone messages. By increasing line and instrument capacities, business enterprises could expand message volume and increase earnings. He suggested, for example, that with transmission speeds of 3,000 words per minute two lines of information could transmit in one day the entire daily volume of 40,000 letters passing between Chicago and New York. He thought the general public, and especially the business community, would favor a telegraphic correspondence system over the mails.

The Associated Press could expand their news reporting services by publishing an entire newspaper simultaneously in several cities. Squier calculated that in an average daily paper containing 12 pages and eight columns per page there were 185,000 words. At 3,000 words per minute an entire newspaper could be transmitted in just over an hour's time.

Squier concluded his paper with the remark that a rapid telegraphic correspondence system operated by a private concern under government contract "could hardly fail to prove of benefit to the people of the United States."[6] He recommended by implication a union of the great telegraphic companies and the Federal postal system along the lines of the British model, but with the American element of free enterprise. The entrepreneurial spirit of the junior officer must have impressed his audience because he obtained generous assistance from the British in proving his system.

The British Post Office Laboratory

Squier's intentions about the commercial development of the Synchronograph are revealed in a letter to the Board of Ordnance and Fortification describing his Synchronograph experiments. He said,

> In the interest of advancing the intelligence transmission service of the world, and especially of the United States, this detailed account of the experiments thus far is submitted for reference to the Postmaster General or such other disposition as the Board may think proper.

In a singular declaration to the Board, considering the recent crisis with Great Britain over Venezuela, he stated, "The system will be presented to the British Postmaster General through Mr. William H. Preece, Chief Electrician of the British Postal Service."[7] The Board's response to this unusual statement is unknown, but Squier pursued the issue.

In early May, Squier visited the Postmaster General of the United States to discuss his project for establishing a telegraphic correspondence system.[8] What else they discussed is unrecorded. Squier probably informed the Postmaster General of William Preece's invitation to conduct Synchronograph trials in London.

A few weeks after his visit to Washington, Squier applied for permission to spend the summer holidays working on experiments in England. While it was not unusual for Army officers to travel in Europe, Squier's visit for promotion of a personal business venture was extraordinary. He notified the Adjutant General that William Preece was placing the British Postal telegraphic system at his disposal for testing the Synchronograph. The Adjutant General agreed to Squier's request as long as the Army incurred no obligations.[9]

During that hectic summer of 1897, Squier and Crehore were invited by Preece to relax for an evening at his home in Wimbledon. There Preece introduced them to the leaders of British electrical science, including Sir Oliver Lodge, Sylvanus P. Thompson, Lord Kelvin, and Lord Rayleigh. Squier's early scientific contacts with British electrical scientists, established first at the 1893 Congress and strengthened at the Postal Laboratory in 1897, portended his future role as an inventor and bridge figure between the United States Army and British science.

In late summer, Squier rendered a preliminary report of his Synchronograph experiments on the British Postal lines. The Adjutant General's response to the elegant lavender stationery of the Grand Hotel on which it was written has not been preserved for posterity.[10] The report recounted how signals generated by their device, installed in the main laboratory in London, were transmitted on a great circular link via York, Glasgow, Aberdeen, and returned to London, a total of 1,100 miles. The Synchronograph-Wheatstone receiver pair attained working speeds of 4,000 words per minute. Comparison of these speeds to those of the Wheatstone transmitter–Wheatstone receiver pair, which was in almost universal use, quickly reveals why the British were so interested in Crehore and Squier's invention. If the Synchronograph could work in excess of 3,000 words per minute, the gain in installed capacity would be a minimum of 12-fold, because the Wheatstone Automatic Transmitter worked at a speed of 250 words per minute under the most favorable line conditions. If the Synchronograph were compatible with short submarine cable operations, the potential savings realizable from increased transmission capacity would be enormous.[11] Over long submarine cables the speed of Crehore and Squier's system, as with any other, would be drastically reduced, for technical reasons. Even so, operating speed was expected to be two or three times faster than contemporary systems. This expectation required further experimentation to verify.

Squier's entry upon the British telegraphy scene illustrated the nature of the dilemma which confronted both Squier and the Army, and of the meaning of Squier's career for the Army. From the time of its mid-century formation, Cyrus Field's great Anglo-American Cable venture had come increasingly under the control of British entrepreneurs. Only one American cable company stood in the lists to compete against a virtual British monopoly of international submarine cables. This one American firm, the Commercial Cable Company, had only two cable landings on North America, whereas the British-dominated Anglo-American had four landings. The vigorous competition to maintain national control of cables was strongly colored by patriotic emotions. The British would have been very interested in obtaining the use of an invention which could double or treble their submarine cable capacity overnight without so much as dispatching one cable laying vessel from port.

As a man of the Victorian era, Squier could well believe that international scientific and business cooperation led to the promised land of peace and prosperity. To some, Squier's aiding the British might seem to place him in an unethical position. Squier's attempts, despite many difficulties, to keep his dual identity as scientist and soldier and his double loyalty to science and the Army were the most remarkable features of his career. The Army was fully informed about the London visit. The easy way out of his ethical dilemmas was to resign from the Army. There is, however, no record of his ever considering that possibility. Indeed, he felt a deep sense of obligation to the Army for his West Point education. It was in his very efforts to maintain his multiple roles and loyalties that he achieved his significance in the Army. They constituted the source of his institutional creativity, which transcended in importance his technical inventiveness.

Squier Joins the Reserve Signal Corps

The American firm, Commercial Cable, and the U.S. Army Signal Corps soon took interest in the Synchronograph. In London, Squier's commercial concerns affected Army work. With rising American interest in the Synchronograph, military considerations shaped further development.

General Adolphus W. Greely facilitated access to the Commercial Cable Company. When Squier returned to Fort Monroe in the fall of 1897, he resumed his artillery researches and his Synchronograph experiments. In the opening months of the New Year, 1898, the country entered a new crisis, culminating in the Spanish-American War. With a war to fight, the Army could ill afford the luxury of laboratories and schools continuing in operation. So Fort Monroe shifted to a wartime footing, its schools and laboratories disbanded. Teachers and students were sent to active field units, but not Squier. General Greely, Chief Signal Officer of the Army, was desperately short of officers because of the war. Knowing of Squier and his work, he asked the Adjutant General to reassign Squier to the Signal Corps Volunteers.[12]

Greely once described himself as "a native American, a man of the masses— non-collegiate in education, without private income, favored by no political influence—whose ancestors for nine generations labored with their hands in New England."[13] He entered the Civil War as a private. After receiving wounds in the Battle of Antietam in 1862, he was promoted to lieutenant. An unusually versatile man, he took charge of the Weather Bureau in 1868, then a part of the Signal Corps, and made it a smoothly functioning unit. He led the Lady Franklin Bay Expedition in 1881 to northern Greenland. Stranded for several years, the expedition was finally rescued in 1884 after scandalous delays. Most perished; the others were close to death. Three years later Greely was promoted to brigadier general and placed in charge of the Signal Corps. When the Weather Bureau was removed to the Department of Agriculture in 1890, the

Signal Corps barely survived the loss of its most important mission. Greely assiduously began new missions in balloons and photography to restore life to the Signal Corps. For this reason he is still remembered as the modern father of the Signal Corps.[14]

Squier was Greely's type of man: energetic, intelligent, somewhat of a loner, and gifted in electricity. Greely hand-picked other members of his command, too, placing great emphasis on their ability to conduct scientific and electrical researches. Lieutenant Colonel James Allen, Captain Samuel Reber, and Captain Edgar Russel were men of the desired stamp. Reber wrote the first Manual of Photography for the Signal Corps' course instituted by Greely at Fort Riley, Kansas. Allen developed the buzzerphone the year before the war. Intended to replace the Morse telegraph sounder in the field, it reflected Greely's fascination with the telephone by adapting the telephone technique to Morse code transmissions.[15] Russel displayed special gifts in practical laboratory research and administration of experimental investigations.[16] Squier was among like-minded officers in the Signal Corps and the quality of his work with them convinced Greely that his services were required in his own office.

Squier and Crehore's anomalous activities received official sanction from Greely.[17] He was convinced that Squier and Crehore's discoveries and inventions could be of "great value to the commercial world, and incidentally to the Army."[18] Since the Army had no cables of its own, advances in signaling techniques had to be implemented by commercial telegraph companies. The Army used commercial lines and cables on a contract basis. Greely felt that the most promising of the Squier and Crehore inventions was a new system of submarine cable transmission, utilizing the Synchronograph as a transmitter and a Siphon recorder as receiver.[19]

Instead of pursuing the photochronograph receiver further, Squier and Crehore chose at this time to emphasize commercial exploitation of the Synchronograph and the use of alternating current on submarine cables. Alternating current appeared to offer advantages to the direct current used in land-line telegraphy, which depended on the use of mechanical repeaters at fixed intervals along transmission lines to amplify the signal. Moreover, while repeaters could not be used on submarine cables, for obvious reasons, and the great sensitivity of Lord Kelvin's mirror galvanometer had allowed direct current to be used for submarine cables, the use of alternating current (AC) for this medium also showed promise.

Military advantages lay more in improvements in submarine cables than in land-line telegraphy. The possibility of attaining speeds of 3,000 to 4,000 words per minute on land lines and 60 words per minute on submarine cables was impressive. Greely had no research facilities suitable for such research. For fiscal year 1898 the Signal Corps budget was a paltry $3,000. By allowing

Squier an unusual degree of freedom, the Signal Corps might benefit from his commercial activity. Squier, in turn, subordinated promising scientific and commercial interests associated with land-line telegraphy to the Army's need of improved signaling devices on submarine cables.

Experiments on Commercial Cable Company's Lines

For obvious reasons the Commercial Cable Company was also interested in their research. In New York, Squier and Crehore carried out experiments on the Commercial Cable Company's Coney Island Cable, which connected New York and Canso, Nova Scotia. The company afforded them every facility in working the 826-mile submarine cable. Several months later they worked the cable with recently acquired commercial quality transmitters in place of their prototype experimental devices. Experiments were also conducted on land lines.[20] The potential in Squier and Crehore's research interested both the Commercial Cable Company and Greely, for different reasons. Economies of high-speed transmission were highly attractive to the Commercial Cable Company, which had nudged its way into international traffic by a vigorous program of rate-cutting. When the first Atlantic cable began operation in 1869 the cost was five dollars per word. The Commercial Cable Company began operations 14 years later and their price-cutting practices reduced the cost per word to less than 50 cents.[21] Greely was surely aware of the personal importance these commercial tests held for his two "brilliant young physicists."

There were two reasons why high-speed telegraphy capability would appeal to Greely. The first was military. More rapid transmission of instructions, orders, and information to troops operating in the field would substantially contribute to greater combat-effectiveness and maneuverability. The second reason was financial. Before the war began, the entire Signal Corps budget was only $3,000. In contrast, communications costs in just one month of the war were $75,000. Telegraph charges for one month alone after the war represented a 300 percent increase in the pre-war annual budget of the Signal Corps. Greely considered support given to high-speed transmission systems would save money and improve military performance. He appeared perfectly willing to allow them to advance themselves in the business world if Signal Corps interests were incidentally promoted.

When Squier was assigned as Chief Signal Officer of the First Army Corps at Lexington, Kentucky, Greely requested the Adjutant General to replace Squier with Reber.[22] Successful high-speed telegraphy trials over the Coney Island Cable clearly influenced Greely's request. Squier relieved Reber of his duties, mustering out volunteer signal companies in Washington. The substantive reason for Greely's accommodating behavior was revealed in his extraordinary request of the Commissioner of Patents to mark Crehore and Squier's patent applications for "Improvements in Telegraphy," covering their use of alternating currents for telegraphic transmissions, "special" according to Patent Office rules. The "special" designation expedited registration of the patents. Greely stated that this matter was of particular interest to the Signal Corps, as put forth in his last report to the Secretary of War, in 1898. Greely planned on the Signal Corps laying a telegraphic cable system throughout America's newly acquired Philippine possessions.[23] Squier saw their installation as the "quickest means of pacifying and civilizing the Philippine Archipelago."[24] Greely observed more cogently that "a system of public order can be maintained or restored with a military force which would be totally inadequate without the advantages of instant communication."[25]

Crehore and Squier Form Their Own Company

The assignment to Washington, and away from field duty, permitted Squier and Crehore to continue their work on high-speed telegraph transmitters. They were also negotiating with Warner and Swasey, the Cleveland industrialists, for financial backing in establishing their own business firm. They kept Greely fully informed of their progress in patent proceedings, experimental work, and business negotiations. In December 1898, Greely conferred with Squier and Crehore when they returned from consultations in Cleveland. They discussed the need for additional tests in which tape perforators would be attached directly to the alternating current transmitter so as to obtain the full benefits of automatic transmission; namely, accuracy, uniformity, and freedom from operator idiosyncrasies. Several months earlier Greely had personally requested the president of Western Union Telegraph Company to permit Squier and Crehore to experiment over the company's telegraph link between Washington and Jacksonville.[26] Greely revealed once again in his action the motive behind his intense interest in supporting a high-speed telegraphic system. He confided to Squier and Crehore his decision to purchase and equip a cable ship which could steam anywhere in the world.[27] Greely's resolve to assume the responsibility of providing worldwide communications for the government represents the significant impact of the acquisition of foreign territories on the Signal Corps. Squier and Crehore's equipment was perhaps a way to realizing Greely's vision of a dramatically increased role for the Signal Corps within the Army through control of electrical communications for the government. Squier's personal and professional prospects had never seemed so favorable.

When Christmas Day arrived, Squier found himself ensconced in the War Department building as assistant to the Chief Signal Officer,[28] although he was still a regular officer in the Artillery Corps. General Greely and the Signal Corps provided the leadership and environment which so thoroughly stimulated Squier's imagination and research. Thus it was that he tried to obtain an appointment in the regular Signal Corps.

Squier Joins the Regular Signal Corps

Attempting to identify and locate all artillery officers after the war, the Commander of the Third Artillery sent a circular to all his units asking for a roster of officers on station. The Commanding Officer of B Battery stated that Squier was on detached duty in Washington under instructions of the Chief Signal Officer and added, "He should be ordered to duty with his battery, where his services are much needed. Has been absent from his battery and regiment for years—never joined, Battery 'B.'"[29] By coincidence, Squier applied the following day for an appointment in the Signal Corps, noting that the recent promotion of Allen had created a vacancy in the regular cadre. When the application reached Greely's desk he appended a note in which he impatiently observed that he had recommended Squier's transfer four or five days previously: "Has the nomination been made? If not, what is the obstacle in the way of its being made?"[30] A week later Squier accepted a regular commission as a first lieutenant in the Signal Corps.[31]

The Crehore-Squier Intelligence Transmission Company

Meanwhile, Squier had made several trips to Cleveland and New York in connection with his telegraph transmitter. Both he and Crehore were busily engaged in soliciting financial backing from Warner and Swasey in marketing their telegraphic devices.[32] Squier and Crehore finally made a formal proposal on 9 January 1899. After a day's thought, Warner and Swasey offered an arrangement whereby they would advance working capital to a new firm, under Squier and Crehore's leadership, and pay salaries and other expenses of the company. In addition they would give their engineering and manufacturing services for one year without remuneration. For these considerations Warner and Swasey would receive one-fourth of the company stock.[33] It was necessary, however, to possess approved and certified patents before incorporation proceedings could start. As previously mentioned, Greely had intervened with the Commissioner of Patents to ask that the applications submitted by Squier and Crehore be marked "special."³⁴ When Squier and Crehore arrived in Cleveland to negotiate the agreement with Warner and Swasey they carried with them a letter of permission from the vice-president of the Commercial Cable Company, George Ward, to conduct future trials of the new cable transmitters over his company's cables.³⁵ On 9 February, Cleveland newspapers carried an announcement of the incorporation of the Crehore-Squier Intelligence Transmission Company. Capital stock was issued with a face value of one million dollars. No mention was made of Warner and Swasey's involvement. Among the incorporators was Harry A. Garfield, eldest son of President Garfield.³⁶

Greely naïvely, perhaps unreasonably, wanted his research and development project moving along a broad front. No sooner had Squier returned from Cleveland, still engrossed in cables, than Greely assigned him to work with Allen in renewing research on radio for the Army. Confident that radio would replace military cables connecting harbor installations, Greely thought radio would also put ships at sea, lighthouses, and Signal Corps stations in touch with one another.³⁷ Before the war, James Allen started the project of developing an Army radio for the Signal Corps. With the demands of the war subsiding, Greely turned once more to the development of radio for military purposes. Pressing rapidly forward, Squier and Allen began successful operation of their radio system between Fire Island and Fire Island Lightship, a distance of approximately 12 miles, by the end of April 1899.³⁸ Regular operation of the Signal Corps system antedated the arrival of Marconi and his radio in the United States.³⁹ Soon Governor's Island and Fort Hamilton, at the entrance to New York Bay, signaled each other by radio. A few months later, a similar radio system was installed in San Francisco Harbor, linking Fort Mason and Alcatraz Island.⁴⁰ Army radio was born.

To further the radio research experience in the Signal Corps, Greely made arrangements for one of his officers to study under Marconi in London. Since Allen was ordered to the Philippines to supervise the laying of submarine cables, Squier went to London.⁴¹ Allen departed on 1 May 1899 and Squier embarked about two weeks later.⁴² On Allen's first cable-laying trip out of Manila Bay, his ship, the U.S.S. *Hooker*, steamed onto Corregidor Reef and sank. When Squier returned in the fall of 1899, Greely told him to take charge of a relief expedition and sail for Manila. Radio investigations were transferred to Samuel Reber. Henceforth, Squier divided his time between his new corporation and the task of preparing a new mission to the Philippines.⁴³ Reaction to news of Squier's reassignment to the Philippines was not long in coming.⁴⁴ Mr. Herbert Satterlee of Ward, Hayden, and Satterlee, attorneys-at-law in New York City, complained to Senator Thomas C. Piatt (New York) of Squier's departure from the States.⁴⁵ He referred to Squier as "the best known electrician in the United States Army and ... one of the few submarine cable experts in this country." Implying that only Colonel Allen's request for Squier's detail to the Philippines and "official etiquette" prevented Greely keeping him, Satterlee concluded with the anxious wish that Squier "be retained where he is for the good of the Service and in the interests of American progress and development in the electric world." Senator Piatt forwarded Satterlee's letter to Secretary of War Elihu Root with an expression of support for Satterlee's wishes.[46] Mr. Harry Garfield wrote a parallel letter to the Adjutant General.[47] Both inquiries eventually came to Greely, who replied that Squier had applied for foreign service several times and that he was the only one who could perform this "most important work."[48] Squier apparently did not rely solely on friends and business associates to delay his departure for the Philippines. In what may contain the seeds of subsequent strained relations between Squier on the one hand, and Greely and other Signal Corps officers on the other, new events worked to the end of postponing Squier's sailing.

Squier received an invitation from Captain C.A. Stockton, President of the Naval War College, to address one of the seminars at the Naval War College in the same week Senator Piatt's letter arrived.[49] There were indications in the letter that Squier had arranged the speaking engagement. Squier told Stockton he had prepared a paper for the Paris Electrical Congress in the fall, to which he was an elected representative of the United States. The paper dealt with the influence of submarine cables upon diplomacy and sea power. Intrigued, Stockton wished to have Squier present it in person.

Squier called the Adjutant General's attention to the conflict between his transfer orders and the invitation, which had priority according to Secretary Root's acceptance of Secretary Allen's January request to allow Army officers to address college classes.[50] Squier also said he favored the "high professional advantages which such a course would offer."[51] Since the lecture schedule placed his presentation on 12 July, he gained several months' reprieve from his assignment to the Philippines. The invitation also led to a change in his original orders, which directed him to leave from California.[52] Greely, accordingly, wrote a request to the Adjutant General to retain Squier in the Washington office for the purpose of supervising construction of the cable.[53] Although Squier had a great deal of work to perform for Greely, he gained several fine opportunities to advance himself professionally and to promote his company.

In May, Squier presented a paper on the performance of his new alternating current telegraphic transmitter before the American Institute of Electrical Engineers at Philadelphia.[54] His paper included the data he collected from his experiment on the trans–Atlantic cables belonging to the Commercial Cable Company during the previous year.[55] The presentation constituted an important event for Squier and Crehore because it was the first public discussion of their new system of telegraphic transmission in actual operation. Their paper received favorable attention. The noted General Electric engineer and president of the Institute in the following year, Charles Steinmetz, praised the paper as the first scientific investigation of the problem of generating power for telegraphy. He considered it one of the most important contributions to the Institute.[56] Michael Pupin declared their work to be the "first scientific measurement of the alternating current elements which come into consideration when an alternating electromotive force is applied to a cable."[57] Arthur Kennelly, whose own professional forte included development of cable transmission theory, stated,[58]

> We have had a theory of alternating current transmission as applied to submarine cables ever since Lord Kelvin's classical work upon that subject, prior to 1865; but I venture to say that we have had no practice, or practical knowledge, with a check upon the reliability of that theory, in published form until today.[59]

Other commentators were all recognizable from previous scientific contacts: P.B. Delaney (of chemical receiver fame), Frederick Bedell, and R.B. Owens.

Squier and Crehore's system differed in several important respects from those in use. The usual cable system employed a battery as its source of power. Intelligence in such a system was transmitted by use of a dot-dash code. Connecting the positive pole of the battery to the cable and the negative pole to earth ground produced a signal designated as a dot. Reversing the connections represented a dash. Letters containing alternate dots and dashes were called *cross letters*, and those composed of successive dots or dashes (the most difficult to transmit), were called *successive letters*.[60] Of these direct current reversing current transmitters, the most popular was the Cuttriss Automatic. Cuttriss held an important engineering post with the Commercial Cable Company and worked with Squier many times during the long developmental phase of Squier's alternating current transmitter. Squier's system was automatic, too, in that a pre-punched tape was fed to the transmitter. A device attached to the transmitter sensed the code through steel brushes sliding against the tape which was drawn mechanically across a rotating wheel. Whenever a hole appeared in the tape, the brush made electrical contact with the platinum wheel beneath it and transmitted a positive half cycle or a negative half cycle, depending on the code. Between each half cycle the cable connection was earth grounded in order to eliminate any residual electrical charges accumulated as a result of the enormous capacity inherent in long cables. Thus, the Cuttriss transmitter delivered a square wave to the cable and the Crehore-Squier system presented a modified sine wave. Herein lay the significant advantage of the latter system.

It had long been known that the greatest power could be delivered by using a simple sine wave, since over long lines only the fundamental harmonic of a more complicated wave emerged. Harmonics differ from the fundamental harmonic by only a simple multiple. All repetitive waves, regardless of shape, can be resolved mathematically into an infinite series of sine and cosine harmonics of decreasing amplitude and increasing frequency. Since only the fundamental harmonic survived at the distant end, all other harmonics represented energy dissipated by the cable, and subsequently lost to the transmission of information.[61] Hence, a sine wave input at the transmitting end of a cable resulted in more energy being received at the distant end than a square wave input could produce at the same working voltages. An even faster working speed could be achieved by tightening the suspensions and decreasing the natural period of the recorder.[62]

There were several other possible advantages to the adoption of alternating current as the source of telegraphic power. Transoceanic cable systems employed very large and expensive shunt capacitors to correct the distorting effects of cables.[63] By working with alternating current, it was theoretically possible to substitute a pair of relatively cheap transformers. Squier tried it and reported in favor of the substitution, although the transformer he used matched imperfectly the line conditions.[64] Battery-reversing circuits also produced undesirably large sparking due to interruption of a high-voltage circuit. By grounding his system between every half cycle when the voltage reached zero, Squier eliminated sparking and avoided excessive wear of electrical contacts.[65]

Another advantage of the alternating system concerned the voltage ratings required to achieve comparable effectiveness of direct and alternating currents. All cables spanning the oceans were carefully and conservatively restricted from operating at high voltages. The limit in effect in 1900 was 50 volts. High voltages led to excessive power dissipation, or heating, in the cable. The heating produced a breakdown of the gutta-percha insulation, an occurrence to be avoided at all costs. Although there existed no direct relation between attainable cabling speed and applied voltage, it was true that obtaining incremental gains in speed required correspondingly higher voltages. The sine wave displayed one peculiar property. Its amplitude was not comparable to the same nominal value of direct current in its ability to cause heating in a conductor of electricity. Its effective value was about 30 percent lower than its nominal amplitude. For example, house current rated at 117 volts has an amplitude of about 166 volts. Because the cable companies measured effective values on the cables, an alternating system could actually exceed the 50-volt limit with a consequent improvement in signaling speed.[66] Squier would only say that his system would "materially" improve cable signaling speed rather than state a definite percentage in improvement, because the operator's skill at reading telegraph tapes played such an instrumental part in speed calculations. Kennelly supported Squier's conservative approach and suggested that a 10 percent increase, while seemingly small, would represent a material advance from the standpoint of finance and traffic capacity. The entire work fascinated him because "not till to-day had [we] available measurements for alternating current transmitters."[67] Satterlee's letter of the subsequent week to the Secretary of War contained more than shallow self-interest or idle claims.

It is necessary to add here that Squier and Crehore were incorrect; AC transmission is not better than DC for submarine cable telegraphy. This would not become apparent until many years later, when better analytical tools had been developed. In 1924, the great electrical engineer Harry Nyquist of Bell Laboratories definitively demonstrated that no improvement is gained by the use of AC, for two reasons: First, Crehore and Squier advocated use of a one-half AC sine wave to represent a unit of transmission (usually a "dot" in a dot-dash telegraphy system). Nyquist definitively demonstrated that the voltage of the received signal, which is the key parameter for submarine cable transmission, is directly proportional to the power transmitted. The power of a one-half sine wave is only 60 percent of the power of a DC square wave of equal magnitude and duration. Second, the use of AC as a "carrier" frequency (like AM radio) on a submarine cable actually reduced the voltage that could be used for the signal component of the modulated AC. To maximize signal reception, the voltage applied at the sending end is the maximum voltage allowed on the cable. This maximum voltage has to be less than the breakdown voltage of the cable insulation. The signal component must be smaller than the AC carrier and, thus, the amount of voltage allocated to it cannot be much more than a third that of the carrier.[68]

Crehore's Career

The Commercial Cable Company was greatly interested in licensing the telegraphy patents that Crehore and Squier had received, and began negotiations with them in 1899. The negotiations failed to reach agreement because, according to Crehore, the $1 million Squier asked for the rights was too high. Crehore later regretted not pressing the issue, but it was too late. In his biography he stated that the money forfeited meant more to him than to Squier, because Squier had his Army salary and Crehore had resigned his position at Cornell to work for their new company.[69] This episode may have soured their relationship, or Squier's frequent absence on Army duty could have been the cause, but their collaboration ceased in the years that followed. As it turned out, Albert Crehore's collaboration with Squier was the most productive of his long career.

Why Albert Crehore changed from an intelligent and resourceful scientist to a laughingstock of the scientific profession remains a mystery. He came up with an ingenious but eventually untenable model of the atom and continued to promote it for the rest of his long life (he died in 1959, at the age of 91). Immediately after Nils Bohr published his first paper on atomic theory in 1913, Crehore published an alternative model of the atom based on classical electromagnetic theory. Bohr's model of the atom hypothesized that electrons orbit around the nucleus and changed orbits by emitting or absorbing energy in discrete quantities (or quanta). Though the Bohr model was in conflict with classical mechanics and electromagnetic theory, it explained all known phenomena associated with atomic behavior, in particular the spectra radiated by different elements of the periodic table. Bohr's model was further refined by the development of quantum mechanics in the 1920s and is now the universally accepted theory of atomic behavior.

When Crehore published his paper in 1913, it was considered a reasonable alternative to Bohr's work, particularly because it was consistent with classical electromagnetic theory. He disagreed with Bohr and did not believe that Bohr's model explained radiation from any but the simplest atoms. In time, the likely validity of Crehore's theory declined as subsequent theoretical and experimental work verified Bohr's hypothesis. Nevertheless, Crehore continued to construct more and more elaborate models to make his theory consistent with experiments. The more evidence that accumulated to confirm Bohr's model, the more books (the last in 1950) Crehore published to justify his own. His books are considered worthless by his colleagues[70] and earned him a reputation as quixotic and unbalanced.[71]

Squier Presents His Work

The Navy's fortuitous invitation to address the Naval War College opened the door to attending the Paris Congress in person and delivering a paper as had been requested.[72] Squier deeply cherished the opportunity. Attendance at a prestigious international meeting and delivery of an important paper were paramount in his mind when he beseechingly informed Greely that "the particular field of my future services is entirely in the hands of the Chief Signal Officer of the Army."[73] Greely consented and wrote to the Adjutant General that Squier would spend the summer inspecting the manufacture of deep-sea cables intended for use in Alaska and the Philippines.[74] But, he insisted, with regard to the invitation from the Naval War College, "Captain Squier will be absent from his pressing and important duties only as long as is actually necessary for the delivery of the lecture on submarine cables."[75] On 12 July Squier presented his lecture at Newport, Rhode Island.[76]

5

The Philippine Cables

The Signal Corps created an extensive network of electrical communications in the Philippine Islands during the years 1900 to 1902. Land lines between government houses, military posts, and military department commander headquarters provided effective communications between all elements of the colonial administration. Squier directed the laying of submarine cables that interconnected the islands. Although he gained valuable practical experience in the Philippines, Squier's prolonged absence from the United States endangered the success of his corporation.

Squier's promotion of his business interests increasingly conflicted with the wishes of his superior, General Greely. His activities in connection with business pursuits led to compromises bordering on dereliction of duty. In a permanent position in the Signal Corps, a technical service, Squier might have successfully institutionalized scientific research and development. But his early years in the Signal Corps were less fruitful in this line than his years at Fort Monroe. Two reasons suggest themselves. First, Greely may have spread Squier's talents and energies too thin. Greely failed to understand the necessity for concentrating in a scientific specialty. As a soldier-scientist of the old school, Greely did not grasp the complexities of modern technologies. He forced Squier into the old mold. In 1898 and 1899, for example, Squier worked on radio, electrical-testing techniques for deep-sea cables, cable-ship machinery, sine-wave telegraphy, and range and position finders. Greely also assigned him to maintain contact with the development of the Langley airplane.[1] Second, military duties related to heading the relief expedition to the Philippines removed Squier from his scientific and commercial concerns. His Philippine duty isolated him. He was unable to pursue his scientific development or aid his company. His economic interests, imperiled by his absence from the United States, exacerbated existing conflicts between his military profession and his commercial activities.

The Decision

In February 1899, President McKinley delivered a special message to Congress on the national importance of spanning the Pacific by a cable under American control. His call for an American Pacific cable produced a flurry of bills in Congress. Some called for granting a government franchise to a private corporation. Other bills stipulated outright government ownership and operation of the proposed cable. In either eventuality, Greely would acquire an important new mission for the Signal Corps.

The decision to lay cables between the islands of the Philippines to facilitate their administration thrust the Signal Corps into a new role. For the first time the Signal Corps was responsible for providing electrical communications in a distant land during peacetime. To meet the challenge would require ocean-going vessels, cable-laying equipment, and cables. Considerable technical knowledge of the mechanical and electrical characteristics of cables was also required.

Greely assigned to Allen the mission of taking the U.S.S. *Hooker*, a ship

George Squier (second from right in the second row) at Camp Atascadero in the Philippines (courtesy CECOM Historical Office).

modified by the Quartermaster Corps to lay cables, to the Philippines and leading cable-laying missions. This was the ship that Greely intimated to Squier and Crehore during Christmas holidays 1898 that he intended acquiring. During the first four months of 1899, Allen was busy with the routine of loading the expedition's equipment on board the *Hooker* and with overseeing the design and manufacture of cable-laying machinery to be installed on the ship. At the same time, in Washington, Squier collected technical data, devised test instruments, and developed techniques for testing the deep-sea cables during the long voyage to the Philippines. Greely's motives in supporting Squier's submarine telegraphy experiments and in acquiring a cable-laying ship were revealed in his evaluation of Squier's services during the preparations for the departure of the *Hooker*. He said that Squier's

> collection of data must be of great value and importance to such officers of the Government as may be charged either with the construction, laying and operations of such cables, or the supervision of the work if done by contract.[2]

Thus, Greely was preparing personnel of the Signal Corps to lay deep oceanic cables and thereby to acquire the responsibility for international commercial, military, and diplomatic electrical communications.

The necessary oceanographic surveys were either complete or in progress. The Navy had surveyed a suitable cable path across the Pacific to Hawaii five years earlier. A few months after the President called for an American cable across the Pacific, the Navy dispatched the U.S.S. *Nero* from San Francisco. Its mission was to survey the ocean floor on a single path from Honolulu to the Midway Islands, Guam, and Luzon. Soundings were also taken from Guam to Yokohama, Japan.[3]

Squier Is Ordered to the Philippines

The U.S.S. *Hooker* left port on 1 May 1899 with Colonel Allen in command of the cable-laying party. After a two-month voyage, the party landed in Manila.[4] On its maiden cable-laying voyage the *Hooker* ran aground on Corregidor Reef, in Manila Bay, and sank. Reaction at Army headquarters is unrecorded, but there must have been some chuckles in Navy offices. Owing to the alert, albeit tardy, action of Allen and his wet party, the cable and other Signal Corps property were saved.[5] With Allen stranded in Manila the only man left behind in Washington with comparable cable experience was Squier. The lot of preparing a new ship and leading a fresh expedition fell to him, when he returned from his summer stay in London.

Meantime, Greely ordered another ship to replace the sunken *Hooker*.

George Squier on the deck of the cableship *Burnside*, in the Philippines (courtesy CECOM Historical Office).

The Quartermaster Corps quickly set about preparing a new ship, one captured from the Spanish and rechristened the U.S.S. *Burnside*.[6] The *Hooker* and the *Burnside* were the first Army cable ships.

On 31 July 1900 the commanding officer of Fort Myer received orders to send a detachment of about ten men to the Army Building in New York City for detail to Squier, who commanded the relief expedition bound for the Philippines.[7] General Greely was extremely anxious that Squier's mission should leave as soon as possible as over a year had passed since losing the *Hooker*. Squier met his detachment for the first time in the second week of August.[8] When they finished loading the final length of 559 miles of cable aboard the *Burnside* they steamed for Europe in late September.[9] Squier sailed with four radio sets[10] specially constructed and purchased at a cost of several thousand dollars, a great sum for the time and circumstances.[11] Greely hoped that Squier would undertake some radio investigations with the electrical engineers and Signal Corps men under his charge. There was no outcome. The sets went unused even in the Philippines, much to Greely's disappointment.[12] Squier inexplicably neglected the radio work which Greely wanted him to perform. Instead, he continued his own work on sine-wave telegraphy. His roles were again in conflict.

Squier's neglect of radio experiments doubtless reflected some resentment over Greely's decision to send him to the Philippines. As early as his Fort Monroe assignment, Squier had articulated what he regarded as the proper sequence of steps in research. The stages he described as ballistic firing, target firing, and field firing can be analogized to basic research, applied research, and field use. He clearly envisioned his contributions being made in basic research. Squier's probable frustration and resentment at being torn from the laboratory, where he strove to stay on the frontier of rapidly expanding technology, and transferred to a desolate location would be wholly understandable. In accepting the assignment, Squier demonstrated remarkable versatility and loyalty to the Army. It must have been apparent to him that his own active participation in science, as well as his business career, were in serious jeopardy. And yet he complied with Greely's decision. Greely's attitude revealed how little the Army understood the institutional dimension of modern scientific research.

On the way to the Philippines by way of the Suez Canal, he took time out to deliver a paper in Paris. The subject of the paper dealt with his commercial sine-wave submarine cable transmitter. It paralleled his Philadelphia presentation. After Paris, Squier sailed on for Cairo, Ceylon, Hong Kong, and Manila, where he arrived on 6 December 1900.

Duty in the Philippines

Two weeks after Squier's arrival in the Philippines, the *Burnside* left Manila with Colonel James Allen in command and Squier, just promoted to captain,[13] in charge of electrical researches and testing.[14] The first cablelaying mission lasted three months.[15] In August 1901, Squier assumed full command of the *Burnside* cable- laying party. During the next nine months he laid 14 new cables in uncharted seas and repaired 13 others. He traveled 9,905 miles in accomplishing the mission and mapped many of the undersea areas along his route for the first time. In all, he supervised the laying of 30 separate cables connecting all the principal islands of the Philippine Archipelago and extending over 1,300 miles. At some points he and his men were obliged to engage in hand-to-hand combat with hostile natives in order to secure a cable landing.[16] The greatest depth reached was off the coast of Mindanao, where a cable descended 1,000 fathoms (6,000 feet) before coming to rest on the ocean floor. This feat placed Squier and his colleagues among the most skilled cable men in the world. Uncharted seas, unlighted coasts, irregular and uncertain currents all added to the difficulties experienced by the captain and crew of the *Burnside*. In all this time, however, Squier failed to develop further the radio sets that he brought to the Philippines.

General Greely was privately disappointed. He publicly stated that the pressure of affairs had made the establishment of experimental radio stations impossible.[17] In his own mind he never received a satisfactory explanation,[18] reporting that it was thought that conditions in the Philippines made the operation of radios expensive and inefficient.[19] Squier's economic interest in cable telegraphy, his busy cable-laying schedule, and possible resentment of having his other careers foreclosed, provide probable explanations for his lack of attention to radio.

Recall to Washington

With effective occupation of the islands won, some business men renewed their efforts to obtain governmental approval for laying a cable to the Philippines. Such a cable would facilitate successful American entry into the profitable China trade. As one of a very few government cable experts, Squier was asked to testify on behalf of a publicly owned and controlled cable to the Far East. At the request of Representative John B. Corliss of Michigan, the Secretary of War recalled Squier from Manila.

During the first months of 1902, hearings were conducted before the Senate Committee on Naval Affairs and the House Committee on Interstate and Foreign Commerce on the question of granting government support for the construction and laying of a Pacific Cable. Several points of importance emerged from the House Committee hearings. All agreed that if a private corporation laid the cable it should clearly be an American firm. In any event, government or private, it should commence cable-laying only with the prior approval of Congress. The Committee majority adopted the view that the construction of cable lines constituted a public utility or franchise held by Congress in the name of the people and not in the hands of the President.[20] The occasion for this assertive declaration of congressional prerogative was a bold move practiced by John Mackay of the Commercial Cable Company.

Ever since President McKinley called for an American Pacific cable in February 1899, Congress considered many bills supporting the effort. In his first message to Congress in 1901, President Theodore Roosevelt called attention most earnestly to the crying need of a cable to Hawaii and the Philippines, to be continued from the Philippines to points in Asia. We should not defer a day longer than necessary the construction of such a cable. It is demanded not merely for commercial but for political and military considerations.[21]

While he admitted that there was

a wide-spread conviction in the minds of the American people that the great corporations known as trusts are in certain of their features and tendencies hurtful to the general welfare.... It is based upon sincere conviction that combination and concentration should be, not prohibited, but supervised and within reasonable limits controlled; and in my judgment this conviction is right.[22]

He advised that either

the Congress should immediately provide for the construction of a Government cable, or else an arrangement should be made by which like advantages to those accruing from a Government cable may be secured to the Government by contract with a private cable company.[23]

President Roosevelt had one company very much in mind. John "Bonanza King" Mackay, the founder of Commercial Cable Company and who, with James Gordon Bennet of the New York *Herald Tribune*, financed the laying of the first American cable, purchased cable in the fall of 1901, in England. He sent a cable ship with the cable on board to Hawaii. Mackay asked President Roosevelt for a franchise when Congress failed to give favorable action on his request to lay a cable without subsidy or other government favor. President Roosevelt thought he had insufficient authority to grant such a request and referred Mackay to Secretary of State John Hay for advice. Following his consultation with Hay, Mackay formed the Commercial Pacific Cable Corporation with a capital stock of $100,000 in September 1901, under the lenient provisions of the Postal Telegraph Act of 24 July 1866.[24] Announcing that he intended to begin laying cable, unless prevented by force, Mackay filed a notice of acceptance of conditions contained in the 1866 law,[25] without the approval of Congress.[26]

Mackay did not have to wait long for congressional response. Representative John B. Corliss (Michigan) indignantly declared that the power to grant a cable franchise resided solely with Congress. In taking the offensive, he charged the Commercial Pacific Cable Company with deliberately violating a provision of their incorporation articles which required that the company be free of foreign control and refrain from colluding with foreign cable firms to regulate rates. Corliss regarded the new firm as being in league with the British-dominated Eastern Extension Company, which had held an exclusive concession from the Spanish crown to operate in the Philippines.[27] Meanwhile, the Eastern Extension Company objected to the presence of Army cables in the islands. The company maintained that the United States was obliged to respect the exclusive concession when sovereignty passed from Spain to the United States. To establish an enforceable claim, the company initiated court action.

Cables that were owned and operated by the Army in the Philippines threatened the British monopolistic control of international electrical communications in Asia. The Eastern Extension Company, a subsidiary of the Eastern Telegraph Company in London, and the Copenhagen-based Great Northern Company together obtained monopoly rights from the Chinese government in the famous Joint Purse agreement of 1896. The agreement formalized a series of informal understandings between the two companies and China about complete control of traffic between Australasia, Japan, China, Russia, and Europe by routes across the north (across Russia) and the south (across India, under the Red Sea, the Mediterranean Sea, and northward to London). Other secret protocols aimed at controlling land telegraphic traffic as well as submarine cable traffic were concluded in succeeding years.[28] The two companies claimed their special privileges until 30 December 1930 when effective international radio transmission systems replaced them. Although fully aware of the massive presence the two companies presented in the Far East, American authorities were innocent of the secret protocols directed against American entry into the Asian telegraph business. Just a few months before the fall of the Philippines, an agreement between Spain and England granted monopolistic privileges to the Eastern Extension Company in the islands and between Manila and Hong Kong, where radiating cables stretched to Australia, New Zealand, and Singapore. It appeared they had successfully checked all possible American telegraphic influence in the Orient. Thus, in 1902, English cable interests became alarmed at the new turn of events. Situated in the Philippines, American traders were now much closer to China and Japan. As the Army interlaced the islands with cables, Congress seriously considered laying a cable to the Chinese mainland. At this point Mackay stepped in, on behalf of English cable companies, so Representative Corliss thought.

Corliss found an ally in the House of Morgan, which controlled the Mexican Telegraph Company. Active in opening South America telegraph communications, the Mexican Telegraph Company unsuccessfully sought many times to win Congressional approval to start a publicly subsidized Pacific cable project. Failing approval of their own plans, the Mexican Telegraph Company advocated a government cable free from foreign interference and control.[29] As the only one offering to lay the cable without subsidy, Mackay's company was the sole alternative to a government project. Roosevelt was already on record as being in favor of permitting a private cable corporation to lay the Pacific cable. Corliss persisted, though, and questioned if Mackay's Commercial

Pacific Cable Company was really American. He insisted on a government-owned cable operated by the Signal Corps.[30] He reiterated the familiar theme developed earlier, that if revenues were to be derived from a Pacific cable they should redound to the benefit of the American people.[31] He estimated potential revenues at between one and two million dollars. Although Corliss carried the majority of his committee, the New York Chamber of Commerce and numerous mercantile associations entreated them to permit private enterprise to handle the undertaking. Into the fray between those wanting corporate sponsorship of the great cable project and those insisting that the public benefit from such a lucrative utility, stepped one of the directors of Squier's firm, Harry Garfield. Corliss's argument for government ownership needed stronger proponents with established scientific and engineering reputations, and, above all, practical experience, deficits Mackay's forces maintained existed in any government attempt to lay or maintain a submarine cable.[32]

Between hearings before the Senate Naval Affairs Committee and the House Committee on Interstate and Foreign Commerce, Harry Garfield asked Secretary of War Root to bring Squier back from the Philippines so he could personally aid the cause of a government cable. He told Root that Squier

> is an ardent advocate of the Pacific Cable and has given the subject most careful attention, extending his investigation in many directions important to the subject in hand ... and I know he advocates a government laid and operated cable.[33]

Greely had already declared that Squier's "previous work in connection with the development of cable engineering has insured him an international reputation."[34] He would have been an impressive and useful witness, but, for some unknown reason, Garfield's request went unheeded until Corliss lent the weight of his office to it a month later.[35] On 30 March the Adjutant General cabled permission for Squier to depart on a leave of absence, effective 17 May 1902, for a period of two months. He must then return to his duties in the Philippines.[36] Squier embarked a week later in Manila and arrived in San Francisco on 17 May.[37] Unfortunately, very little is presently known of Squier's activities during his leave of absence. Most certainly he advised John Corliss and just as certainly he threw himself into the work of his corporation.

Two weeks before his authorized two-month leave was to expire, Squier asked the Adjutant General for an extension of one month.[38] Failing to obtain approval, Squier asked the Acting Chief Signal Officer to be released from further duty in the Philippines and assigned to Alaska in connection with radiotelegraph installations.[39] Representative Corliss joined the others and wired Root with a request to extend Squier's leave one month: "his meritorious service justly deserves this favor, and I trust public interest may permit."[40] Squier resubmitted his request for extension, stating that General Greely had promised an extension in an exchange of letters in February and that, with respect to his mission in the Philippines, the "system of cables was successfully laid and the routine of their maintenance well organized before my departure."[41] On the same day, Garfield appealed to the Secretary of War to extend Squier's leave, due to expire in just three days, in order to advise on the use of his sine-wave telegraphy system by an American corporation engaged in constructing and laying a Pacific cable. Who was this company? The only one reported to be engaged in laying a cable was the Commercial Pacific Cable Company.

Commercial Pacific was interlocked with the Commercial Cable Company and the Postal Telegraph Company through George C. Ward, who served as vice-president and general manager of the two Commercial companies, and Clarence Mackay, son of the Bonanza King and an officer in both Postal Telegraph and Commercial Cable companies, which his father had founded.[42] Squier had had close working relations with Commercial Cable and its proprietors during his Canso and Trans-Atlantic Cable trials. He might have had an arrangement with them. On the other hand, Corliss, Mackay's avowed foe in Congress, had intervened several times with the Army to gain Squier's presence and assistance. Garfield and Corliss's joint appeals mean something significant in light of Garfield's tone of urgency:

> The question of Captain Squier's presence here is a question of such importance to both the public and to private interests that it is impossible to speak of the one without involving the other.[43]

Perhaps they intended to mount their own Pacific cable project as a bona fide American firm, with Corliss's blessing. In any event, little more is recorded from present sources to develop what transpired and what was planned. The possibilities for serious conflicts of interests were legion. Squier completed his leave and returned to his Manila post.

Completion of the Philippine Tour

Soon after Squier returned to the Philippines, as Greely disapproved his request for an assignment to Alaska, he assumed a new position as Superintendent of Telegraphs for the Government Telegraph System. The Signal Corps had earlier transferred military control to Governor General William Howard Taft's civil government, so the government system carried all civilian and military telegraph messages. Squier held the post until his reassignment the following June. In the meantime he acquired the additional responsibility of Superintendent of the Manila Telephone System. Despite considerable indulgence in granting Squier his desires, Greely appears to have quashed plans involving a trans–Pacific cable and the Crehore-Squier Company. Events in Washington washed over whatever hopes they might have entertained.

In December 1902, the President granted the Commercial Pacific Cable Company non-exclusive permission to lay a Pacific Cable from San Francisco to China via Honolulu, Midway, Guam, and Luzon.[44] Roosevelt's nod to the Commercial company became invested with irony as the years passed.

The effect of the foregoing events was to transfer unwittingly America's cable to the Far East into the hands of a British-dominated corporation. Two decades later Clarence Mackay revealed for the first time that the British-controlled Eastern Extension and Great Northern firms owned, respectively, 50 percent and 25 percent of the capital stock of Commercial Pacific Cable Company. Only by entering into secret arrangements with these two foreign companies could Commercial Pacific obtain a part of the lucrative Far East communications business. American communications firms finally broke the British monopoly with the initiation of reliable international radio broadcasting.[45] One week before Squier reached San Francisco in July 1903 to take up his new post as Chief Signal Officer of the Department of California, Mackay's company landed their cable in the Philippines.[46]

6
Founding of the Signal School

The Army displayed especial ineptitude in planning for operations before and during the Spanish-American War. To ensure adequate forethought in providing for military defense, Secretary of War Elihu Root recommended a series of far-reaching reforms. Preeminent among them was the establishment of a General Staff headed by a Chief of Staff. Congress approved his recommended reforms in early 1903, despite objections to their Prussian antecedents.[1] In November, Root directed the establishment of a U.S. Army War College as an instrument for educating officers of flag rank in the new command structure and as a center for progressive change. The college, in time, assumed more of the planning functions of the General Staff, becoming the principal source of war plans and information about foreign armies through its subordinated attaché system. Professional education for lower ranking officers, suspended during the war, recommenced in 1901 in five service schools located at Fort Leavenworth, Kansas.

The Leavenworth schools, under the direction of one commandant, trained company- and field-grade officers in the employment of combined arms at the regimental level. The interaction these schools afforded officers of the several combat corps was one of its special merits. The addition of a Signal School, in 1905, marked two developments: first, the enhanced prestige and sense of value the Signal Corps acquired throughout the Army in the postwar years with a consequent amelioration of traditional army hostility toward the technical services; and, second, the gradual exposure of the Army to a wider circle of influence. Isolation of the Army from civilian society did not effectively end until the First World War, but the first steps taken toward integrating the Army in the larger society occurred as a result of the new educational arrangements created by Root. The process of integration involved increased cooperation with the technical services which were attuned to current trends in the technology of their civilian counterparts. Squier played an important part in this process. He came to Fort Leavenworth from two years at San Francisco. During that period he served as Chief Signal Officer to General Arthur MacArthur, commanding officer of the Department of California. Squier established Benecia Barracks Signal School outside of San Francisco, carried on his telegraph business, and experimented with the conduction of electrical currents in living plants and trees. General MacArthur was well pleased with Squier's services and communicated his pleasure to Greely. Greely replied that "personally [he was] very fond of him" and informed MacArthur that he had designated Squier to take charge of establishing the Signal School at Fort Leavenworth in the fall.[2] As assistant commandant he promoted joint training of Signal Corps and combat units. One cavalry colonel visiting Leavenworth as a special inspector took note of Squier's cooperation with line organizations and the speed and reliability of signaling methods. Formerly, the colonel confessed to General Allen, he preferred orderlies to convey his messages. Now he would recommend their use only for distances of less than a few hundred yards. He recommended assigning another Signal Corps company of troops to Fort Leavenworth.[3] Squier also supported closer cooperation between the Signal Corps and combat units in the use of aeronautics for military purposes.

Left: George Squier at Fort Leavenworth c. 1900 (courtesy CECOM Historical Office).
Right: George Squier in civilian dress, 1905 (courtesy CECOM Historical Office).

For a period of time, in 1906, line officers actively sought the establishment of an aeronautical facility at Leavenworth. At a mass meeting of the faculties a resolution favoring the location of a balloon plant and school at Leavenworth was passed. Interest in balloons and airplanes flourished in at least this one important segment of the Army some time earlier than is credited at all in aviation histories. Squier's ideas and policies on closer cooperation between the technical services and the combat arms met with evident approval from the schools' commandant, Brigadier General Charles B. Hall. He described Squier as "probably the most competent officer in his Corps and one of the most accomplished officers in the Army."[4]

In his position at Fort Leavenworth Squier entered upon a new phase of his career. He seemed much less interested in his own scientific discoveries, especially after the failure of his and Crehore's company.[5] He was eager to teach and to foster new ideas and institutions in the hope of thereby opening the military to the new technologies of flight and communications. Out of his own disappointments he sensed how different institutional arrangements might have supported and fostered his own scientific career and provided the Army more effective access to new technologies.

The Signal School, Fort Leavenworth

The Signal School differed from its predecessor branch schools. Organized by War Department General Order No. 145, series 1906, as an Army school, it was charged with participating in the mission of providing training in combined arms operations. Other schools of the Signal Corps trained men and officers for signal duties alone. For the first time, line officers of the Army were exposed to theoretical and practical applications of electrical communications in classes conducted by officers with matchless experience around the world in radio, telegraphy, and submarine cables (as well as visual-sighting devices). Combat officers now had an opportunity at genuine participation in developing military doctrine which would realize on the battlefield the full potential of modern technologies—electrical communications and aeronautics—controlled by the Signal Corps. From now on, no General Greely would find it necessary to impose his technology upon the Army. Indeed, all subsequent relations between the Army and the Signal Corps may be viewed as a continuing struggle by the combat arms to "capture" the latter's technology.

Combat officers at Leavenworth recognized the value of radios and balloons, imperfect though they were, and supported closer association with the Signal Corps, even to the point of recommending that the Signal Corps be henceforth regarded as part of the line of the Army. Root's intention to reduce the independence of bureau chiefs, disapproved by Congress, was finally achieved by the mid-1960s with the disestablishment of the Signal Corps. Radio and other forms of signaling had simply become too important for one semi-autonomous bureau chief to be in a position capable of thwarting the wishes of combat officers and their commanders.[6] Combat officers saw radio as crucial to their survival and demanded that electrical communications be placed under their control. The centralization of command was well under way.

The very origin of the Air Force, as an organization independent from the Army, is likewise rooted in the Fort Leavenworth experience. The great hostility, bitterness, and disaffection which accompanied the separation of aviation from the Signal Corps in 1918, in the midst of the war, might also be viewed as a "capture" by combat officers of a technology which had become vitally important to the successful outcome of battle. The capture process began at Fort Leavenworth rather than a decade later, as the time is customarily assigned in Air Force histories.[7] Squier paradoxically tried at Leavenworth to promote joint development of aeronautics; it was from him that control over aviation was wrested a little over a decade later.[8]

To delineate the mission of The Signal School, general regulations were published within a year of its founding.[9] As head of the school, Squier's most important duties included making recommendations on the general regulation. The entire charter of future professional education for technical and line officers was at stake. He included as part of the school's educational mission a requirement to instruct students in military aeronautics.[10] When the general order appeared in print, Squier immediately directed a recommendation to the School Secretary that the school obtain modern balloon equipment in order to fulfill the requirement for instruction in military aeronautics. The Commandant approved and recommended favorable action.[11] Allen chose, however, to locate the new equipment at Fort Omaha, citing considerations of personnel and money as his reasons.[12] This reply failed to satisfy the cavalry and infantry officers, not to mention the signal officers. A mass meeting of all the Academic Boards, composed of assistant commandants and instructors, convened in early February 1907, to discuss the Chief Signal Officer's endorsement and formulate a reply. Squier's voice is clear in the record of the proceedings:

> The subject of military aeronautics has passed the experimental stage and is a permanent part of the equipment and instruction of foreign armies. Recent marked advances in the construction of air machines have added greatly to the military interest in the possible development and the use of this auxiliary as an adjunct in war.[13]

This position represented the sense of the Boards which recommended Fort Leavenworth rather than Fort Omaha as the site of a new gas-generating plant and balloon house. Among the reasons cited for preferring Leavenworth over Omaha was one which reveals considerable insight into the early problems of aeronautics and, subsequently, aviation. The Boards, composed of line officers and Staff College officials, argued that

> at Fort Omaha only the technical training necessary for the actual manipulation of the balloon can be attempted, while at this station, its tactical use with the three arms of the service can also be studied.[14]

Indeed, the eventual separation of the Aviation Service from the Signal Corps might be viewed as an affirmation and a concrete realization of the priority of the latter sentiment over the former. In contemporary jargon the sentiment is expressed in the phrase "The mission of the Air Force is to fly and to fight, and don't you forget it." The experience at Fort Leavenworth, even though not specifically referring to airplanes, should have served to put the Signal Corps on notice.

Allen missed a splendid opportunity to gain combat arms support for his efforts in the development of aeronautics by establishing the balloon facility in a location far removed from a center of influence and power. Squier's letter received an almost unprecedented 18 endorsements, traveling twice to Allen's desk. Signal Corps' leadership in subsequent years would have felt blessed to secure a fraction of the support they spurned at Fort Leavenworth. Squier became associated with advocacy of aeronautics as a result of this remarkable correspondence. Allen eventually compromised and allowed the balloon to go to Fort Leavenworth, assigned the gas plant to Fort Omaha, and sent Lieutenant Frank P. Lahm, Paris Exhibition prize winner and expert aeronaut, to Fort Leavenworth.[15]

Staff and Curriculum

As commanding officer of the Signal School, Squier also held the position of Assistant Commandant at Leavenworth. He reported to the Commandant of the Infantry and Cavalry School, Signal School, and Army Staff College. Captain Saltzman from Benecia Barracks, California, enrolled as a student and soon became one of the instructors. Captain William (Billy) Mitchell, newly arrived from Alaska where he assisted in establishing the Signal Corps telegraph and cable system, joined Squier as an instructor in the Department of Signaling and commander of the Signal Company attached to the post.[16] He later achieved fame as a "crusader for air power" during the First World

War. He was the subject of a famous court-martial, which questioned the permissible limits of vigor in pursuing his advocacy of air strength.

The course of study embraced three departments: Signaling, Signal Engineering, and Languages. Students learned theoretical and practical lessons in optical, acoustical, and electrical signaling in the signaling department. Saltzman taught from Swoope's *Lessons in Practical Electricity*, Maver's *Wireless Telegraphy*, and Root's *Military Topography and Sketching*.[17] Squier lectured on alternating current electricity; control systems; laying, operating, and maintaining submarine cables; installing and operating gas and oil engines; military aeronautics; photography; and topography. He also established a laboratory in the basement of Sherman Hall, where instruction and independent investigations were conducted. He recommended doubling their size in the next year.

He instituted the practice of semi-monthly journal meetings, borrowed from his graduate student days at Johns Hopkins University. He led discussions on topics of interest in the professional literature, both military and scientific, and, eventually, on matters of current interest to the Chief Signal Officer. Students were required to present verbal reports, abstracts or written précis on articles of importance for general discussion and criticism. These meetings, held less frequently under Squier's successor, became the notable Leavenworth Technical Conferences. Leavenworth earned a reputation as the "intellectual center of the Army."[18] Squier also used his journal meetings to examine issues which he later raised with the Chief Signal Officer, who, in turn, sent questions to Squier for examination by his students. By far the most pregnant issue they treated was the portent of aviation to military strategy and tactics.

The Study of Balloons and Airplanes

Traditional Air Force histories direct criticism toward the War Department, Board of Ordnance and Fortification, or the Army as a whole for failing to maintain an active interest in aviation after the failure of Samuel Langley's "aerodrome" to sustain successful flight. Like most anti-establishment commentaries borne out of revolutionary fervor and long-restrained frustration, they contain broad generalizations. In their tendency to trace an approved lineage of pioneers who laid the foundations for the contemporary institution, they overlook important contributions from outside sources. General Curtis LeMay, Chief of Staff of the Air Force from 1961 to 1965, for example, considered the origin of his service occurred with the issuance in 1910 of the Army's first aviator rating to Benjamin Foulois, who later rose to the rank of Air Force brigadier general.[19]

Many of these historians view the Army in an unfavorable light. According to Goldberg, editor of a history of the Air Force, later re-published as a manual intended for ROTC cadets throughout the country, the "U.S. Army had not kept up with the greatly accelerated aeronautical activity in Europe after 1900.... Not until after 1907 was the Army sufficiently impressed by the great new aeronautical developments to take the first step toward building an air arm."[20] He claimed that news of the Wright Brothers' flights was met with "disbelief, skepticism, and disinterest."[21] When, in October 1905, the Board of Ordnance declined to formulate any requirements or specifications for an airplane until the Wrights produced such a craft capable of flying and carrying an operator, the government ceased to communicate with the Wrights for another 18 months, according to the authoress of another official Air Force history.[22]

The same feeling pervades other accounts of the period.[23] The Army was too large and diverse an entity to make the object of such complaints. Researchers of the period have overlooked the "intellectual center of the Army," where Root expected a more professional and progressive military would evolve. The questions of aeronautics and aviation received extended consideration in Squier's journal meetings. The accomplishments of the Wrights were well known and discussed in the Signal School.

Squier spent Christmas holidays, 1905, in New York, where he kept a membership in the University Club. Staying on for most of January, he attended a talk by his old friend Alexander Graham Bell, a Regent of the Smithsonian Institution, who confirmed the general results of the Wrights' airplane trials. For many people, not just the Army, there existed considerable doubt about the authenticity of claims emanating from Dayton, Ohio. The public was even confused about the difference between a "flying machine and an airship," a confusion the Wrights exploited to maintain secrecy about their invention.[24] Squier penned a note to Greely in which he said he expected to return to Fort Leavenworth in a few days and that "he may decide to stop over at Dayton, Ohio, unofficially to learn what I can of the facts."[25] Squier enclosed a clipping from the Sunday *New York Journal-American* (21 January 1906) concerning the French purchase of the Wrights' patent rights in Europe. Greely penciled a reminder on the letter to inform Captain Fournier, as "he recently spoke to me about visiting Ohio to investigate a flying machine."[26]

Squier found passage between the scientific and military communities facilitated by his unique qualifications. This augured well for military aviation, on two counts. First, his position at Fort Leavenworth gave him unusual opportunities to raise the issue of aeronautics and aviation for general discussion. He was respected by the Commandant, line officer, and was listened to with care by the faculty.[27] Second, with solid friends in the scientific community, he could obtain information for the Chief Signal Officer as a reputable engineer and scientist without being obtrusive. This ability became doubly important with the departure of Greely in early 1906, when he succeeded MacArthur as commanding general of the Pacific Division. General James Allen, the new Chief Signal Officer, did not have Greely's prestige or international scientific reputation. With Greely's reassignment, only Squier, of all the officers in the Signal Corps, possessed those qualities.

The development of military aviation might well have been different had Greely remained at his old post, but in 1906 aviation could hardly have held the same importance in Greely's mind as maintaining and operating the Signal Corps' worldwide communications system, a system that included the world's first commercial radio link, the longest submarine cable in the Western Hemisphere, and tens of thousands of miles of telegraph line strung through tropical jungles and across arctic wastes. For this task Allen had more than qualified himself to be an effective leader in combat, and an efficient administrator in peacetime. Nevertheless, Squier's insatiable scientific curiosity made him the most likely candidate in the Signal Corps for pressing upon Allen the need for more positive advocacy of aeronautics, and, later, aviation. His enthusiasm for science and his knowledge of scientific societies and their members stood in strong contrast to the reception of the first reports of flying activities in the office of the Chief Signal Officer in 1902.

Captain Virgal, a French diplomat, wrote Greely a personal note in May asking if he could give information about the Wright Brothers' experiments and their apparatus described in a communication, he had learned, to the Western Society of Engineers.[28] Russel was asked to look into the question of a flying machine invented by the Wrights and found nothing upon which to base a satisfactory reply. Greely then wrote to H.W. Fisher of the Standard Underground Cable Company at Pittsburgh, inquiring if there was such a Society of Western Engineers and if he could give him any information on the Frenchman's query.[29] A month passed and no reply arrived from Fisher. Finally, Greely asked Russel again to investigate the question.[30] It is unknown if Captain Virgal ever received his answer. Greely's surprising ignorance of the Western Society and, apparently, of Octave Chanute, who presented the illustrated lecture, is mysterious considering his personal friendship with Samuel Langley, who knew Chanute well.[31] Squier's eclecticism, in contrast, led him to consider virtually anything of scientific interest and to develop professional contacts with scientists of different disciplines. One of his chief contributions to the Signal Corps, and later the Army, was an ability to open Army windows to the outside world.

Since Squier directed the selection of topics for consideration during journal meetings, existing records of the papers delivered reveal his own thinking. Aeronautics was prominent. Shortly after Lieutenant Wieczorek delivered a lecture on "Balloons and Flying Machines," Squier forwarded the manuscript, along with his personal comments, to Allen. Although the lecture has gone astray, Squier's observations are preserved. He urged Allen's attention to Wieczorek's remarks on the tactical and strategic effects of airplanes. He also emphasized the importance of reports contained in Kansas City papers of the Wrights' experiments and included clippings on the uses of balloons in English and Bavarian armies.[32] He stressed the importance of balloons and airplanes in his own lectures before the assembled school. Regarding maneuverability and practicality of aircraft for army purposes he stated:

> Although balloons, both captive and dirigible, have amply justified their use as an adjunct in war, yet it is to the attainment of a successful flying machine heavier than air, capable of carrying considerable loads, and producing a speed of thirty to fifty miles an hour that the military student at the present moment looks with keen concern. In this connection it should be stated that the first practical flying machine is an established fact and is the creation of the Wright Brothers of Dayton, Ohio. Each of the Wright Brothers has made numerous flights over their testing field near Dayton, sometimes at an elevation of about eighty feet and at other times passing close to the earth at only about ten feet from the ground. They have been able to circle over the field of operation and even to describe in the air the figure eight, thus demonstrating their perfect control over their apparatus, both in the vertical and horizontal directions. They have succeeded in remaining continuously in the air for thirty-eight minutes, and then descended only on account of the exhaustion of fuel supply. The velocity attained was about thirty-seven miles per hour. The machine has not only sustained its own weight, but has also carried a man, and a gasoline engine weighing 240 pounds, exerting a force of twelve to fifteen horse power, and in addition an extra load of fifty pounds of pig iron. The apparatus complete with motor weighed not less than 925 pounds, while the supporting surfaces consisted of two superposed aeroplanes each measuring six by forty feet, so that the machine as a whole had a flying weight of nearly two pounds per square foot.[33]

Were, indeed, the French and British the only ones who had had representatives in Dayton, as claimed by Hennessey?[34] Was all of "official Washington" *really* surprised at the Wrights' performance as late as 1908, as suggested in a popular history of aviation?[35] Squier's description of the Wrights' airplane and its performance capabilities were not remarks gleaned from newspaper clippings. The Wrights were particularly cautious about disclosing such details, especially after August 1905, when further flight trials were suspended for almost three years lest someone observe secret features of their design.[36] It is clear that Squier had either visited the Wrights in Dayton, as he suggested he might in his note to Greely, or he obtained confidential information from his scientific and engineering colleagues, Alexander Graham Bell or Octave Chanute, because of his confident and detailed account of the Wright airplane.[37]

Aviation was an achievement that had knowledgeable sponsors within the Army, contrary to traditional histories which view President Roosevelt's intervention in 1907 as crucial to the ultimate incorporation of aviation within the Army. Goldberg stated that Roosevelt "had to intervene to bring the War Department and the Wrights together."[38] Hennessey said that the President "ordered" Secretary Taft to investigate the Wright claims after they had been brought to his attention by prominent members of the Aero Club of America.[39] In later years, Squier claimed to be a "pioneer member" of the Aero Club. To what extent he was involved in these representations is highly problematical, but not improbable considering his style of action and his recommendation written in late 1906:

> If nothing else had been accomplished in this direction, the performance of the Wright Brothers in the United States should be sufficient to cause the War Department to undertake the further development, and ascertain the possibilities and limitations of such a weapon of warfare. The Aero Club of America is doing pioneer work in aeronautics, and it would be of great advantage to the War Department if a similar organization among National Guard members of State Signal organizations was effected to encourage the development of the military side of the subject at summer maneuvers.[40]

His statement contains two key elements of subsequent developments: the War Department was prevailed upon, by highest authority, and the Aero Club of America played the principal role in promoting the cause. More of this aspect of Squier's activities in aviation as they affected policy in Washington follows in the next chapter.

7
Origins of Army Aviation

The story of the Aeronautical Division of the Signal Corps, formed on 1 August 1907, two weeks following Squier's arrival from Fort Leavenworth, has been a neglected chapter of military aviation history. Some authors have seen its formation as a product of external pressures, stirred by accounts of balloons and airplanes publicized by the Aero Club of America, and the personal intervention of President Roosevelt. Such a view of outsiders alone prodding the Army into action must be modified by what happened at Fort Leavenworth. Similarly, the relationships Squier established with the Aeronautical Division and the Smithsonian Institution suggest some appropriate revision in the traditional, but substantially true, accounts of the period. Military historians have focused on military institutions and neglected vital relations with other governmental scientific bodies which were as deeply interested with the advancement of aviation as the Signal Corps. If the scope of investigation is sufficiently broadened to encompass the Smithsonian and the Washington Aero Club, a small group of individuals, an "invisible establishment," materializes. Each member of the group had his own reasons for joining with the others in promoting government-sponsored research and development of aviation. They usually reflected the special interests of the groups which they represented.

Squier vigorously promoted Army interests in aviation before public gatherings and in print. John F. Victory, first employee and later the Executive Director of the National Advisory Committee on Aeronautics, called him a "spark-plug" for aviation.[1]

The Army Orders an Airplane

Squier arrived in Washington in July 1907 to assume his new post as Assistant Chief Signal Officer under General Allen. There is no document recording his participation in the decision to establish an Aeronautical Division within the Signal Corps, although its first chief, Captain Charles Chandler, said that Squier recommended the action.[2] Such a recommendation would have harmonized completely with his advocacy of aeronautics at Fort Leavenworth.

Two weeks later, in mid-August, the champion balloonist Lieutenant Frank P. Lahm, was given a three-month sick leave in France with the understanding that he would investigate aeronautical progress at British and German aeronautical centers.[3] On his own initiative he added Belgian and French aeronautical activities.[4]

Meanwhile, General Allen assigned Squier the task of preparing a thorough study of the entire field of airplane, dirigible, and balloon theory and practice. The research material came from military attachés in Paris, Berlin, and London.[5] Later, the Commissioner of Patents was also asked to furnish patents issued during the past 17 years pertaining to dirigible balloons, airplanes, helicopters, and other flying devices.[6] Squier conducted his study while temporarily assigned to the Department of the East, headquartered at New York City, where he could confer easily with engineers and Aero Club officials. Before he left for New York, he and Allen made plans to bring the needs and promise of aeronautics before the public in the hope of securing outside support for their budget request in Congress.

One of the most important events in their stratagem for developing public support was the dispatch of the Signal Corps aeronautic detachment to the Jamestown Exposition, near Norfolk, Virginia. They demonstrated a passenger-carrying captive balloon and contributed toward completion of an aerial glider designed by Israel Ludlow, an early member and official of the Aero Club, who took special interest in heavier-than-air machines.[7] This stratagem constituted a significant exception to the traditional assertion that outside forces alone prevailed upon the Army to develop aeronautics. That the stratagem was intended to promote military aeronautics was clearly revealed several months later. In commenting on a newspaper announcement that Israel Ludlow, director of aeronautics at the Jamestown Exhibition, stated that "a determined effort will be made (this winter) to get Congress to appropriate $500,000 for aeronautics,"[8] Squier said:

> This looks as though we were on the right track, and that by the time Congress meets we may not have to fight the battle alone to obtain funds for government aid of military aeronautics, but very probably can secure the combined assistance of such an organization as the Aero Club of America.[9]

When General Greely saw Captain Thomas Scott Baldwin's dirigible fly at an air meet in St. Louis the same month, its performance made such an impression upon him that he asked the War Department to give the Signal Corps $25,000 to purchase a small non-rigid dirigible balloon.[10] It was done. A hopeful step forward in creating public enthusiasm for army aeronautics had been taken but it appeared that Chandler and Allen were willing to confine their attentions to dirigibles and neglect airplanes.

When the Board of Ordnance and Fortifications asked Allen about the military usefulness of airplanes, he and Chandler, who prepared the answer, rendered a revealing reply. They declared the airplane unsuitable for military purposes and recommended no appropriation from Congress for the purchase. Basing their opinions on several factors, they argued that the rapid improvements in electrical communications obviated any need of airplanes to carry messages. For reconnaissance, the airplane was even less practical once the prone operator was exposed to a wind stream greater than 30 miles per hour, making it quite impossible to use field glasses, maps, or to draw sketches. And despite contrary claims made by the Wright Brothers, they had no evidence to suggest that airplanes could carry a passenger "in addition to the engineer." Even its utility as a vehicle for dropping bombs was considered dubious. They calculated that a bomb dropped from an altitude safe from ground fire would possess an accuracy of no better than one-half mile of the target. Their ideal of a military "flying machine" was one "which can stop over a certain point and sight down with a plumb line."[11] They also expressed objection to reported aerodynamic instability and mid-flight gasoline engine failures which would "undoubtedly cause it to turn sharply, upward or downward, and dive like a kite, with fatal results to the operators." Requirements for clear open spaces and, indeed, even a launching rail made machines of the Wright Brothers' type unsuitable for field service. The flying machine Allen apparently favored was one constructed on the helicopter principle, able to ascend without horizontal motion. Many inventors were working on that principle, but years would be required for its perfection. Meanwhile, they recommended reliance upon a fleet of dirigibles.[12] Allen and Chandler were not alone in their attitudes. Skepticism about the Wrights' claims was widespread. There is nothing to indicate, however, that Squier shared their doubt.

In late October or early November, the Aero Club of America held an exhibition of motors in New York. Patent Commissioner Moore and General Allen may have attended with Squier. The reliability of motors in aircraft was one of Allen's known concerns. When Squier returned to Washington, he requested information from all companies displaying motors. The Aero and Marine Motor Company of Fall River, Massachusetts, sent a brochure, but they expressed uncertainty whether the Signal Corps intended their use for dirigible balloons, airplanes, or helicopters.[13] Glenn Curtiss said he was willing to take contracts for building "any type of flying machine" using new engines capable of producing 100 horsepower.[14] When the Automobile Show came to New York in early November, Squier and W.R. Kimball, a consulting engineer in the city and designer of helicopters, attended to see if auto manufacturers had motors capable of worthwhile adaptation to aerial navigation.[15] Although six-cylinder, water-cooled engines (they were usually four cylinder) appeared to be the most popular, he favored four or more cylinder, air-cooled, two-cycle engines for aeronautical use because every stroke was a working stroke. He thought they would be operated for only short periods, and that engine temperature would not become troublesome. He noted that great advances had been made in developing reliable ignition systems so that now they should be able to draw satisfactory specifications.[16] A few days after the show, Kimball showed Squier some of his helicopter models which he promised to make fly with men if government support could be arranged. Squier's activities in New York were directed toward drawing intelligent and reasonable specifications for the power plant.

By mid–November 1907, correspondence between Squier and Allen became so heavy Allen found it necessary to advise George P. Scriven, Chief Signal Officer of the Department of the East, that, henceforth, Squier was authorized to communicate directly with his office in order to "avoid encumbering your files with letters on the subject."[17] It is quite possible, however, that Allen wanted greater confidentiality.

An M.R. Hutchison wrote Admiral George Dewey, head of the Navy, about valuable scientific and engineering data compiled by Peter Cooper Hewitt, an electrical engineer of New York City. Among Hewitt's various scientific contributions was the development of mercury-arc lamps. Hutchison said Hewitt's greatest contribution lay in preparing data on dirigibles, based on several years of secret investigations. Hutchison explained that Hewitt was willing to share his information confidentially with the government. He declared Hewitt's information would make possible the construction of dirigibles capable of carrying several tons of cargo at 50 to 60 miles per hour.[18] Dewey immediately contacted Allen, who was obviously impressed by Hutchison's opinion of Hewitt's work.[19] Allen knew Hewitt and had spoken to him about specifications being prepared for an army dirigible balloon. Hewitt expressed dissatisfaction with that sort of airship. He favored some unexplained heavier-than-air machine. Allen asked Squier to learn from Hewitt what type of aircraft he had in mind and to explore with him what specifications would be reasonable for airplanes. His call was to remain confidential.[20] Squier complied and reported another visit to the A.M. Herring's laboratory, where he viewed a light gasoline engine designed for airplanes.[21] Squier also reported a meeting arranged by the Adjutant General of the department with a Mr. Harper, who had a special engine and airplane design to show him.[22] He told Allen that the "further one goes into this matter, the deeper one gets, and every day now brings forth something new."[23] Allen expressed doubt as to whether Congress would support individual inventors in their researches.[24] After several more meetings, Squier returned to Washington to help prepare airplane specifications.[25]

Wilbur Wright told his friend Octave Chanute that he expected to meet with General Allen and members of the Board of Ordnance and Fortification on 5 December 1907.[26] As an experienced aerodynamicist, Chanute was skeptical of its outcome because he considered the Board and General Allen to "have only very crude ideas on aeronautical matters. They are just as likely to jump down on the wrong side as to remain on the fence."[27] They may well have been ignorant of the extensive preparations made by Squier in New York for drawing intelligent and reasonable specifications, as the Aeronautical Division purposely kept their actions confidential.[28] Historians have been just as innocent of the elaborate study that Allen directed Squier to perform. It is generally held that the specifications were written as a result of the meeting mentioned in Wright's letter to Chanute. Orville Wright himself said as much later.[29] The mistaken impression is left that the Signal Corps had made no preparations of their own in drawing the specifications. Tentative specifications were forwarded to Crozier with the comment that they had been discussed with Wilbur Wright.[30]

The final specifications involved the advice and counsel of many other engineers. Two days after the meeting Russel and Squier sent copies of the tentative specifications for comment on their practicability and reasonableness to Alexander Graham Bell, Captain T.T. Lovelace of the Aero Club of America, A.M. Herring, Peter Cooper Hewitt, Israel Ludlow, W.R. Kimball, and the Wright Brothers.[31] After discussing the tentative specifications with other members of the Aero Club, Herring replied that their consensus was that "a practical machine could be had but bids could not be drawn from any of the most advanced workers with specifications as at present."[32] Kimball remarked that the specifications were not impracticable but "probably too severe to attract the conservative element of manufacturers in view of the newness of the art."[33] Hewitt, on the other hand, suggested that the minimum speed of 40 miles per hour should be raised to 50 miles per hour—or even 60.[34]

The question of a suitable motor, however, still worried Allen and Squier. Squier wrote Columbia University Professor F.R. Hutton, President of the American Society of Mechanical Engineers, requesting a private conference on the subject of airplane motors, which he described as "the key to the solution of the problem."[35] With conferences concluded and replies received, General Allen advertised the final specifications to an incredulous public on 23 December 1907.[36]

The *New York Globe* attacked the War Department, charging that "nothing in any way approaching such a machine has even been constructed (the Wright Brothers' claims still await public confirmation)."[37] Editors of the *American Magazine of Aeronautics* criticized Army officials for preparing specifications incapable of being fulfilled and predicted that no one would enter a bid.[38] The overseas public was equally unbelieving.[39] Despite such doubts about the wisdom of War Department officials, 24 bids were received in the War Department before the February 1908 deadline and two contracts were let.[40] A.M. Herring, one of the contractors, failed to produce an aircraft and withdrew after several extensions of the deadline. The Wrights remained as the only contractors still in the race to produce an acceptable airplane. When Orville Wright arrived at Fort Myer, Virginia, near the end of August for his acceptance trials, Squier greeted him. As head of the Aeronautical Board, Squier was charged with supervising the trials.

Army Trials of the Wright Airplane

Orville wrote his brother that the acceptance committee seemed very friendly. He also judged them as strict in their expectations of Captain Baldwin's dirigible, which was tested just before the airplane trials.[41] Dr. Alfred Zahm, a local professor of engineering and president of the Aero Club of Washington, asked Wright to take up lodgings at the Cosmos Club, where many of Washington's most prestigious scientists met.[42] It has been said that more important decisions were made in the parlor of the Cosmos Club than in the halls of Congress.[43] Comfortably settled among scientific colleagues, Wright began the frustrating task of preparing his airplane motor for the trials. On 3 September, with difficulties stemming from poor gasoline, hot bearings, and defective magnetos at long last corrected, he lifted off the ground for a one-minute flight around the Fort Myer field.[44] After a week of practice solo flights, Orville took Lieutenant Frank Lahm for a six-minute flight, six times around the field.[45] The feature of dual flight distinguished the Army trials from all previous public flights.

Ecstatic newspaper accounts told of the excitement created among spectators who had come out to Fort Myer from Washington for the first public demonstration of a Wright airplane.[46] Squier declared to the assembled reporters: "This is a big moment, boys. This is an event of world importance. Don't forget that."[47] Reporters explained to their readers that Lahm was the

> first army officer in the world to fly in an aeroplane, a type of airship which military experts say will come into general use in all the armies of the world before many years have elapsed.[48]

7. Origins of Army Aviation   87

Top: The Wright brothers' plane at the Fort Myer trials, 1908 (courtesy CECOM Historical Office). *Bottom:* Wilbur Wright (in straw hat) timing a flight; George Squier is at extreme left (courtesy CECOM Historical Office).

Three days later Orville carried Squier aloft for a new world's record of a nearly ten-minute flight with two men. On landing, the reporters engulfed them,[49] and Squier exclaimed:

> That was bully. It is the most exciting sport in the world. The thing that amazes one is the easy manner in which Mr. Wright controls the machine. It is remarkable. I'd have to exhaust the list of descriptive adjectives if I started out to describe the sensations of the trip.

Turning to Orville, he said, "I want you to keep that record standing for a couple of days at least. Will you do it?"[50] Squier's transparent excitement was notable. With Allen in Europe, he served as the President of the Aeronautical Board and as Acting Chief Signal Officer. To the reporters, his declarations were those of a responsible military expert. It is likely that he is the expert referred to above who predicted the popularity of airplanes in future armies.

When questioned if he thought a dirigible air battleship carrying armament, as proposed by one British inventor, was possible, Squier remarked: "That is extremely hard to say. Now that Mr. Wright is accomplishing such wonders in the air almost anything in the aeronautical line seems possible."[51] Secretary of War Luke Wright, on the other hand, expressed skepticism on the present practicality of airplanes. He said:

> I can't see that these aeroplanes are going to be especially practical just yet. They are remarkable in that they represent the actual conquering of the air, but until they are still further developed I do not think that they will be of much service from a military standpoint. They might be of some slight use in scouting but they are in the experimental stage just at present.[52]

Squier's predictions for the military use of airplanes seemed restrained, cautious, and unimaginative by comparison to today's realization of the potential of aircraft, even of its realization just ten years later. For his own time, considering his official position and the inclinations of his superiors, his attitudes were progressive and unorthodox. The public became more and more excited

Sign at Fort Myer showing that George Squier was the second passenger of the Wright Brothers (courtesy CECOM Historical Office).

with each passing day, witnessing new aerial feats and records. And then the feared event occurred.

On the 17th of September Orville took up his third passenger, a young aerial enthusiast, Lieutenant Thomas Selfridge. A mechanical failure caused the airplane to go out of control and plunge to earth. Orville was injured and Selfridge, who had designed four of Bell's planes and worked with the Aerial Experimental Association for nearly a year, lost his life.[53] Squier was sorely affected during the short time following the accident while Selfridge still lived and attempts were made to assess the extent of Wright's injuries. A *Washington Post* reporter told his readers that Squier walked up and down the porch with his hands behind his back and his Panama hat pulled down over his face, a "picture of dejection." He reported Squier's concern: "It's frightful It's frightful! Just at the moment of success, too."[54] The next day the Aeronautical Board investigated the cause of the accident with the assistance of Lahm and Chanute. Squier was absent during the investigation. He later approved their findings and forwarded their report to the Adjutant General.[55] Squier also sent a wire to Allen, in Paris, simply stating Selfridge's death.[56] On the following morning he ordered a steel stake driven in the ground where Selfridge fell to his death to mark the location of the world's first military aviator casualty.[57] When Squier commanded America's air forces about a decade later he directed that Selfridge Field, Michigan, be named in honor of the first military aviator to lose his life. Despite the loss of Selfridge, Squier and the Aeronautical Board were enthusiastic about the future of airplanes.

Orville told his brother that Squier said, "This is just the beginning of business with us. He says the U.S. is going to be our best customer."[58] Squier's enthusiasm was evident in the large number of photos, accompanied by appropriate narratives, he sent to military commanders, school commandants, newspaper editors, and magazine editors.[59] In public, however, Squier carefully maintained a discreet impartiality since Herring, who had won another delay, was due to present his airplane to the same board for acceptance trials.[60] Squier's optimism about obtaining a sizeable congressional appropriation to further the future of American aviation was tempered by the impact of Selfridge's death.

That Squier and other airplane enthusiasts had hoped successful trials would spur Congress to sponsor American aviation was evident in a letter from Ernest LaRue Jones, the editor of *Aeronautics*, to Squier. He expressed regret over the loss of Selfridge and the adverse affect it would have upon aviation, and "for your appropriation—considered sordidly."[61] The Jamestown strategy to promote authorization for purchase of an Army dirigible of the previous year had seemingly failed for airplanes. Nevertheless, Squier replied with conviction that "the great subject of aeronautics will press forward and that it has achieved such a tremendous headway that at present no power on earth can stop its progress."[62]

Squier's role and statements during the Wright trials of 1908 reveal him as the probable source of support for airplanes during the previous fall when Allen and Chandler adopted a clear policy before the Board of Ordnance and Fortification in favor of dirigible balloons for military purposes. If any doubt existed about Squier's sympathies they were dispelled by his answer to Lester French, editor of the *Journal of the American Society of Mechanical Engineers*, on the relative importance of airplanes to dirigible balloons. Squier stated:

> This office regards the dirigible balloon of equal promise with the aeroplane and believes that both types of and combinations of them should be developed side by side, each having its own particular field of usefulness. The dirigible balloon will probably be the burden bearing machine of the future, while theoretical considerations indicate that the aeroplane type, in single units, at least, is limited to smaller tonnage and probably high speed.[63]

In strong contrast to Allen and Chandler's concept of static employment of dirigibles in warfare, Squier suggested a dynamic role for the airplane:

> From a military standpoint the general development may be compared to past naval development in that the dirigible balloon will represent the battleship of the future, while the aeroplane will represent the fast cruiser or torpedo boat for scouting purposes and as a fast destroyer.

Squier's firm confidence about the future of military aviation found unusual expression in recruiting literature prepared during the Wright trials. In 1908, Army enlistments were seriously decreasing. To assist the Adjutant General in preparing a new booklet to be issued to all recruiting officers, each of the branch chiefs was asked to prepare a manuscript on the activities of their organizations with a view to attracting young men to the Army. Squier submitted a remarkable manuscript, which advertised opportunities for a "limited number of specially selected men" to secure training and experience in airplanes.[64] Although the plane crash ended any chance of Wright's airplane completing its acceptance trials at Fort Myer, Squier eagerly wanted an airplane for the Army.

In early October, Squier submitted a suggestion to the Aeronautical Board for yet obtaining acceptance of the Wright airplane in 1908. Wilbur had another model in Europe, where he acquired public accolades and honors for his aerial accomplishments and technical achievements in conquering the air. Squier proposed that the Board allow Wilbur to fly for Allen, still in Paris, and thereby complete the acceptance trials.[65] But the Board rejected the suggestion and stipulated new trials for the following year.[66] At this critical juncture in obtaining support for military development of the airplane, Squier commenced a period of intense publication about the airplane for military use. His culminating effort was a major paper read before the American Society of Mechanical Engineers at their annual meeting at New York, in early December.

The Status of Aeronautics, 1908

Allen, Squier, and Lahm went to New York for the annual meeting of the American Society of Mechanical Engineers.[67] Lahm presented a popular lecture on aeronautics, illustrated by slides of Army Dirigible No. 1 (the one built by Captain Baldwin) and the Wright airplane trials. So avid was the interest in aeronautics that attendees filled the large auditorium of the Engineering Societies Building to overflowing, the first time any presentation had more than filled the auditorium since the building opened.[68] It was Squier's paper, however, that received the serious attention for its technical treatment of the subject of aeronautics and its possible application to military problems. Aeronautics was considered a subject due the special consideration of engineers as, according to the editor of *Engineering News*, the methods of the engineer had been substituted for those of the inventor in the development of airplanes. American and English editors alike appreciated the solid engineering treatment Squier brought to the subject. *Engineering News* called it the best general review of the progress to date that has appeared anywhere.[69] The British journal *Flight* considered it so significant that they published it unabridged. They found the paper

> well calculated to dispel such lethargy as there is in the attitude of the public towards aeronautics; and we therefore feel that it is deserving of the greatest degree of publicity that can be given to it.[70]

Squier indeed intended to direct professional attention to the problems of flight and to stir public interest in its commercial and military possibilities.

What made his paper substantially different from previous expositions, and from attitudes current in the Aeronautical Division, was the equal treatment he accorded airplanes and dirigibles. He reviewed the extensive development (primarily military) of British, French, German, and American dirigibles and airplanes. His information on dirigibles came largely from Lieutenant Lahm's report submitted to Allen in early January; Lahm dealt exclusively with dirigibles and completely ignored airplane developments.[71] That part dealing with airplanes reflected Squier's own influence in internal Signal Corps policy making. Historians of airpower have sought positive statements by early Army officials on the use of airplanes for obtaining control of the air and delivering ordnance to targets beyond the enemy lines. Generally they have been disappointed in their search in the period before the First World War. It is in this regard, therefore, that Squier's 1908 paper is particularly notable.

As no one used airplanes for anything but sport and amusement in December of 1908, any discussion, then, would necessarily be theoretical. Squier tried to present the broad outlines of military aeronautics and aviation. Unlike his colleagues who gave primacy to dirigibles, he declared that both types of aircraft would be developed—each for a different purpose. Maintaining that Euclid's square-cube theorem connecting the volumes and surfaces of similar figures (volume is proportional to the cube of the major dimension of an object; surface area is proportional to the square of the major dimension) placed a severe limitation on the weight that airplanes could carry, he said dirigibles in the future would be used to carry heavy loads. Airplanes would take advantage of their possible great speed through reduced plane surface area and increased power.[72]

Dirigibles and airplanes together comprised his elements of airpower in a form recognizable by contemporary airpower theorists. He assigned three major roles to dirigible and airplanes: strategic, tactical, and reconnaissance. To the comparatively small dirigibles he assigned the tactical function of bombing bridges and supply depots close to the mobile army or coast defenses. The heavier dirigible would carry larger burdens of explosives and operate far behind enemy lines against his bases of supplies; his dry docks, arsenals, ammunition depots, principal railway centers, storehouses and his navy. The airplane could fulfill a reconnaissance role and certain other unusual missions. Urgent messages could be dispatched at high speed, important commanders ferried into critical locations to lend the power of their personalities to the tide of battle or personally direct artillery fire, and raids made into the adversary's capital city for the purpose of capturing the very leaders of an enemy government.[73] Squier thought that placing responsible leaders of state in immediate and personal danger would act as a deterrent to aggressive behavior.[74] Thus, he argued, the introduction of aerial navigation introduced a third dimension to fighting between nations and negated the importance of national frontiers as natural lines for fortification.

Squier warned that, under the new conditions posed by addition of a third dimension to warfare, national boundaries no longer afforded definite limits to military movements. His thinking was deceptively modern. He omitted an important element of modern airpower doctrine introduced by two of the basic theoreticians of airpower, General Giulio Douhet, and America's leading early exponent, General Billy Mitchell. The element which Squier failed to address was intimidation and terrorization of the adversary's civilian population. Massive destruction of noncombatants and unfortified areas would clearly have been rejected as criminal. He asserted that airplanes and dirigibles provided the power unknown in previous wars for achieving great victories, but cautioned that they must also "tend to produce results with the minimum loss of life."[75] The sentiment of the world demanded, he thought, that the aim of military force was to capture and not to destroy:

> It may be said that the consummation of military art is found in maneuvering the enemy into untenable situations, thereby forcing a decisive result with a minimum loss of life and treasure.[76]

And airpower became the instrument, as Squier predicted, by which such untenability might be achieved. His statement of the purpose of military power sounds contemporary:

> We have arrived at a conception of the principle of an efficient army and navy, not to provoke war, but to preserve peace, and it is believed, that, following this principle, the perfection of ships of the air for military purposes will materially contribute on the whole, to make war less likely in the future than in the past.[77]

This possible use of airpower, he thought, was evidence that the world was becoming more humane with each passing year.

While Squier's thoughts about the role of airships in warfare and commerce sound commonplace now, they were regarded as radical, bordering on irresponsible, when he presented them.[78] The editor of *Engineering News* criticized Squier for declaring that the airplane was to play an important role in future warfare.[79] Several months earlier they had taken issue with government promotion of aircraft, and the intervening events gave them no cause for changing their position, notwithstanding Squier's claims.[80] Warfare was a serious business. Certainly, they argued, no more vulnerable machine existed than the airplane. It was slow-flying and incapable of being armored due to severe weight limitations. Squier's assertions about airships bombing supply bases and naval ships were dismissed as unrealistic since modern artillery could easily knock down even high-flying craft, and since explosives had a limited radius of destruction. Even the enthusiastic British journal *Flight* said the inclusion of actual specifications for an airplane lent "an unmistakable strain of reality."[81] *Engineering News* considered Squier's assurance of a military and commercial future for aeronautics to be potentially mischievous. They advised potential investors and inventors to remember that aerial navigation has more scientific than commercial promise.[82]

Although Squier may have seemed daring, he did have one important constraint placed upon his public utterances. On 10 March 1908, the United States acceded to the declaration of the Second International Peace Conference, held at The Hague in the fall of 1907. Of all the first-class powers only the United States agreed to "prohibit, for a period extending to the close of the Third Peace Conference, the discharge of projectiles and explosives from balloons or by other new methods of a similar nature."[83] Such a treaty obligation imposed serious limitations upon federal officers when discussing publicly the military potential of aircraft. Nevertheless, Squier assumed the heavy responsibility of openly declaring the inadequacy of prohibition. Were actions required in the interest of self-defense to be proscribed? If it was permitted in time of war to sow the seas with mines, why disallow dropping aerial mines? And if dropping aerial mines against enemy shipping was permitted, on what grounds should fortified places on land be exempted?[84] Although Squier excluded the bombing of civilian populations as part of legitimate warfare, he clearly argued for a concept of airpower considerably in advance of his contemporaries and containing all the essential elements considered necessary to fulfillment of air-war missions, namely, strategic bombing, tactical support, and reconnaissance.

Squier's paper was immediately recognized as more than a popular presentation on aeronautics.[85] One of its most valuable features (from a scientific standpoint) was a comprehensive aeronautical bibliography. Squier commenced the research himself from Aeronautical Division holdings, acquired in part from foreign attachés, and the Aero Club of America library.[86] In early November 1908, with Allen still in Europe, Squier obtained the help of his friend Alfred Zahm to undertake the compilation of the bibliography for inclusion as an appendix to his December paper.[87] This bibliography may have formed the basis for aeronautical bibliographies subsequently published by the Smithsonian Institution under the direction of Paul Brockett.[88]

Squier deposited a copy of the paper with the Smithsonian and sent additional copies to foreign armies and aero clubs, senators and congressmen, and adjutant generals of every state.[89] This one paper, and Squier's experience gained through flights in balloons, dirigibles and airplanes, established him as one of America's authorities on aeronautics. Secretary Walcott of the Smithsonian soon sought his advice on ways of promoting the scientific study of aeronautics within the United States.

The "Invisible Establishment" of Aeronautics

In December, at Alexander Graham Bell's suggestion,[90] the Regents of the Smithsonian Institution authorized a Langley Medal to be awarded for specially "meritorious investigations in connection with the science of aerodromics and its application to aviation."[91] When Bell wrote Walcott, he suggested Squier and Chanute as obvious members of a committee to select the first recipients of the Langley Medal.[92] Unsure of who other members might be, he suggested John Brashear, Charles Pickering or Simon Newcomb.[93] Walcott replied with a request that Bell himself serve on the committee as he wanted only "men of known competence in the subject of aerial navigation."[94] Invitations were sent to Squier, Chanute, Brashear, and James Means, editor of the *Aeronautical Annual*.[95] Walcott designated Chanute chairman of the committee.[96] Walcott told Lodge that he had selected gentlemen "of known competence in this science, whose judgment it would be absolutely safe to follow."[97] From the beginning, the committee divided over interpretation of the term "investigations" in the Regent's resolution and the extent to which it affected the eligibility of the Wrights' accomplishments for recognition.

What at first must have seemed to the Regents a simple task of selection became stalled on the question of whether the medal should be conferred for achievements in pure science or in the applied science of flying.[98] A close friend of the Wrights, Chanute held the pure science side of the dispute and asked why, if the Regents meant otherwise, would the resolution use "investigations" rather than "results" or "achievements."[99] Somewhat later Squier himself confessed his misunderstanding of the Regents' intentions to Chanute.[100] From the very beginning, Bell, Lodge, and Walcott favored the selection of the Wrights.[101] Squier, Means, and Brashear joined them.[102] In addition to the matter of interpretation, Chanute bore resentment over the Wrights' concealing their work in the hope of large profits.[103] Now the Regents intervened and informed the committee of their opinion that "the Wright Brothers were preeminently entitled to be the first recipients of the medal on account of the notable work in the science of aerodromics."[104] Several weeks later, the controversy having ended decisively, Chanute transmitted the committee's selection of the Wright Brothers to Walcott.[105] He had actually informed the Wrights only a week after the Regents indicated their choice.[106] The presentation of the medal was delayed by almost a year because of difficulties in striking the medal[107] and finding a time when the Wrights were unoccupied with patent suits.[108]

Meanwhile, Walcott asked Squier to chair another committee. Marvin McFarland of the Library of Congress Aeronautics Division, the editor of the Wrights' papers, identified a small group of men who were personal friends of Langley and who felt committed to justifying Langley's importance, if not priority, in the achievement of powered flight. Prominent among this group were Walcott, Greely, and Bell.[109] Known as the "friends of Langley," they exercised sympathetic influence over their subordinates in composing a sizeable lobby of individuals committed to opposing a favorable outcome for the Wrights in their infringement suit entered against the Herring-Curtiss Company, Glenn H. Curtiss, and the Aeronautic Society of New York, in August 1909.[110] The group also assumed a positive, constructive role in promoting government support of aeronautics. In its expanded form, the group included Zahm, Squier, Curtiss, Bell, and Walcott. They constituted an "invisible establishment" of aeronautically minded men who shared a common devotion to science. They represented governmental or quasi-governmental scientific bureaus, attained high standing in their professions and their bureaus, and belonged to Washington s most prestigious scientific club, the Cosmos Club.[111] (Paul Brockett was very much a part of the establishment's activities, as Walcott's secretary, although only meager information exists about his life and career. He deserves more study than he has received.) The invisible establishment tried to foster the scientific study of aeronautics. While the Aero Club of America would not have opposed such a goal, they concentrated their efforts on flying achievements and public promotion of flying as a sport, in the hope of interesting Congress and the Government in encouraging aviation. It was the invisible establishment that strove toward the founding of a national aeronautical research laboratory.

A Proposal for a National Aeronautical Research Laboratory

Despite the tragic ending of the Wright trials in September 1907, Allen retained the hope that Congress would appropriate $500,000 so the Signal Corps could take up study of aeronautics "in a serious way."[112] Within six months, expectations of receiving real assistance from Congress waned. Allen explained that the main retarding factors in the development of aeronautics in the United States were lack of a commercial market, the unavailability of a place where aviators could practice flying, and the need of a good engine.[113] Without aeronautical appropriations, the Signal Corps could not realistically proceed with a scientific and engineering study of flight. Although the upcoming Wright summer trials might have helped Allen's cause, the Secretary of War had already decided that reduction of expenses in the War Department enjoyed a higher priority than support of aviation.[114] By mid–1909, Allen lost all hope of accomplishing anything in aviation in the next year. Meanwhile, Walcott, still engaged in the Langley Medal project, directed Brockett to produce a bibliography of aeronautics.

The most competent man, in Walcott's opinion, to review Brockett's work, report on its technical merit, and its suitability for publication, was Squier. Thus, in April 1909, Walcott asked him to form a committee to examine Brockett's manuscript.[115] A month later, Squier and Zahm recommended publication.[116] The award of the Langley Medal and preparation of aeronautical bibliographies were important initial steps in fostering the scientific study of aviation in the period following Langley's secretaryship. But the Smithsonian had insufficient funds to support the scientific study of aviation itself and was forced for the present to rely upon the Signal Corps for the necessary funds.

Allen asked again for $500,000 for fiscal year 1911, with the understanding that the Government did, indeed, intend to take up aviation.[117] He noted some foreign governments were spending up to five times his requested amount in the belief that dirigibles and airplanes were absolutely essential adjuncts to a modern army. Allen stated that $500,000 represented a minimum level of support and should not be further reduced. He advised complete abandonment of aviation support, pending further consideration, rather than accept a reduced appropriation.[118] Even the successful Wright trials in July, when the Aeronautical Board, chaired once more by Squier,[119] accepted the world's first military airplane, failed to stimulate sufficient congressional support for an appropriation. Nearly two years elapsed before an appropriation was made for aeronautics.

Public enthusiasm was high, however, over the potential of flight. Numerous invitations flowed into the Chief Signal Officer's office for him or officers of his staff, notably Zahm or Squier, to address meetings on aeronautics. Allen's uniform reply betrayed his extreme disappointment over the Congress's refusal to support aeronautics. The reply was reminiscent of his handling of the balloon controversy at Fort Leavenworth several years earlier, insofar as it failed to recognize the importance of patiently building a large and effective constituency capable of influencing those responsible for appropriations. After a flurry of publications and activity reaching over a period of two years, Allen suddenly terminated any further public contacts. When the Franklin Institute in Philadelphia requested either Zahm or Squier to address their winter meeting, Allen replied that, as little practical work could be accomplished by the Signal Corps in aeronautics, he had decided to decline, without exception, all requests for outside lectures.[120] Allen sent the same reply to requests from the Society of Arts of the Massachusetts Institute of Technology, the Aero Club of Pennsylvania, and the American Association for the Advancement of Science."[121] Even the formidable Glenn Curtiss requested Squier's detail to the first International Aviation Contests to be held in America, at Los Angeles. Allen rejected Squier's assignment and sent Lieutenant Paul Beck, an infantry officer with a checkered career (on detail to the Signal Corps) who was stationed at the Presidio in San Francisco.[122] Although Congress refused to appropriate money for aviation, Allen still had the potential support of the Board of Ordnance and Fortification. Allen's termination of public engagements did not signal his complete lack of interest in aviation. He made one last effort, in 1909, to obtain financial support for airplane developments. Board of Ordnance and Fortification allotments for aviation had provided a total of

$46,000, of which only $30,000 was expended due to cancellation of Herring's contract. Allen twice asked the Board for permission to use the balance of $16,000 for experimentation and the purchase of other types of airplanes, mainly the Curtiss types, since Congress had made no appropriation for fiscal year 1910.[123] When the Wrights initiated their patent infringement suit in court, Allen promptly returned all $16,000 to the Board with an expression of doubt as to "whether in the immediate future it would be advisable to purchase additional flying machines."[124] Allen's last attempt to secure support for aviation collapsed with the Wrights' legal action.

Allen's efforts to develop support for an airplane program within the Army foundered in the Wrights' litigation. Nearly everyone in aviation turned against the Wrights.[125] Chanute's fears were realized. The Wrights appeared to care more about money than about common sharing of their knowledge for the sake of science. In an age acutely sensitive to big corporations, the specter of monopoly loomed before everyone's eyes. It seemed to threaten the growth of powered flight in America. Squier earlier remarked to Orville Wright that the government would be their best customer. The patent suit served, however, to retard any further military investment in aeroplanes for several years and to inhibit development for another eight years.[126]

As a consequence of Allen's inability or disinclination[127] to pursue further funding for aviation, Squier's link with the Smithsonian became more important.[128] Delayed nearly a year after announcement of the recipients, the Langley Medal presentation ceremonies offered members of the invisible establishment an opportunity to discuss the need for aeronautical research.[129] Wilbur Wright declared that there was a great deal of work to be accomplished, enough to keep a "large number of investigators busy for a lifetime."[130] Walcott thought most investigators would be concerned with the manufacture and sale of airplanes so the Smithsonian might become involved in further aeronautical investigations "if there was something in the field of abstract science."[131] The next few months saw a change in Walcott's attitude regarding a proper role for the Smithsonian to play in aeronautics.

In mid-April, Walcott asked Bell to consult with Squier and Zahm on the conduct of aerodynamic experiments that would advance the development of aeronautics.[132] The three readily came to agreement that America needed a National Aeronautical Laboratory, supported at government expense, to investigate all scientific questions related to the problem of flight. Squier proposed the establishment of a committee, representing the various departments of the government currently conducting aerodynamic research, to guide research policies. He suggested the following departments and appropriate agencies within them: Army, Navy, Commerce and Labor (Bureau of Standards), and Agriculture (Weather Bureau).

Proper coordination of work was the principal concern. A similar situation existed a half decade earlier in the radio field. Difficulties between departments were then resolved by the creation of a Board on Wireless Telegraphy, composed of representatives from the Army, the Navy, and the Weather Bureau.[133] Squier suggested forming a parallel board for coordinating government aeronautical research under the leadership of the Smithsonian, because of its "commanding position ... in the scientific work of the government."[134] Only government sponsorship could assure a level of support essential to a laboratory in which the fundamental problems of aerodynamics could be investigated. Without the promise of a market, private concerns would avoid the heavy investment of basic research. The interdepartmental committee should address basic research as its first priority. Squier believed that the report of such a committee "could not fail to carry great weight with Congress in securing an appropriation."[135] Bell considered that a laboratory, under the direction of the Smithsonian, could investigate specific matters at first and later cover the entire field of aeronautics.[136] The paradigm of such a committee and laboratory was the newly established British Advisory Committee for Aeronautics.

For at least a year the office of the Chief Signal Officer had investigated foreign aeronautical laboratories. Allen requested the General Staff to obtain all possible information on the aerodynamic laboratory at Koutchnif, Russia, and shortly afterward on the British Advisory Committee on Aeronautics.[137] Squier sent Lord Rayleigh's report on the British Aeronautical Committee's activities to Bell and Walcott.[138] As a result of his conferences with Zahm and Squier, Bell suggested that Rayleigh's report might well form the basis for an American research organization and program.[139] Walcott responded with an enthusiastic well-done and promised he would seek an interview with the President on the subject of appointing the recommended committee.[140] Although effort produced no immediate results, it did lay the groundwork for the subsequent creation of the National Advisory Commission on Aeronautics,[141] organized along the very lines suggested by Squier, who drew upon the models provided in the Inter-Departmental Committee on Wireless Telegraphy and the British Advisory Committee on Aeronautics.[142]

8

Radio Over Telephone Lines

Squier's long hiatus in research and invention ended with his assignment to a major new research effort aimed at developing improved military radios. His unique research in combining the engineering of line telephony and radio initiated an entirely new branch of communications engineering. Although telephone and telegraph companies saw little immediate commercial application, Squier's system of multiplex telephony provided a concrete demonstration of the practicality of multiplexing voice messages, i.e., the simultaneous carrying of two or more messages in the same or opposite directions on a single-wire circuit. Telegraph systems were long capable of multiplex operation. Squier accomplished the same feat for voice conversations by directing radio frequencies along telephone wires and detecting them at the receiving end in the same way that radio engineers detected their signals, i.e., by using a crystal detector or DeForest's new vacuum tube. Dubbed "wired wireless" by Squier, the system later became known as carrier current telephony.

Squier's resumption of research in electrical communications was abetted by three significant events. First, in 1908 one of the country's leading electrical communications firms, the Stone Telegraph and Telephone Company, disbanded. Several experienced and talented engineers were forced to leave the company's laboratory at Cambridge; among them were the "Stone Trio." Each of them subsequently made their own contributions to the development of electrical engineering. The Stone Trio consisted of George H. Clark, Frederick A. Kolster, and Ernest R. Cram. Clark, the first of Stone's "Three Musketeers," as Stone himself referred to them, found outside employment first. The Navy hired him as a radio engineer with the imposing title of Sub-Inspector of Wireless Telegraph Stations for the Bureau of Equipment.[1] Kolster worked several years as an assistant in Lee DeForest's laboratory in New York. In 1911, Edward B. Rosa (pronounced Ro-zay) director of electrical researches at the Bureau of Standards, hired Kolster to investigate difficulties in radio engineering originating from industry.[2] Cram, the last of the Musketeers, obtained the civilian position of Radio Engineer with the Signal Corps. He remained with the Signal Corps for 25 years. Clark commented that "for a number of years the 'Stone Trio' were very closely coupled with their country's official wireless procedure."[3] Stone himself resumed his former profession as an independent consultant and expert in patent causes in New York.

The second and third events concerned money and laboratory space. In 1908, the Signal Corps applied for a radio research appropriation of $30,000. While radio communications were widely recognized to have a promising future, their practical military applications in 1908 were severely limited. The Navy showed a similar interest in the practical applications of radio to their operations in the establishment of the U.S. Naval Radiotelegraphic Laboratory in the new Bureau of Standards laboratory building. Soon thereafter the Chief Signal Officer requested the Secretary of War to procure laboratory space for the Signal Corps at the Bureau of Standards.[4] Director Samuel Stratton willingly assigned space just down the hall from the naval laboratory to Squier and his assistants. For nearly a quarter of a century, these two laboratories represented the only government agencies conducting their own scientific and standards research in the Bureau of Standards.[5]

The Decision to Renew Radio Research

Dissatisfied with commercial radio telegraph sets available on the 1907 market, the Signal Corps began designing and fabricating their own radios in a small laboratory in Washington, D.C.[6] By mid–1908, Signal Corps technicians were achieving success. Some 46 radiotelegraph sets were assembled by laboratory workers and distributed to field units.

Despite these gratifying advances in radiotelegraphy, interest focused upon radiotelephony as a more effective means of military communication.[7] Messages received in voice offered distinct advantages over those in Morse Code. Voice transmissions could eliminate or reduce the need for special training of operators. Voice transmissions certainly permitted faster reaction to orders. Allen therefore initiated a period of practical testing of several commercial radiotelephone sets.

For almost a year, several radio manufacturers demonstrated the performance of their sets. At the end of the test period, Signal Corps officers recommended rejection of all the radios.[8] Most radios of the time used the phenomenon of a spark, or an arc, generating radio waves. Manufacturers featured a variety of ways of producing and controlling the spark, but essentially all radio equipment depended on the favorable electromagnetic properties of the spark. Electromagnetic or radio waves created by the spark traveled through space without the aid of wires. However, its broad-band transmission possessed an undesirable property. As sparks produced numerous radio frequencies, receivers in the vicinity were subjected to serious interference.[9] Even when attempts at narrowing the band of frequencies succeeded, the band itself was unstable, moving to and fro in the frequency spectrum. It did not have to move very far before causing interference in nearby receivers tuned to neighboring bands. Another unfavorable property of spark transmitters was their low power. Signal Corps requirements demanded radios capable of transmitting between 50 and 100 miles. The best equipment tested reached only a tenth as far. Low power also led to a form of interference created by natural electrical disturbances in the atmosphere, unsurprisingly called atmospherics. Such considerations as ruggedness, portability, and ease of operation, naturally counted as well. An ideal transmitter of voice communications generated a single frequency, high-power, stable wave in the radio frequency range, which extends from about 13,000 cycles per second (called Hertz; to 10,000 Megahertz). Spark transmitters failed to meet these requirements.

Only one other electrical device of the time could theoretically satisfy the requirements of the Signal Corps. It was the alternator, a generator that produced an alternating current. Squier thought highly of the alternator because it could produce a stable wave, of one frequency. Transmitting a single stable frequency did away with the problem of radio interference associated with broad-band transmission. One notable alternator of the time, developed by Reginald Fessenden, could generate a very low radio frequency signal of 50,000 Hertz. But low power, bulk, and susceptibility to atmospherics ruled out Fessenden's alternator as a source of power for field radios. Therefore, Allen and Squier advised the Board of Ordnance and Fortification that wire lines remained the only practical method for transmitting voice communications.[10]

Nevertheless, there were no apparent reasons for believing that radiotelephony could not ultimately be as important as wire telephony, especially in coast defense applications. So Allen proposed developing new apparatus as soon as funds became available.[11] Allen suggested to the Chief of Coast Artillery that he submit an estimate of $30,000 for development and purchase of radiotelegraph apparatus to be installed at each of the radiotelegraph stations in the headquarters of the seven artillery districts in the United States. Allen informed the Chief of the Coast Artillery that tests performed in his office showed that "within one year practical wireless telephone equipment for operation up to distances of 50 to 100 miles will be available."[12] It was late in the year to make the request, but he believed that a year's delay could not be afforded. The Army Appropriation Act of March 1909, therefore, included an authorization of $30,000 for Signal Corps research in radiotelephony.[13]

When the money became available on 1 July, Allen placed Squier in charge of the research project.[14]

When Squier began his experiments, reports were circulating about a new alternator built by E.F.W. Alexanderson. It produced an alternating current with a frequency of 100,000 Hertz. Squier later described the appearance of the Alexanderson high-frequency alternator as having "exactly the same effect at that time as the introduction of a new telescope would have for the astronomer." He said he realized he possessed a "new electric telescope in his hands with which he could explore the new electric heavens."[15] With a part of his recent appropriation, Squier purchased one of Alexanderson's new alternators from the National Electric Signaling Co., Fessenden's firm.[16] With the prized Alexanderson alternator in his possession, he sought a new laboratory in which to conduct his experiments.

A New Laboratory for Radio

For the first time since 1895 at the Johns Hopkins University, Squier reentered the life of a researcher in a civilian laboratory. In 1909, he established a new research laboratory at the Bureau of Standards, located outside Washington in new facilities. Since 1900 the Bureau of Standards had grown in the image of Samuel W. Stratton. Accepting an invitation in 1899, Stratton left his professorship in physics at the University of Illinois and took the position of Inspector of Standards in Washington. With the encouragement of the Secretary of the Treasury for whom he worked, Stratton drafted a bill for a National Standardizing Bureau and began organizing arguments for congressional hearings to come.[17] Stratton's proposals found wide support in the scientific and business communities. The United States was the last of the major Western powers to institute a standards laboratory in accordance with scientific principles and staffed by scientifically trained men. International commerce and science both necessitated a high-grade laboratory for setting standards. Congress responded by creating the National Bureau of Standards on 3 March 1901.

One of the most urgent needs in the matter of standards concerned research on electrical matters. So one of Stratton's first duties as director was finding an outstanding man to plan and establish a division of the bureau devoted to electrical researches. He hired Edward B. Rosa from a physics professorship at Wesleyan University. Rosa had studied electricity under Henry Rowland at the Johns Hopkins University, graduating in 1891. Of the staff of the electrical division, almost half were graduates of Hopkins. Rosa and others may have known Squier from their student days.

With 13 subdivisions, the bureau was soon one of the largest and most complicated branches in the government.[18] In February 1903, the bureau was transferred to an already-outsized Department of Commerce and Labor. The bureau was still located in downtown Washington when its new physical laboratory was scheduled for completion. In fact, it was finally ready for occupancy in late 1904, two years behind schedule. About four years later the Navy asked Louis W. Austin to come to the bureau to investigate certain applications of radiotelegraphy. Austin left a teaching post in physics, at the University of Wisconsin. He also counted in his experience two years at the German *Reichsanstalt*, the larger German equivalent of the Bureau of Standards. Austin stayed on for the next 24 years as director of the U.S. Naval Radiotelegraphic Laboratory.

Austin was already ensconced as director when Allen asked for space at the Bureau for the Signal Corps. By the end of 1909 three laboratories at the Bureau of Standards worked on radio. No conflicts existed, however, since the bureau laboratory, under J. Howard Dellinger, investigated high-frequency radio waves and the two service laboratories worked on generation and detection techniques of low-frequency radio waves.[19]

Complete radio stations were installed at the bureau and in the old Signal Corps laboratory, located at 1710 Pennsylvania Avenue. The two laboratories, situated about seven miles apart, were specially connected by a commercial telephone line. Two arc transmitters, which produced frequencies in the 400,000 Hertz to 500,000 Hertz range, were also installed. These frequencies were close to the highest frequencies attainable at the time. No long-distance transmissions were attempted; instead, they sought first a complete understanding of the practical problems of operation.[20] Alexanderson himself was dispatched from his firm to direct the installation of his alternator in the basement of the building at 1710 Pennsylvania Avenue.

Multiplex Experiments

Squier combined the radio and wire engineering specialties in one stroke by connecting a radio to a wire circuit. No loss of radiation occurred, i.e., the signal was guided entirely along the wire instead of through it, where electrical losses take place (the physicists call it "skin effect"). The frequency range he explored had hitherto been uninvestigated. The new Alexanderson alternator made careful examination of that region of the radio frequency spectrum possible. He used a triode vacuum tube as a detector, probably the first person to use one in connection with speech or telegraphic transmission over physical circuits.[21] DeForest himself seems not to have contemplated such a use for the triode vacuum tube.[22] Squier showed that a number of radio devices could be used on wires. No one before him had demonstrated the application of so many radio devices in telephony. Many, and most notably the triode, became essential in wire telephony. Their combined use made it possible to send more than one message on the same wire at the same time—the definition of multiplex communications. Squier termed his system of combined radio and line telegraphy "wired-wireless." The first successful operation of his unique system took place in September 1910, about ten months after his investigations began.

Modifying the Alexanderson machine to produce satisfactory waves for all frequencies between the very end point of human hearing at about 20,000 Hertz and 100,000 Hertz, a region where it was not originally designed to operate, posed the most vexatious moments for Squier, his assistants, and Alexanderson.[23] At last they succeeded in obtaining a full range of satisfactory frequencies from the alternator.

The first phase of the project consisted of pure physical research. No one before Squier had investigated the frequency band between 20,000 Hertz and 100,000 Hertz. Apparatus capable of generating stable waves of one selected frequency in that entire region were simply unavailable before the Alexanderson alternator. Spark and arc transmitters customarily emitted frequencies above 200,000 Hertz, while telephone circuits employed audible frequencies below 5,000 Hertz.

In need of a long telephone cable to test his ideas and to obtain certain electrical measurements of a cable carrying high-frequency currents, Squier leased a seven-mile line from the Bell Company in Washington.[24] During his experiments, the cable also carried normal commercial telephone traffic without interruption. Of foremost interest electrically were the resonance and attenuation tests. Graphing the resonance data produced a curve which showed the frequencies that registered the strongest received signal strengths. The attenuation curve indicated which frequencies arrived at the receiver with the weakest signal strength.

The resonance and attenuation curves Squier obtained were convincing evidence that voice signals in the radio frequency range could be carried on telephone lines. Moreover, by using tuning circuits it would be possible to send multiplex messages at different radio frequencies along one wire. The experiments Squier developed on the Washington telephone line tested the principle on a duplex-diplex basis. (In the terminology of the day, derived from submarine-cable communication, duplex meant two simultaneous signals traveling in opposite directions on the wire; diplex, to two simultaneous signals traveling in the same direction on the wire.) His achievement was unique.[25]

On the afternoon of 29 September 1910, General Allen witnessed the official test of Squier's multiplex telephony system conducted between the laboratories at 1710 Pennsylvania Avenue and the Bureau of Standards. Squier described the occasion in his notebook:

> General Allen first inspected the receiving station at 1710 Penn. Ave. and the arrangement was such that he could listen over each circuit separately or listen over both circuits at the same time by holding the two receivers, one to each ear.
>
> At 1710 Penn. Mr. Rufus R Bermann Assistant in Wireless Telephony Signal Service U.S. Army had charge of the receiving circuits and also acted as one of the four operators required for this test.
>
> At the Bureau of Standards George O. Squier Major Signal Corps and Mr. Ernest R. Cram Expert in Wireless Telephony served as operators for the test and Master Signal Electrician. John F. Dillon Signal Corps was in charge of the operation of the high frequency generator. General Allen after completing his tests at the receiving end of the line 1710 Penn Ave proceeded to the Bureau of Standards and made similar tests.
>
> The talking tests were severe in character and included miscellaneous conversation both ways on the battery telephone circuit [commercial] while other speech of the same and of different character was being transmitted by high frequency circuit to 1710 Penn. Ave. A whistling test was also made and tunes unknown, e.g., sent unannounced from the Bureau over the high frequency circuit were identified on the high-frequency receiver at 1710 Penn. not having been heard on the other receiver and the identifications repeated back to the Bureau on the battery telephone circuit. The severe tests conducted during the afternoon were completely successful.[26]

To have sent and received many more signals over the same wire would have required sharper tuning circuits. Squier's system relied on resonance tuning, which allowed maximum signal current to flow for selected frequencies. The number of frequencies passed, however, was so large that they overlapped and interfered with the information signal carried in neighboring multiplex channels. To avoid inter-channel interference, it was then necessary to maintain large spectral distances between channels. Not until the introduction of George A. Campbell's wave filter in 1917 could the full potential of multiplex telephony be realized. The wave filter was superior to resonant circuits for providing sharp tuning. A brilliant research electrical engineer and mathematician for the American Telephone and Telegraph Company, Campbell designed filters which permitted highly selective tuning of a few frequencies or a band of frequencies. The filter strongly attenuated all other frequencies, thereby permitting the relatively dense packing of channels without mutual interference, each channel having its own wave filter.

Preference for Multiplexed Wires

There were several important military advantages to a multiplex system. Transmission of messages directly from transmitter to receiver along a wire permitted considerable secrecy. Radios, on the other hand, broadcast information for all to receive. More important, though—and this was the notion that motivated his research—was the simple need to make one wire carry more than one message at a time. Large armies, with many separate command posts in the field, relied heavily upon information and orders carried over lines of communication. If one wire could perform the labor of several wires, a considerable increase in efficiency, speed, and effectiveness in military command and control of large bodies of troops could be achieved.[27]

The Patents

The military aspect of the invention appealed so strongly to Allen that he took complete charge of the invention and immediately instituted patent proceedings.[28] He was evidently happy at the outcome of Squier's research. In less than a year's time his organization had made a valuable contribution to electrical communications with their congressional appropriation. It was an accomplishment that would attract favorable attention from professional and business groups and, ultimately, from Congressmen. Allen attached such importance and urgency to obtaining protection for the government that a Patent Office examiner was assigned exclusively to the Signal Corps headquarters to draft specifications and prepare the applications.[29]

This procedure contrasted strongly with the approach adopted ten years earlier. Allen's taking charge of the patent proceeding reflected a change of attitude in the Army toward inventors in uniform, deriving in part from the influence of the Root Reforms. In another time, Squier boldly bargained personally with the Board of Ordnance and Fortification. In 1910, Allen intervened. He ensured government protection for what he believed would be a major advance in electrical engineering by securing the professional services of the Patent Office. While the government obtained more protection, the soldier-inventor received less assurance of reward and uncertain freedom to exploit his invention. Squier took legitimate pride in the fact that he had founded an entirely new branch of communications engineering, namely, multiplex telephony and telegraphy, based on the transmission of radio waves along wire.[30]

The first patent appeared on 3 January 1911. He acknowledged in it that previous attempts at multiplex telephony had been tried, but they all failed because they omitted a radio detector or equivalent electrical device. The presence of a detector was essential because the rapid oscillations of a radio wave left the telephone receiver unaffected, i.e., a telephone receiver served as a suitable detector for low frequencies but not for high frequencies. Thus, an essen-

Demonstration of multiplex telephony in Squier's office on March 23, 1922, 12 years after the invention that made the modern telephone system a reality (courtesy CECOM Historical Office).

tial part of his invention was the incorporation of some form of integrating detector capable of transforming radio-frequency waves into oscillations capable of utilization by a receiver.[31] He had experimented with many types of detectors, and found the triode vacuum tube "an exceedingly useful type, for it requires very little attention in the way of adjustments."[32] His second patent differed from the first in that it specified placing various sources of high-frequency current in series with the line rather than across the line.[33] He added the caveat that, although he had discussed only telephonic messages, the system was perfectly adaptable to the simultaneous transmission of telegraphic messages.[34] The third patent covered a ground connection, which he considered an important addition to the science of telephony. Low-frequency battery telephone systems omitted ground connections because strange noises in the receiver usually occurred. The arrangement required the addition of another wire and additional expense. The addition of two small capacitors made possible the use of silent-earth circuits.[35] One prominent engineer thought this feature one of the system's notable advantages.[36] The fourth patent further rearranged resonant circuits on the line for maximum sensitivity to incoming signals.[37] All four patents were dedicated to the public and attended by great fanfare and acclaim.

General Allen told the press that the unrestricted use of this method was available free of charge to all people of the United States. He described multiplex as the culmination of Squier's life work. Allen commented that

> no one connected with the service challenges his unquestioned right to have retained the invention for his own use and profit, and the patents were issued in his name. That he chose to give them freely to the public gives an added and romantic interest to the story of a really important invention.[38]

John S. Stone, a noted inventor and consultant declared "a new art has been born to us."[39] *Telephony, Telegraphy, and Wireless* magazine stated it was the "greatest advance made in electrical communication since the introduction of the telephone itself."[40] Frank B. Jewett, Chief Engineer of American Telephone and Telegraph, traveled to Washington to see it in person.[41] The Franklin Institute at Philadelphia appointed a committee of distinguished engineers to investigate Squier's research and results. Elihu Thomson, of the General Electric Company, John Stone, and George W. Pierce, of Harvard, studied all aspects of Squier's multiplex invention on a visit to his laboratories in Washington. They reported to the Institute:

> These quantitative experimental results characterize Major Squier's work as a distinct contribution in the field, separating it from the prior art in which, so far as pertains to multiplex telephony through line wires, no such useful results could be predicated; proving for the first time the practicability of the invention on a commercial scale, and giving good promise of an actual commercial use of the principles involved for multiplex telephony over wires.[42]

Since Squier was away on the Mexican border, serving as Chief Signal Officer of Pershing's Maneuver Division, S.G. McMeen, a nationally prominent consulting telephone engineer, read his paper to the annual meeting of the American Institute of Electrical Engineers in Chicago.[43] Although the paper was well received, the highly influential Frank B. Jewett wondered about the applicability of Squier's system to commercial purposes. He described Squier's experiments as indeed beautiful and his invention as an extraordinary laboratory apparatus. McMeen responded that Bell's own invention was nothing more than that when he had completed the telephone. Nothing fundamental had been added to telephone engineering since 1900, in his opinion, and Squier's work should be considered as a fundamental contribution.

Jewett's principal substantive objection dealt with expected attenuation of high-frequency signals over long-distance telephone lines.[44] He considered that such great attenuation would obtain that the system would be commercially unfeasible. He said:

> Further, a few simple computations will show that the attenuation of current at the high frequency which must be used in Major Squier's system is enormously greater than the attenuation at the frequencies which go to make up ordinary speech. The phenomenon of current attenuation in long telephone circuits is not different in character for frequencies of from 200 or 300 to 3000 cycles per sec. and the same attenuation formulas can be used.[45]

In fact, attenuation formulas applicable to low frequencies were inapplicable to the frequencies Squier introduced on the line. The state of knowledge for those frequencies and their behavior on wires was undeveloped. Jewett's error was a serious one in the light of subsequent engineering experience, especially with the advent of vacuum-tube amplifiers.

Jewett's unfavorable judgment of Squier's multiplex system retarded its acceptance. Many engineers, steeped in the conventional engineering wisdom of low-frequency transmissions, agreed with Jewett's judgment:

> I do not say, of course, that it may not be feasible to apply it commercially in certain localized cases where there are special reasons for desiring an additional circuit without the necessity for running additional wires. This could hardly be considered as a general commercial application, for the purpose of extending the range or efficiency of telephonic communication.[46]

Jewett added: "As yet I have not been able to determine that the research, beautiful though it may be from a physical standpoint, possesses any great commercial value of possibilities."[47]

Probably unanticipated by both himself and Squier, Jewett's criticism was the beginning of a long, fractious dispute that ended in a suit brought before the United States Supreme Court. Squier's multiplex patent was the key element in the vast expansion of the American Telephone and Telegraph Company that began in the second decade of the 20th century. In his criticism, Jewett quoted some calculated attenuation figures for high-frequency transmission of signals over commercial telephone circuits indicating such high projected attenuation that the power required to transmit these signals would be enormous. Jewett, in his remarks, raised other issues as well, negating the commercial applicability of this work, including possible problems of interference, the signaling required, problems with transmitting high-frequency signals in the presence of loading coils, for example. His criticism was not only technically wrong; it was contradicted shortly afterward by the actions of the company for whom he worked, which gained great commercial success by exploiting Squier's invention. The actions which AT&T took to incorporate carrier multiplexing into its telephone plant were recently analyzed by the distinguished Mischa Schwartz, Charles Batchelor Professor Emeritus of Electrical Engineering, Columbia University. In his final judgment Professor Schwartz states that the available evidence strongly indicates that AT&T's use of carrier multiplexing was based on Squier's invention, despite the claims of ATT&T executives to the contrary.[48]

Squier was conscious of the novelty of his position. He said that he had always to proceed from the standpoint of the radio engineer, instead of the telegraph or telephone engineer.[49] A large part of the invention consisted of looking in a fresh way at two seemingly separate disciplines.[50] Despite Jewett's criticism, inventors adapted Squier's fresh way of looking at the problem of multiplexing a line. Charles Culver, for example, conducted a series of "guided-wave" telegraphic experiments during 1913–1915 over telegraph lines of the Chicago, Milwaukee, and St. Paul Railway and the Delaware, Lackawanna, and Western Railway. He sent no telephone messages because he lacked a power source capable of emitting a single, stable frequency. According to Culver, the nature and cost of the power unit used by Squier further delayed practical application of the Squier system of multiplex.[51]

Several months after the American Institute of Electrical Engineers' meeting, Squier returned to Washington. His promising radio research was cut short by a new assignment.[52] The policy that officers return to the field after four years in Washington was also a part of Root's reforms. Root adopted it to reduce the deadwood at headquarters and to extend staff experience to more officers.[53] But the policy also inhibited the development of specialized scientific personnel and, once more, thwarted the progress of one of Squier's ideas from invention to innovation.

9
Science and Syndicate

Squier spent four remarkable years in London as military attaché to the Court of Saint James. Although he lacked the usual social credentials of an attaché, his reputation in electrical engineering provided valuable diplomatic access to the scientific stratum of British society.

Squier continued his promotion of scientific research in the U.S. Army from afar. He arranged a series of joint radio experiments between the British Association for the Advancement of Science and the U.S. Signal Corps. He also served as an important source of information on the military progress of aviation in Europe. When the Smithsonian Institution reopened Samuel Langley's aeronautical laboratory, Squier served on one of the laboratory committees. He sent printed material on European aviation to the Smithsonian and the U.S. Army and assisted American aeronautical experts in meeting British builders, designers, and aviators.

Squier's attempts to sell his multiplex patents entangled him once more in conflicts of interest. When a House of Commons committee on imperial radio affairs asked him to testify about the Marconi system of radiotelegraphy he found himself in a dilemma. He did not care for the system, but he was also negotiating with the Marconi Company over the sale of his multiplex patents. Squier also formed a partnership with the largest manufacturer of cable terminal equipment in the world, Muirhead and Company, to develop a military radio based on his multiplex principles. One of their first potential customers was the U.S. Army Signal Corps. Squier's association with Muirhead may have led to the formation of a large syndicate of electrical communications interests dominated by Lee DeForest. Other American firms, such as Western Electric and American Telephone and Telegraph Company, with whom Squier also sought or concluded patent purchase or lease agreements, might well have wondered about Squier's impartiality as an Army officer.

Squier's New Assignment

Squier received notification in March 1912 that he had been selected to assume the position of Chief Signal Officer of the Central Division, headquartered in Chicago.[1] Then, in a sudden change of plans, Allen recommended Squier to Chief of Staff Leonard Wood for assignment to London as the American military attaché.[2] Wood immediately wrote Brigadier General A.L. Mills, President of the Army War College, who supervised the military attaché system, to say that it seemed "to be an excellent recommendation."[3] Such a recommendation from Allen was curious because the Signal Corps was traditionally short of officers. The loss of one so senior in rank and experience would not have pleased Allen. Yet he acquiesced in the desires of the Chief of Staff. Even Squier was astonished.

George Squier in Manchester, England, 1915 (courtesy CECOM Historical Office).

Evidence for an explanation about the sudden change in Squier's career points to Wood's personal intervention. Walter Hines Page, Ambassador to London, reported in his memoirs that Squier's posting to London reflected the Army command's increasing concern that progress in the application of science to war in Europe would leave American military professionals far behind. Page said Wood was determined to change the character of the attaché post at London. Wood wanted to send a serious soldier of proven worth in scientific investigations who could "survey the great armed camp which Europe had become, and derive from it such lessons as would promote the military preparations of the United States."[4] Service attaches before the First World War belonged to what was usually referred to as "society" or the "governing classes" of the Western world. They ordinarily came from the so-called best families and regiments. The man Squier was selected to replace, Major Stephen Slocum, was "notable in the service as being the proud nephew of one of the chief heirs of Mrs. Russell Sage, who was supposed to gild his career beyond the dreams of avarice."[5] Neither Squier's family background nor military service fitted him for the assignment, except his scientific attainments and Wood's desire to take diplomatic advantage of them. Wood's action in sending an attaché for the purpose of obtaining information on the military applications of science and technology was highly unusual, if not unprecedented, in the hundred-year history of military attaches.

In an apparent gesture of compromise to Allen, who may have objected to Squier's absence, is a handwritten note by the Adjutant General on the letter of notification stating that "this officer has been informed that his detail will not exceed two years."[6] Colleagues often considered a returning attaché as spoiled for following his military profession in the field.[7]

Squier Arrives in London

Squier arrived in London as a member of a large American delegation to the Third Radio Telegraphic Conference, held June 8, 1912.[8] Delegates gathered with the loss of the *Titanic* in April still fresh in their minds. The United States Congress responded rapidly to the disaster by passing an act requiring auxiliary power supplies and mandatory monitoring of specified emergency frequencies.[9] These key features were included in the American recommendations for an amended radiotelegraphic convention.[10] The majority of delegates believed that requiring all vessels to carry radio equipment exceeded their authority. They did accept unanimously, however, a resolution calling for universal installation of radio equipment on certain classes of ships.[11] Four days before the conference ended, Squier presented his credentials to the King at Buckingham Palace.[12]

The Marconi Scandal

Squier assumed his duties at a time when the famous Marconi Scandal was just becoming public. The crisis broke when the British government announced plans to contract with the Marconi Company to build an imperial system of radio stations. Parliamentary forces opposed to the government plan demanded a full public review of the case. They questioned the integrity of high government officials involved in the contract negotiations and the government claims for technical superiority of Marconi radios.

On the question of technical superiority, a select committee of the House of Commons asked several American radio experts for their opinions. Squier was among those asked to testify.[13] Meanwhile, one of the leaders opposed to authorization of the Marconi contract, Henry Norman, called upon Squier privately, without knowledge of the government, to obtain his opinion against confirmation of the contract.[14] Because the entire issue had created such heated public debate, Squier at first declined to appear before the select committee for fear of taking public sides in a domestic controversy. Moreover, Squier felt that his testimony would be undesirable to the British government, owing "to his views of the Marconi system in general."[15]

He also doubtlessly felt a need to be circumspect about rendering public opinions concerning the Marconi system. Only a few months previously he had entered into negotiations with the Marconi Company for sale of his foreign multiplex patents for 100,000 pounds sterling. Although the Marconi concern did not pick up the option to buy, future negotiations were still possible.[16] While the Marconi Scandal had a destructive effect upon Marconi and almost cost the Prime Minister his position, some positive scientific initiatives resulted from the hearings. They were in the area of radio wave studies.

International Radio Research

Little scientific work was undertaken before 1912 on the effect of radio waves upon the aurora borealis (and vice versa), the behavior of radio waves in the arctic and tropical regions, and the effect of solar eclipses upon radio propagation. The British Association for the Advancement of Science (BAAS) appointed a Committee for Radiotelegraphic Investigations to study these various problem areas. Dr. William Eccles, a close friend of Squier, was appointed Secretary. Another friend, Sir Oliver Lodge, was selected Chairman. One evening during supper, Squier said that the Signal Corps was in a position to assist the BAAS (on behalf of international cooperation in science) in completing their observations of radio phenomena.[17] The necessary arrangements were made and, for the following year and a half, Signal Corps officers in Alaska, Hawaii, and the Philippines kept daily records for the BAAS. They reported on the effect of the aurora borealis upon signal strength and reception. Their reports refuted some of Eccles's data that radio transmissions had an effect upon auroral phenomena.[18] During the solar eclipse of 21 August 1914 all the southern stations of the Signal Corps made observations of radio strays and weather.[19] This joint radio project was continued. Squier's close association with men like Eccles, Lodge, and Rayleigh did indeed change the character of the London attaché post, as Wood desired.

European Aeronautics and the Smithsonian

During the same time period, Squier supported Smithsonian efforts to advance the scientific study of aeronautics. In early 1912, members of the invisible establishment, led by the Secretary of the Smithsonian, moved to gain support for a national aeronautical laboratory. President Taft appointed a commission in December 1912 to study the question and make recommendations to Congress.[20] The bill drafted by the commission failed to pass the lame-duck session of Congress in early 1913.[21] Rather than rely on the incoming Wilson administration to succor aviation, Walcott, also a member of the Taft Commission, obtained permission from the Smithsonian Regents on 1 May 1913 to reopen Langley's laboratory.[22] With a view to making the laboratory a permanent institution and providing guidance according to the idea suggested by Squier in 1910, Walcott formed a joint interdepartmental and civilian advisory committee with President Wilson's approval.[23] Zahm, recorder of the committee, became director of the laboratory. The first meeting was held on 23 May 1913.[24]

Two weeks later Zahm asked Squier to serve on the Sub-Committee of Collection and Correlation of Aeronautical Information, which Zahm chaired.[25] The other member of the sub-committee was Paul Brockett. This channel represented one of the means by which European aeronautical information reached the Smithsonian. By special arrangement between the Army, Navy, and the Smithsonian, aeronautical information received by one party was made available to all. By end of summer 1913, Zahm, Brockett, and Squier collected 100 volumes, 400 pamphlets, and 59 periodicals. Publications were purchased or obtained in exchange for Smithsonian publications.[26]

The leadership of the Aerodynamical Laboratory also relied on firsthand inspections of European laboratories and factories. Zahm notified Squier in July that he and Naval Constructor Jerome C. Hunsaker would arrive in London in early August and told him that he hoped to be able "to see the people and establishments that will best serve us in our mission."[27] When they returned to Washington, Hunsaker conferred with Lieutenant Colonel Samuel Reber, head of the Aeronautical Division. That no Army representative accompanied Zahm was indicative of the difficulties aviation people were experiencing within the Signal Corps. Hunsaker's presence reflected a profound respect for research laboratories in the Navy and the Navy's forthright movement into growth situations that offered promise for Navy control.[28]

Reber also received aeronautical information from the War College through its attaché system. Much of what he borrowed came from the *London Post*. From the lists of books, journals, and articles listed by the War College, the Chief Signal Officer borrowed approximately five items per week on behalf of the Aeronautical Division. Although most articles were of a secondary nature, some first-account analyses were transmitted, e.g., a descriptive article of the Dunne Biplane. Quite possibly an account of Squier's visit to the 5th International Aero Exhibition in Paris (at no expense to the government) also made its way to the Aeronautical Division.[29]

Squier reported discussions of airplane employment doctrine held at the Royal Service Institution, paying particular attention to one conducted by Colonel F.H. Sykes, Commanding Officer of the Military Wing of the Royal

Flying Corps. Sykes predicted the appearance of airplanes capable of flying 100 miles per hour and having a 700-mile radius of action. He thought machines would be made silent and their visibility decreased. He foresaw a useful military role for airplanes beyond reconnaissance and rapid intelligence reporting. By flying low over parked ammunition and supply wagons to "sow explosives amongst them," enemy military operations could be interrupted. Squier approvingly reported Sykes's attitude that mobility and readiness for instant action at any time represented the "essence of being" of a Flying Corps.[30]

Although topics on training, organization, and employment doctrine dominated the lists, considerable attention was also given to aerial tactics and lessons of the Balkan Wars, a subject of keen interest to all military fliers. So important were the combat uses of aircraft in the Balkan Wars that the Assistant Chief of Staff considered assigning Squier to the Constantinople attaché post, in addition to his London post, when the resident attaché, Major John R.M. Taylor, became temporarily disabled.[31] Besides secondary accounts, personal analyses, and training material, Squier sent instrument designs and official reports of British Aeronautical activities to the War College Division of the General Staff and the Chief Signal Officer.

Squier sent the Chief Signal Officer a copy of a British government publication entitled "Military Aeroplane Competition, 1912, Report of Judges Committee." The report contained valuable graphs of engine and aircraft performance. Because the performance of airplane engines was still not well understood, the report was especially popular among aviators.[32] Multiple copies of the "Report of the Advisory Committee for Aeronautics, 1911–12" and the "Memorandum of Naval and Military Aviation, 1912" were forwarded directly to the Chief Signal Officer.[33] Special aeronautical recording devices developed at the Royal Aircraft Factory were brought to the attention of the War College Division.[34] Sometime later he sent a set of six confidential blueprints provided by Brigadier General David Henderson, Director General of Military Aeronautics in the British army. Just what they represented is unrecorded, but clearly their presentation to the American military attaché represents an unusual policy of encouragement to American aviation.[35] In addition to sending aeronautical information to the General Staff and the Smithsonian Institution, Squier pursued his private commercial interests more vigorously than at any other time during the previous decade.

London Inventions and Enterprises

Within a year after arriving in London, Squier found a laboratory in which to continue his research in multiplex telegraphy and telephony. Muirhead and Co., the world's largest manufacturer of cable terminal equipment, such as receivers, transmitters, relays, and recorders, opened their laboratory to Squier. They took keen interest in his development of a compact, portable radio receiver capable of handling both telegraphic and telephonic transmissions by simply connecting the input terminals to a telephone line. By December 1913, Muirhead became convinced of the military promise of Squier's Field "Wired-Wireless" Receiver and purchased the sole right to manufacture the radio set.[36] Squier was to receive 20 percent of gross sales.[37] When an article and photographs of Muirhead's new military radio receiving set appeared in the January 1914 issue of the London *Electrician*,[38] Major Wiedman, Director of the Army Signal School, requested that the Chief Signal Officer purchase two receivers for testing.[39] The Chief Signal Officer agreed to buy two sets and test them in the Washington laboratories.[40] Later in the year Western Electric Co. entered into an agreement to market the field military radio receiver worldwide for 15 percent of net sales.[41]

In June 1914, Squier and Muirhead brought out a version of his radio suitable for installation in private residences. Squier exhibited his new version of the radio receiver at the Annual Conversazzione of the Royal Society.[42] It could be used on regular antennas or commercial telephone wires without disturbing existing services. By proper selection of electro-magnetic constants, sharp tuning of the antenna and the oscillatory circuits could be obtained in steps over the wide range of radio frequencies from 300 meters to 3,000 meters.[43] Variable coupling made the set suitable for either radiotelegraphic or radiotelephonic reception. Inconvenient and unsightly outside antennas were obviated by utilizing part of the vast network of telephone lines as an antenna. It was found that the horizontal orientation of telephone lines minimized the extent of static noise introduced to the receiver, since vertical antennas seemed to be more sensitive to atmospheric noise.

With Muirhead and Western Electric handling the military version of his radio receiver, Squier tried to renew interest at the American Telephone and Telegraph Company in developing the receiver as an adjunct to telephone subscribership. A short time before Squier left the United States in 1912 negotiations were started with AT&T for the purchase of some of his foreign patents covering "Wired-Wireless." When Squier left for England the negotiations were suspended until such time as the merits of the invention could be investigated to the satisfaction of AT&T. From December 1913 until September 1914, the Marconi Company made vigorous efforts to acquire control of Squier's foreign patents, as they were the only "mother" patents in radio not controlled by one of the major radio corporations. Finally, Squier broke off negotiations. In August, Squier approached John Waterbury, a director of AT&T and fellow delegate to the Radio Telegraphic conference, with a proposal that AT&T acquire the property rights in the receiver patents, then pending in his name in the United States, England, Russia, and Germany, when issued.[44] Again, in the following months, he suggested that AT&T handle, in combination with Western Electric and the Western United Telegraph Companies, all his radio patents covering multiplex transmissions. So sure was he of their ultimate success he offered to take his remuneration from what they turned out to be worth in actual practice.[45] Squier pressed Waterbury for a decision, expressing the opinion that after two years of waiting he was entitled to a definite answer, yes or no. John J. Carty, Chief Engineer for AT&T, advised Waterbury that after several conferences on the subject he could not propose an arrangement that he considered advantageous to Squier or to AT&T. Carty wrote that he and his engineers had very high personal regard for Squier, and wished him well, but thought that the company should follow a policy of watchful waiting.[46]

In early January, DeForest wrote Squier about combining his radio telephone and telegraph patents with DeForest's, for "a very strong situation could be built up in this way, to our mutual benefit."[47] While Squier considered this proposal, he indulged his old attraction to alternating current telegraphy over landlines and submarine cables.

Alternating Current Telegraphy

By 1915, it was assumed that developments in cable telegraphy had reached finality and that progress would henceforth occur in radio.[48] The appearance of new receiving devices designed by Alexander Muirhead, S.G. Brown, and E. Huertley, capable of increasing signaling speeds by 40 percent, offered new promise for cable communications.[49] For all the effort devoted to improving receiving devices and modifying cable characteristics, relatively little attention had been devoted to improvement of the transmitter.[50]

Squier's revival of AC theories for working submarine cables presented before the Physical Society of London found an enthusiastic audience.[51] Charles Bright, a well-established cable expert, remarked favorably upon Squier's "ingenious theory" for using AC generators on ocean cables. (This was, of course, ten years before Nyquist proved that the use of an AC carrier did not improve transmission speed or accuracy.)[52] Bright was unaware, though, of Squier's commercial involvement with Muirhead, who brought Squier's theory into practical working.[53] W.A.J. O'Meara, Engineer-in-Chief of the British Post Office, declared that Squier's new viewpoint provided a strong argument

> for the abandonment of the arbitrary square-topped wave impulse hitherto used for telegraph signaling purposes, and for the adoption in its place, both in the case

of land lines and in that of submarine cables, of the more scientific, and therefore more efficient, pure sine-wave impulse.[54]

Another authority on submarine cables, H.W. Malcolm, recognized "that the uninterrupted single phase alternating current of sine-wave form was the last word in bridge balance," i.e., smoothness of operation.[55] He would undertake a theoretical discussion in a leading journal to help advance Squier and Muirhead in their work. Squier told Walcott that experiments on his system were actively going on at the Eastern Telegraph Company's laboratory in London. He confidently expressed his belief that cable telegraphy had a "brilliant future ahead ... and depends on substitution of AC [alternating current] engineering for present existing methods."[56] Squier was doing his best to turn belief into reality. In June 1915, Squier assigned one-half interest in his alternating current telegraphy patent to Muirhead and Co.[57] A month earlier he had formed a joint alternating current telegraph company with Muirhead. Squier received 1,000 pounds sterling in cash and 15,000 pounds sterling in fully paid shares of the new company.[58]

It appears that Squier sought a broader base of support for his new system than one company could provide, albeit that Muirhead was the world's largest. In September 1915, a Sir Robert A. Hadfield, British industrialist and member of the Royal Institution and the Royal Society, informed Sir Henry B. Jackson, First Sea Lord and noted radio expert, that Squier had accomplished the successful application of alternating current to cable work after several years of experimentation at Muirhead's and that Squier "was quite willing to allow the use of his discovery without any charge to the British Government."[59] The very next day Squier received an invitation from Jackson to discuss the principles of his new cable system. Squier gathered from Jackson's conversation that the system should be introduced from the top down instead of going directly to the engineers of the different companies. Jackson showed great interest in the system and promised to bring it to the attention of the proper people at the top.[60]

Meanwhile, Squier tried to persuade Wilkins, Manager of Muirhead, that their partnership should be reorganized in order to provide larger coverage for their telegraphic system throughout Europe and the British Empire. For unknown reasons, Mr. Wilkins objected to Squier's proposal to join forces with A. Baxendale. Perhaps it was because Baxendale handled the European business affairs of Lee DeForest. In any event, Squier told Wilkins:

> The history of most good patents is found to be, that it takes so long to get them ready for the market, that they are just about developed when the patent expires. I have seen this occur a dozen times, and don't propose to have it occur in this case if I can help it.[61]

9. Science and Syndicate   121

A week later Squier engaged Baxendale as his agent.[62] Baxendale agreed to represent Squier's interest in alternating current telegraphy on the continent and throughout the British Empire.[63]

Their prospects brightened in the new year, 1916. In January, Squier exulted about the breadth of his English patent. He told Wilkins:

> If I should not have done this work, all we should have got would have been an ordinary instrumental patent describing the apparatus, whereas, as our English patent now stands, it does not make the slightest difference what apparatus we use.

Thus, he felt confident in having developed another patent in the field of alternating current signaling on cables. He confided exuberantly to Wilkins that he had more than 80 patents, so "I have learned a few points about the drawing up of a good patent."[64] His work received additional attention following his remarks before the Royal Institution in the same month.[65] Soon after, in accordance with Jackson's promise, arrangements were concluded between Secretary of the British Post Office and Squier for a full test of the new telegraph system.

During the months of February and March 1916, the system was tried in the Experimenting Room of the Engineer in Chief. The signals were all of equal length, one-half cycle, while dots, dashes, and spaces were represented by different amplitudes. Changes in amplitude were achieved by varying the amount of resistance in the local circuit. A gold-wire relay inserted between the receiving relay and the printing mechanism converted the signals at the receiving end into linear dots and dashes suitable for reading by Morse operators. The trials compared speed and smoothness of operation of Muirhead's system and the one in use by the Post Office, which used reversing battery polarities. A definite advantage in speed was found in the case of the sine-wave system. Subsequent tests, performed to determine if the improved signaling speed was due to the wave-form or some other system feature, revealed that the use of continuous waves alone accounted for an increase of about 20 percent in speed. Although some improvement in duplex balancing occurred, the Post Office engineer thought the alternator might better be replaced by a battery and some unspecified auxiliary apparatus.

The dramatic increase in speed, of course, provided considerable promise.[66] Cables were burgeoning with messages in 1916, strikingly unlike the situation when Squier and Crehore first introduced their telegraph devices 16 years earlier when cables stood idle for hours on end each day.[67] Squier wrote DeForest:

> The London telegraph world is finally waking up and I believe we are on the eve of tremendous advancements, particularly in land-line and cable telegraphy, which I regard as totally out of step, from an engineering standpoint with the other branches of electrical science as it exists today.[68]

While his demonstrations of sine-wave telegraphy were proceeding apace, Squier suggested to DeForest that they join forces through the agency of Baxendale to develop multiplex telephone and telegraphy in Europe.[69] DeForest replied with the statement that "the new syndicate will take up the development of wired wireless in Europe, for I believe it has a wide field."[70] But Squier had no opportunity to personally pursue the exploitation of his patents in Europe, for a month later he was notified of his relief from attaché duty in London, effective at the end of April.[71]

10
Secret Missions

When World War I broke out in August 1914, British Foreign Secretary Edward Grey determined to continue the conciliatory policy toward the United States adopted in the 1890s as a result of the Venezuelan crisis. Preservation of American friendship became even more important as it had become apparent that hostilities would last for a greater period of time than anyone had initially anticipated. As a man of great popular prestige and influence in the cabinet, Grey regarded the friendship of America "as England's most vital interest of all."[1] British policy in the early months of the war reflected Grey's own evaluation of the importance of American friendship.[2]

Among the cabinet members, one rivaled Grey in popularity and influence. Of very different political convictions, a soldier, and a stranger to England by reason of long military service in the East, the Secretary of State for War, Lord Kitchener treated other members of the cabinet with a mixture of military contempt and apprehension. According to Lloyd George, Chancellor of the Exchequer, all Kitchener's colleagues were frankly intimidated by his presence, because of his repute and enormous prestige among all classes of the people outside. A word from him was decisive, and no one dared to challenge it at a cabinet meeting.[3] Despite his power and influence, Kitchener hesitated to reveal military secrets before the cabinet, whom he regarded as a group of strangers.[4]

On the fundamental policy of maintaining friendship with the Americans, however, Kitchener shared Grey's conviction. But he would pursue the policy in his own way and on his own terms. One significant instance of his willful policy of seeking American benevolent neutrality, if not a military alliance, on his own initiative during the early months of the war, was the exclusive privilege he granted Squier to go with the British army in the field. Kitchener literally sneaked him out of England for an extended visit to the front lines in France. Kitchener granted him full access to any information he desired, any location he wished, and any individual he cared to interview. He authorized him to draw unlimited gold coin from any British army paymaster. His three trips in the course of the war before American entry were unique among attaches. Squier's diplomatic value to Walter Page, the American Ambassador in London, became so marked that he personally requested the Secretary of State to retain Squier beyond his two-year tour. Squier's scientific and military reputations and confidences materially assisted the Ambassador's efforts' to glean information from the wartime capital. Favors shown Squier indicated Kitchener's determination to make the United States a military ally.

George Squier as U.S. Military attaché in England, 1914 (courtesy CECOM Historical Office).

Squier Goes to the Front

In August 1914, Squier was in London on borrowed time. In late 1913, a new administration committed more to non-involvement in Europe than military preparedness governed the United States. General Wood decided against succeeding himself, and the new Chief Signal Officer, George Scriven, was strengthening his control over the Signal Corps. In late 1913, Scriven initiated Squier's reassignment with the Secretary of War, Lindley Garrison, stating that Squier stood at the head of the list for duty in the Philippines.[5] That his attaché, who had such high standing with the embassy and in the community, should be sent to the nether regions, as far from Europe as possible, disturbed Page. He immediately telegraphed both the Secretary of State and the Secretary of War requesting that Squier's tour be extended to a full four years.[6] Although his reluctance and pique were evident, Garrison extended Squier's tour one year, but emphatically retained the freedom to relieve him on 1 January 1915.[7]

Shortly after the war began in Europe, the War College held a series of meetings to formulate policy concerning requests from American attaches abroad to observe the war.[8] They decided attaches were more urgently needed at the capital of a belligerent than on the front lines, however, lower ranking officers might be sent as observers. Significantly, both Aviation Section and Signal Corps representatives were omitted from proposed officer assignments as observers to various corps or arms of the belligerent forces, a policy about which Reber repeatedly complained to Scriven.[9] Military attaches were notified by cable and letter to remain at their stations, to obtain valuable information on war movements, and to assist ambassadors or ministers.[10]

The War College transmitted word of their prohibition by cable in response to Page's information a few days earlier that the British were willing to permit Squier and two other officers to accompany the British army in the field. Squier telegraphed details to the War Department, and affirmed his intention of leaving when an opportunity presented itself.[11] The War College then reiterated its position to Squier, expressly forbidding him to visit the front.[12] Squier replied in a lengthy communication that all the military attaches had been expecting to leave for the front for the past three weeks. Since their names had been personally approved by the King there was no diplomatic way of withdrawing from the party of visiting attaches without furnishing adequate reasons through the State Department. He concluded decisively that in case "I do not hear from you and the party leaves for the front in the meantime I shall go with it. The Ambassador has of course seen the message. He approves it and recommends it."[13] Chief of Staff W.W. Wotherspoon, Wood's successor, lamely replied that the Secretary had no objection to Squier's visit to the front. The order was based on the idea that the Ambassador would need his services in London.[14]

All of this pre-planning and anticipation seemed to be of no avail when the British canceled all plans to permit attaches of neutral countries access to the fighting zone. Even military officers of the two principal allies, Russia and Japan, languished in London while awaiting permission to accompany British troops in the field.[15] The Russians had most cause for impatience since their attaché, Lieutenant General Yermoloff, was the doyen of the Military Attaches Corps in London. All governments made appeals to the British Empire to allow their attaches to observe the fighting, but none was honored—officially. It seemed the War College policy would be observed after all.

On 11 November 1914, the same day the King opened Parliament, Lord Kitchener, of Khartoum fame and newly appointed Secretary of State for War, sent word to Squier early in the morning that he wanted to see him as soon as conveniently possible. Squier arrived at 10:30 a.m. in full-dress uniform and was received by Kitchener alone in his office. Kitchener explained that he had conferred with Generals Joffre and French about a visit from Squier to the British sector of the front lines, and they approved. He wished Squier to "simply disappear" from London and go to French's headquarters, by way of Paris, in the north of France.[16] Page's biographer reported that Kitchener had suggested Squier's temporary detachment from the embassy in London to become the military attaché in Paris.[17] Kitchener's invitation to Squier and subsequent plans were made without consulting the Foreign Office.[18] Page approved the project. An ardent spokesman on behalf of American intervention on behalf of the Allies, Page sent no information of the project to the State Department until its successful conclusion.[19] Kitchener dispatched Squier to the front lines when the presence of military attaches of other neutral powers would not have been permitted for a moment.[20] Kitchener gave orders that none but Squier was to be allowed in the zone of the British Expeditionary Forces in France from the outset.[21]

When Squier passed through Paris he picked up two young American officers, First Lieutenant F.W. Honeycutt, 3d Field Artillery, and First Lieutenant Carl Boyd, 3d Cavalry, to take to General French's headquarters for a few days so they might see something of professional advantage. About Kitchener's subterfuge (the art of which he learned well after 40 years of service in the East), Honeycutt remarked,

> It is rich to think of that Russian Lieutenant General and five Japanese Officers back in London, and Colonel Squier off on his little hunting trip in Scotland. They will explode, I am afraid, when they find out about it.[22]

Kitchener's call put the American military attaché on the front lines, a privilege withheld from all other attaches.[23]

Kitchener gave Squier special permission to keep a personal diary of his travels in France. He committed to it a wealth of information concerning supplies, messing, artillery, airplanes, personalities, motor vehicles, combat practices, cavalry, general staff practices, and troop morale. The tone was sometimes quaint; the result perhaps of professional detachment or naïveté. Many expected a conflict of short duration. The brutalizing processes of war were slow to begin. Gentlemanly codes of deportment prevailed in the early days of the war: e.g., when a German airplane was downed near Indian troops, the Commanding General invited the two enemy officers to tea at his headquarters before sending them to the rear as prisoners of war.[24] Everywhere Squier went he was met with enthusiasm and genuine expressions of friendship earned during the several years before the war. Honeycutt testified to Squier's friendship with Sir John French, head of British forces, and Sir David Henderson, Chief of the Royal Flying Corps. He recalled that, on one occasion he saw Squier greeted by these individuals and remarked:

> He certainly seems to be in "solid" with them. There was unfeigned pleasure shown by all of his friends whom I saw him meet, they seem to act as if they couldn't do enough for him. It was quite simple to see how important the personal element is in this observing game.[25]

Beyond the personal element, of course, was the earnest wish of many British military leaders that the United States enter the war.[26]

Squier's diary was unique. Unlike other attaches and observers, Squier had complete freedom of movement and inquiry. Kitchener gave orders to show him everything. Many military leaders in the field realized, by late 1914, that the war would be protracted. A war of attrition and a severe test of national economies was at hand. They handled the American military attaché in such a way that a clear diplomatic signal of British military desire for American participation would reach Washington. Squier himself thought that the real purpose of this visit was to enable the American Army to learn firsthand the details of managing modern warfare.[27]

Squier's own army experience had ill-equipped him for the function he imagined to be performing. Aside from brief service in the Artillery Corps, Squier's entire career was spent in technical service. Even his combat arm experience was atypical of the line officer's duty. Doubtless there was military value in his observations, particularly on the unexpected usage of ordnance, the use of tactical and strategic air warfare, and the implications of armored warfare. Although the principal value of the secret mission lay in the diplomatic signal intended for Washington, the diary did provide eye-witness accounts of frontline action as it unfolded. Kitchener's policy denied these observations to all other military observers from neutral nations.

Airplanes in Action

One of Squier's preoccupations throughout his tour on the front lines concerned gathering information on the uses of the airplane in battle. Squier observed the common airplane missions of reconnaissance, the suddenly important artillery spotting, and the beginnings of tactical and strategic bombing. No number of airplanes, he found, could satisfy Artillery Corps personnel's demand for assistance in marking and identifying targets.[28] Radio was definitely preferred over systems of marking, which relied mainly upon Very pistols shooting red and green lights or searchlight signals to indicate long or short and left or right of target. The reception of signals by radiotelegraph sets installed in airplanes first took place in the American Army in 1911.[29] In the following year airmen directed artillery fire from airplanes.[30] Although the limited range of transmission and weak power of the radiotelegraph sets restricted its general usefulness, radio had one clear advantage over other means of signaling from aircraft: an observer in an aircraft could transmit enemy positions and targets regardless of visibility. Noting that airplanes had not yet achieved a destructive threat equal to the other arms, Squier reported that bombs they released from airplanes killed people every day, but "such things are considered as trifling incidents." He observed that one's whole perspective and scale of values changed in modern warfare and, yet, the "old habits persist and I have seen several officers wearing monocles, [while] living in caves and dug-outs and ankle deep in mud."[31]

The revolution in the military uses of aircraft was nevertheless underway. No one questioned the indispensability of aircraft, and Squier took copious notes of conversations held with lower echelon commanders as well as with such men as Sykes and Henderson. Indeed, aeronautics had become so important that Henderson was transferred from command of the 1st Division I Corps back to his old position as Chief of the Royal Flying Corps.[32] Squier's notes supplemented observer forms supplied by the War College and returned to them upon completion of Squier's tour.[33] Old acquaintances of Squier, Henderson, and Sykes spent hours discussing with him the changing role of aviation in the war. The specifics of these conversations were not reported, but Squier did suggest a tactical bombardment mission for aircraft. He expressed puzzlement over the Germans' failure to attack Le Havre, where the British unloaded and stored provisions for their army in France. He observed that hitting the gasoline stores alone would cripple the British military effort.[34] Squier said that sending aircraft to bomb truck parks, munitions dumps, and port facilities would have dealt the British a serious blow.

One unexpected experience in conducting military operations on such a grand scale abroad was the greater difficulty in maintaining an adequate supply of ammunition to the front than in supplying adequate levels of troop rations. Demand for food by a given number of troops is quite predictable, but expenditure of ammunition, a function of when battles occur, proved to be more difficult to anticipate.[35] Squier emphasized the wholly unanticipated demand for ammunition, which threw off staff calculations of all major armies. Ammunition parks were particularly vulnerable to attack by airplanes and were carefully guarded by newly developed anti-aircraft guns of novel design.

Anti-aircraft guns came under the operational control of recently created units directly responsible to Corps commanders. Young officers sought service with the new units since they offered opportunities for conspicuous service. When Squier arrived, only a few of the guns had appeared in France. He had heard of them in London and relished the chance to see one at close range. His own artillery experience at Fort Monroe hardly encompassed such considerations as firing against targets projected upon a large hemispherical surface. British ordnance engineers at Woolwich Arsenal fashioned the first anti-aircraft guns by adding a second recoil cylinder to a standard 13-pound gun used by the British Horse Artillery and mounting it upon a special vertical pivot carriage. The carriage, in turn, was placed upon the chassis of a London motor bus. Maximum firing elevation was 60 degrees, but the gun could be rotated and fired around full circle. Their psychological impact upon German pilots was remarkable, and zones protected by their presence were soon free of hostile aircraft.[36] Although anti-aircraft guns never attained much accuracy, the German pilots wasted little time in retiring from the area. Squier found other novel adaptations of the London bus.

A Mobile Army in Buses

Three fleets of London buses, consisting of 50 buses each, were used to quickly transport 1000-men battalions, complete with field equipment, to points on the line in need of reinforcement.[37] One fleet was kept in constant readiness at General Headquarters, St. Omer. On occasion, all three fleets combined to carry as many as 4,000 men to the trenches.[38]

Thus began the symbiotic relationship between motorized war and modern methods of communication, which increasingly characterized World War I and subsequent wars. The London buses and Paris taxis foreshadowed the growth in armored warfare. Fascinated by such imaginative realization of military mobility, a concept long heralded in the mobile army and made possible by electrical communications and motor vehicles, Squier took special delight in finding a scientific field laboratory carried about in a truck commandeered from an English millionaire.

Science at the Front

Dr. Sydney Rowland, a professional bacteriologist with the Lister Institute of London, outfitted his mobile laboratory in eight days at a cost of 1,500 pounds sterling. The 40-horsepower truck was completely lined inside with aluminum sheets and stocked with every piece of scientific equipment necessary for research. A picture of Lord Lister hung on one wall. Rowland told Squier that his was the first bacteriological laboratory to accompany an army in the field. He spoke highly of the military medical work of Gorgas, Reed, and other American medical officers. When Squier told him of his personal friendship with Gorgas, that he had recently accompanied him to Oxford to receive a DSc, and that he also occasionally spent a weekend with Sir William Osler at Oxford, Rowland opened up and told of his many research activities in the field, feeling assured of a sympathetic listener.[39]

Two research activities were of special interest. The first practical benefit of maintaining a mobile field laboratory occurred when several cases of typhoid fever were detected in one of the transportation depots located near St. Omer, General French's headquarters. Rowland drove to the transportation detachment and took urine samples from all the men. He found one member of the detachment to be a "carrier" without displaying any symptoms. The man was removed at once and no further cases developed. Rowland advised Squier to tell General Gorgas that "he should require every recruit who enters the American Army to be examined bacteriologically to see if he is a 'carrier.'"[40] He personally hoped that every recruit could be administered the anti-typhoid serum treatment, but British public opinion would not allow it.

A second instance of research concerned an investigation of the soils of northern France. Wounds in the war had led to a large number of reported cases of gas gangrene and lockjaw, far larger than in previous wars and beyond all expectations. Indeed, the large incidence of such cases provided the necessary support for Rowland's research. Rowland found that the fertilization of French soils with manure over many years had produced an incredibly rich medium for the growth of bacteria. If any dirt entered the wound, no matter how slight, of an injured soldier, he would inevitably become infected. Rowland verified his conclusions about the dangerous nature of the soils by a simple experiment wherein he inoculated a guinea pig with a few drops of ground water from the trenches. Just 18 hours later the pig died. A post-mortem examination revealed gangrenous conditions similar to human wounds from the trenches.[41] Rowland prepared a paper on his research which reported, for the first time in history he believed, on a piece of research laboratory work accomplished with an army in the field, written and published at the front. Squier sent a copy of Rowland's "A Report on Gas Gangrene," Field Laboratory, G.H.Q., 11 November 1914, to the War College in advance of his official report.[42]

Squier curiously devoted relatively more attention in his diary to the meeting with Rowland and to Rowland's scientific work than to fields of science in which he was expert. He was clearly fascinated with the notion of scientific research aiding directly the war effort at the front. Like himself, Rowland was an individual scientist engaged in military duties in a combat zone. They may have found themselves to be kindred souls. In any event, Rowland's contribution to the British fighting capability through scientific research, which was undertaken within earshot of the fighting, impressed Squier. Scientific research in support of a war effort need not be a rear-echelon activity.

General Staff Organization

On the second day of the new year, Squier left Paris and arrived back in London late in the evening.[43] Four days later, on 6 January 1915, Ambassador Page informed the Secretary of State for the first time that his military attaché had just returned from seven weeks at the front.[44]

The Ambassador's approval of Squier's trip to the front may be viewed as part of Page's well-known lack of impartiality in the European war. So fervent was Page's advocacy of the rightness of Britain's cause that he participated in drafting British replies to messages sent from Washington with which he disagreed.[45] Even Colonel House, wandering about Europe seeking peace for the President, and widely known to be sympathetic to ending the war by adding American military muscle to the cause of the Allies, found it necessary to dilute the strong pro–British advocacy of Page's dispatch to the President about the proposed blockade of Germany.[46] Page regarded Squier's diary as important evidence in support of his case for American aid and showed it to House.[47]

For months, Page had tried to impress President Wilson and his Secretary of State, Robert Lansing, with the desperately brutal situation in Europe. He felt that American offers of mediation were mockeries.[48] Page saw in Squier's diary a way of relating the slaughter and brutalization from the standpoint of a reliable American observer. Page was convinced that specific approaches to ending the conflict were futile: by comprehending the need for forceful American action the carnage could be stopped.[49]

Squier's Report to the War Department

Squier treated each branch of the British army separately in his report. Appended to each section was a detailed account of organization, an item of extreme interest to American General Staff planners. Indeed, it is certain that one of Squier's special subjects for investigation was the operation of the General Staff, now being subjected to the actual trial of war. Wood's personal identification with the General Staff concept in the American Army made its evaluation in a theater of operations significant for contending factions. While it seemed that Wood had succeeded in impressing his will upon the Army with the premature retirement of Adjutant General Ainsworth, resentment against Wood by certain members of Congress and the bureau chiefs persisted. He also encountered severe criticism for his cooperation in establishing an informal army reserve in collaboration with the American Legion, a policy vigorously opposed by President Wilson, who saw in it an attempt by Roosevelt and Wood to cause him political embarrassment. The national preparedness mood of the country, of which the reserve scheme was a manifestation, also threatened to jeopardize the President's neutrality policy by revealing an intention to enter the war. It stands to reason that one of Squier's purposes during his visit would be discovering as much as possible about the wartime operation of the General Staff system.

This innovative, yet untested, institutional change in national armies required careful, informed observation. Squier's own experience in founding the Signal School, a part of the General Staff system in the United States, and in serving as assistant Chief Signal Officer qualified him uniquely to act as General Macomb's poobah. The year before the war, Brigadier General William Crozier, then Chief of the War College Division, described Squier as "a careful student of the effect of modern means of observation and communications upon tactics."[50] Squier saw the "good and the bad just as it came, day or night."[51]

He described the division of the General Staff in its two general branches, Operations and Intelligence. Officers assigned to the operations branch devoted their attention exclusively to planning and executing military operations. Invariably, he noted, the men in Operations possessed advanced staff training, usually obtained from the Staff College in England and supplemented by experience in the South African war or outlying frontier expeditions. The operations orders they prepared were drawn up with the same precision and care one would expect to find in time of peace at an institution for staff training. In the case of a general bombardment by a corps, the amount of ammunition for each caliber, the exact times of shelling commencement, their durations, and recommencement were specifically prescribed. Office work in support of operations, he said, should be held to a minimum. Squier observed that a "modern war is largely a business proposition and it is simply a question of efficiency."[52] Thus, any means of improving communications between command elements were welcome.

One mode of communications that impressed Squier the most was the automobile. Clearly, he thought, the day of prancing steeds for commanding generals and their chiefs of staff was past. With the exception of General Sir James Willcox, commanding the Indian Corps, no general or chief of staff traveled about on horseback. Each corps commander used the Rolls-Royce provided him. Each automobile cost about the same amount of money as one round of ammunition for the new 15-inch guns of the *Queen Elizabeth*, then engaged in bombarding the Dardanelles. The cost was well worth the enormous improvement in command effectiveness and efficiency.[53]

The automobile also served the needs of a special system of liaison officers utilized by both the French and British armies. Each of the hand-picked officers in the system possessed combat experience, accepted by all corps commanders and the commander in chief as *persona grata*, and consistently displayed tact and solid judgment. They amplified orders and instructions issued from General Headquarters for the corps commanders and, in turn, personally informed the commanding general of situations existing in the corps areas. In addition to their role of adding a personal dimension to the chain of command, they acted as King's Messengers between officers of high rank and command. Knowing several of the liaison officers from London, Squier spent part of his time traveling with them on their rounds. He found that it was rarely apparent where a commander selected his headquarters. He explained that "in these days of aeroplane bombs," commanders chose modest homes instead of grand villas or hotels.

George Squier with Captain Le Maitre, Aviation Corps, French army (courtesy CECOM Historical Office).

Squier paid close attention to the influence the airplane exerted upon combat strategy and tactics. That part of Squier's report devoted to aviation contained instructions for the identification of airplanes, descriptions and comparisons between French and German dirigibles, official Admiralty statements on air raids on the Belgian coast (during February 1915), and tables of organization of Aeroplane Squadrons and Flying Depots. Under separate cover he dispatched ten copies of the new *Provisional Training Manual* of the Royal Flying Corps to the War College Division.

He left no doubt of the importance of aviation to the conduct of future wars. Like most military men of his day, Squier considered the use of the airplane for strategic and tactical reconnaissance as indispensable. Going beyond that conviction, however, he stated that it was becoming more evident each day that the airplane, as a direct means of attack for specific military purposes, namely, tactical bombardment, was a weapon of great moral and military value.[54] He advised that, if the war continued, one could expect to see many repetitions of the famous Royal Flying Corps raids against the Belgian coast. Disclaiming a desire to indulge in prophecy, Squier maintained "nothing is more certain than that the era of war in the air is upon us, whether we like it or not."[55] Ever mindful of research possibilities, he suggested one affirmative benefit from the war would be "the development of the science of aerial navigation at an unprecedented pace."[56]

A Conduit of Information to the War Department

In 1915, Squier told Walcott of his warfront visit and said the "scale on which aircraft is being developed here makes all our efforts appear trifling in comparison."[57] Walcott replied with the hope that Squier would deposit his notes on the rapid advances being made in aviation because of his exceptional opportunities for observation in the fighting zone.[58] The most spectacular of these advances were improvements in speed of climb, maneuverability, Fokker's synchronized through-the-propeller machine gun, and Georges Guynemer's automatic gun, which fired several types of projectiles through the hollow hub of a propeller.[59] At a deeper, more fundamental level, were the evolution of specific types of aircraft for specific military purposes (scouting and artillery fire control, pursuit, and combat), the application of research and development to large-scale production of aircraft, the use of radio to coordinate aircraft activities with other combat arms, the evolution of aerial-warfare doctrines, and the techniques for training flyers.

While Walcott was aware of what opportunities Squier obtained to observe aviation developments, the head of the Aviation Section, Samuel

Reber, seemed uninformed of Squier's secret mission and reports. Reber complained to Chief Signal Officer Scriven on repeated occasions that no military observers with aviation experience had been detailed to either side in the European conflict. Noting that "practically no reports of any value" covering the operations of the flying corps of the various armies or their organization and equipment had been received, he urged that at least one member of the Aviation Section be assigned as a military observer.[60] Acting on Reber's recommendations, Scriven made several unsuccessful attempts to have a billet created for aviation observers among the list of observers being prepared by the General Staff.[61] Reber's curious attitude is made more difficult to understand because Squier's report was distributed to Scriven, Wood, the Chief of Engineers, the Chief of Ordnance, the Commandant of the Field Artillery School, the Assistant Secretary of War, and the Secretary of War.[62] Unless Scriven kept Reber ignorant of Squier's report, or of its many appended manuals and publications, Reber's attitude is difficult to reconcile with the large amount of information being received by the War College Division from London. Squier sent large numbers of books, Army Orders, periodicals, Acts of Parliament, proposed budgets, war photographs, regulations, dispatches of war leaders, tables of war equipment, and aeronautics publications on medicine, ordnance, signaling, railroads, awards of medals, technical work conducted in the National Physical Laboratory (a copy of which was sent directly to Scriven), work of the parliamentary committee on Aircraft Insurance, treatment of prisoners of war, night flying, Zeppelins and other enemy aircraft, sailing instructions, maritime pilot lists, Zeppelin raid accounts, tropical diseases, horse breeding, national and international law in time of war, and command orders from various military establishments. When asked to place emphasis on obtaining original combat photos, Squier eventually found a valuable collection of 300 photos taken by an American chauffeur to Major General H. dB. DeLisle.[63] The chauffeur, Frederick A. Coleman, was one of those rare individuals permitted to take photographs in the war zone. His personal annotation of each one added greatly to their worth. Squier sent a duplicate set to the War College Division.[64]

Among the books Squier sent to the War College Division was one entitled *The Officer Training Corps and the Great War*. Squier appended a note stating that General Wood was interested in the subject and recommended that one copy be forwarded to him.[65] Some other works were the *Aviation Pocket Book* by R. Bolase Matthews; *Aircraft in the Great War* by Grahame-White; *The Art of Reconnaissance* by Brigadier General David Henderson; "The Development of the Aeroplane" by Dr. R.T. Glazebrook, who presented his paper before the Aeronautical Society of Great Britain; and, *Aircraft in Warfare* by F.W. Lanchester, a member of the British Advisory Committee for Aeronautics.[66]

Squier also reported on radio matters. Early in the war a Lieutenant S.C. Hooper, United States Navy, assigned to duty with the United States embassy in London, prepared a detailed report on the organization of the British radio service. Of special interest in his report, which Squier forwarded to Washington, was a thorough exposition of British communications intercept operations. The Marconi Company operated eight or ten stations for the Post Office. They did nothing but "listen in" for messages, signals, and press news from the enemy and neutrals. Each station monitored a wave length (or range of frequencies) and copied everything heard. Message destinations were associated with specific transmitter sites, interconnecting links, power, tactical call letters, working schedules, and operating frequencies. Squier discussed modes of working tactical units in detail, including a description of message format.[67]

Plans to relieve Squier from his London post by 1 January 1915 were canceled.[68] Page's regard for Squier was respected. The Secretary of War designated Squier as an observer with the English forces and stated he would remain in London as long as he had a "good prospect of going to the front."[69] Unknown to the Secretary and the Chief of Staff, Squier was already at the front. When he returned, Lord Kitchener intimated that other trips would follow if word of his previous one was kept secret. Squier was retained in London on the chance that more trips to the front would follow.

Squier Returns to the Front

Sometime before the end of May 1915, the Supreme Military Command decided to permit only observers from Allied nations.[70] As of 1 March 1916 the British would allow no observers to see anything of military operations in France, with the exception of Squier. The French had allowed one American medical officer to visit a front-line receiving hospital. The British subsequently invited the Americans to send a detail of medical officers to England for conferences with Sir William Osler.[71] When some observers were finally permitted in the combat zone in the fall of 1916 they were company grade officers. The Allies did not desire high-ranking officers studying operations.[72] Their greater experience made them more informed, and potentially more valuable to the enemy side.

Sir Edward Grey arranged Squier's second visit through Lord Kitchener in October 1915.[73] Accompanied by two American Army captains, Squier visited Le Havre for two weeks. They were allowed to see every element of the base. Because Le Havre served as a principal supply base for British military operations in France, it offered a useful model for the study of logistical problems associated with trans-oceanic shipping.[74]

Squier's last trip dealt with the administration of an army in the field. He and his small party of observers studied the operations of the Adjutant General at his headquarters in Rouen. His command consisted of 70 officers and 1,800 clerks, who kept the war records on 1.2 million soldiers. The entire establishment operated in the Bishop's Palace, attached to the Cathedral of Rouen. The Adjutant General gave them every opportunity to obtain detailed information on record-keeping in time of war. When Squier sent his report to the War College Division he attached the observers' reports, a complete set of forms used by the Office of the Adjutant General, and a manual dealing with registration and conduct of correspondence in military offices.[75]

All in all, Squier made three trips to the combat zones in France.[76] Each visit concerned itself with an important phase of conducting military operations in the field. Walter Page's biographer, Burton Hendrick, asserted that if the Army had acted on the information dispatched by Squier it would have had a "policy and a program, if not an army, on the Declaration of War in 1917."[77] He was personally convinced that Squier's reports never saw the light of day once they arrived at the War Department, especially those dealing with aviation. Reber's complaints about the absence of aeronautical information from Europe would support Hendrick's contention. Even Miles was refused permission to see his own reports after his return from Russia.[78] In April 1916, Squier received word that he would be returned to Washington. Army aviation had given a poor showing during the recent Mexican-border operations against Pancho Villa. Everyone agreed that reorganization was in order. Young aviation officers were resentful of their leadership and becoming difficult to manage. General Macomb, Chief of the War College Division, regretted transferring Squier from London where his services were valuable to the entire government, but as he stated,

> The immediate necessity for strengthening the Aviation Section ... and of having an experienced officer fully conversant with the latest developments in this new art available for duty in the War Department, are of paramount importance.[79]

11
The Biggest Thing of the War

Squier returned to the United States, caught up in the fever of national preparedness. Leonard Wood's and Theodore Roosevelt's insistent urgings for strong national action on military expansion finally took root in the emotions of the American people.

It soon became apparent that achieving a state of military preparedness would be difficult. The United States was undeveloped or heavily dependent upon one of the belligerents in every area of industrial science related to aeronautics, chemicals, glass, and guns. Agencies for coordinating military needs with industrial capabilities were non-existent. Facilities for scientific research within the military services were improving, but still unequal to a program of national preparedness.

Squier's arrival provided the Army with one man who possessed some of the needed contacts in industry and professional science. The need for scientific research and industrial cooperation in aviation surpassed all other areas. In 1916, aviation was still regarded as an experimental enterprise.

Nine months after Squier set foot in New York he was promoted from Lieutenant Colonel to Brigadier General and appointed Chief Signal Officer. Within his charge were the entire aviation and communications missions of the Army. The assignment recalled Greely's use of Squier as a one-man technology brigade. His responsibilities spanned the military uses of radio, submarine cables, telegraph systems, cryptography, photography, meteorology, rockets, guided missiles, airplanes, devices for clandestine signaling, and psychology. He oversaw a collection of operational, technical, and research duties which were beyond the ken or competence of any one man. The impossible magnitude and scope of the assignment dramatized the extent of the Army's failure in accommodating scientific research within the military. The war in Europe was widely viewed as a battle between scientists. The American Army was unprepared for that kind of war.

Squier tried several approaches in obtaining scientific assistance in developing aviation. This chapter describes his use of the National Advisory Committee for Aeronautics, a cooperative advisory agency, to establish aviation policy, marshal scientific and industrial resources in aeronautics, bring peace to the aviation industry, and provide coordination with other governmental agencies interested in aeronautics. The next two chapters illustrate other arrangements Squier explored for institutionalizing scientific research and development within the Army.

Squier faced division within the Aviation Section of the Signal Corps, opposition in the General Staff, and encouragement as well as harassment from the general public on the question of aeronautics. In ways that were highly questionable from the standpoint of military protocol and destructive of his reputation, Squier forced aviation upon the Army.

Brigadier General George O. Squier, 1917 (courtesy CECOM Historical Office).

The Aviation Section

Squier found the Aviation Section in a deplorable state of affairs upon his return.[1] He must have had mixed feelings about his assignment. Until mid-April he thought he was returning to Washington to head "Government Telegraphy, where I shall have good opportunity for progress."[2] Now, the maelstrom of discontent in the Aviation Section caught up with Squier and banished his chances for further personal activity in the active work of science. Unfair and unrealistic administrative regulations depressed morale of officers in the Aviation Section.

Flying officers felt dispossessed and disenfranchised. In testimony before a special committee of the General Staff, one young flying officer, First Lieutenant H.A. Dargue, eloquently pleaded the aviators' case. After eight years of Army Aviation and appropriations of $1 million, he complained, there were only 23 qualified pilots and not one unit fully equipped for field service. Nothing short of separation from the Signal Corps would satisfy the dissidents. Dargue complained that, even if airplanes should be properly regarded as most suitable for the service of information and reconnaissance, the Signal Corps had failed to develop a single radio set for airplanes. He resentfully claimed that all experimental work was accomplished by fliers and appropriated by the Signal Corps, which took credit for the achievements.

Widespread objection existed among pilots to the exercise of command by senior Signal Corps officers. Dargue said that 21 of the 23 fliers in the Aviation Section felt "that officers of the line of the Army are more competent to man this branch of the service, this really being a branch of the line of the Army."[3] In a defiant gesture of unwelcome to Squier, then in mid-Atlantic on his return from London, Dargue asserted on behalf of the 21 officers that "the Signal Corps, so far as we have been able to find out, have no officers who have shown themselves competent to take this command."[4]

If the aviators—all junior officers—felt oppressed, their seniors felt equally tormented. Chief Signal Officer George Scriven held against the aviators a long list of grievances which he nursed for several years. He complained about their youth and inexperience, deficiency in discipline, and lack of proper knowledge of the customs of the service and duties of an officer. He condemned them for resorting to insubordinate acts to further their ambitions in forming a separate aviation arm:

> Selfish [substituted for "unscrupulous efforts," which Scriven lined out in his draft manuscript] desire for personal aggrandizement on the part of officers has tainted and obstructed many efforts to obtain beneficial changes in conditions in the Army.[5]

Scriven's most fervent complaint against the aviators concerned the Goodier case. Lieutenant Colonel Lewis E. Goodier's son was seriously injured while in flying training at San Diego. In trying to have the flying school commandant ousted, the Colonel wrote critical letters to friends throughout the Army about questionable "brother officers." Rumors of disaffection and insubordination became so widespread that the Army preferred charges against Goodier, who held the important post of Judge Advocate of the Western Department. Goodier's court-martial found him wanting in the proper regard of his office relative to junior officers.[6] In separate findings by an Army investigative committee, recommendations to censure Samuel Reber, Chief of the Aviation Section, for disrespect to a coordinate branch of government, lack of business method, and lack of loyalty to General Scriven, his superior, were sent to Secretary of War Newton D. Baker. Secretary Baker concurred and added a personal censure of Scriven for failing to supervise discipline within the Aviation Section. Reber was relieved of his command and Captain William (Billy) Mitchell took temporary charge.[7]

Army aviation was under attack from the public as well. Its notable failure during the pursuit of Pancho Villa, in March 1916, stirred official efforts to obtain more modern aircraft for the First Aero Squadron and to reorganize the Aviation Section. Of their first flight from Columbus, New Mexico, to Casas Grandes, Mexico, the commanding officer reported that, of eight airplanes, one turned back because of engine trouble; three became lost and made night landings in Mexico, one of them cracking up; and the remaining four were forced down by darkness, short of their destination. On the following day one pilot was unable to attain altitude enough to cross the Sierra Madre. After carrying mail and dispatches and performing reconnaissance duties for a month, there were only two planes left, neither of which was fit for further service.[8] Against this background of internal strife, external criticism, and poor record of achievement, the public press speculated about Squier being the natural man to take charge of the Aviation Section. His scientific reputation, familiarity with the European war, and early involvement in aviation affairs seemed to qualify him for his new assignment. The press welcomed the announcement of his appointment. One magazine hailed him as America's Air Ace Extraordinare, the one who would finally move Army aviation forward.[9]

That Squier's entire career was technically oriented and that his last command was in the Philippines one-and-a-half decades earlier escaped attention. Some felt, though, that the shortage of senior officers in the Signal Corps left no other choice possible. Major General Joseph O. Mauborgne, Jr., Squier's wartime assistant, said in later years that Squier was simply the only man the Signal Corps could assign to the problem.[10] His administrative abilities were most at question. General Scriven appraised Squier as "a very zealous energetic officer especially suited to broad schemes and given to broad views. Not especially a good routine officer but a very valuable man for the service."[11] Still, on the basis of Squier's reputation in science, General M.M. Macomb, Chief of the War College Division, advised the Chief of Staff that Squier was "an officer of great ability and is believed to be the one best fitted to undertake the work of the Aviation Section."[12]

Squier plunged into his new duties. He had a constituency of well-wishers in aeronautical clubs, the Smithsonian, and the National Advisory Committee on Aeronautics. Within the Army it was different. In the Aviation Section itself he had vexed relations with subordinates to overcome. Even the General Staff seemed unsympathetic to the problems that were now Squier's to tackle. For example, when Squier rushed an equipment manual into print without obtaining the approval of the General Staff, he soon heard from Army Chief of Staff Hugh Scott. Without the equipment manual, air squadrons could not draw supplies from the Ordnance or Quartermaster Corps. General Scott took the matter to the Secretary of War, who directed Squier to explain by what authority he approved and published the equipment manual.[13] Squier was in a dilemma. Without a manual, air squadrons could not receive the necessary supplies to fulfill the planning criteria that "an aero squadron upon receipt of orders for field service, should be equipped with a sufficient quantity of supplies to maintain it for at least six months."[14] Scriven replied that the error in publishing the manual without approval was due to "inadvertence."[15] A General Staff officer reviewing the issue admitted that the manual in question had indeed been marked tentative, but Squier's actions were a clear violation of standing orders. Nevertheless, the manual should be approved for further distribution.[16] Over three months later the Ordnance Corps was still unable to supply the items for aviation requested by the Signal Corps.[17] Just as in 1909 Squier decided to build his aviation activities on the foundation of governmental science located at the Smithsonian Institution, namely The National Advisory Committee for Aeronautics.

Early NACA-Army Relations

On 4 March 1915 Congress authorized the creation of an Advisory Committee on Aeronautics.[18] Dupree called it the "capstone of the federal establishment of science."[19] Borrowing a trick from John Wesley Powell of the Geological Survey, Walcott engineered passage of its organic act as a rider to the Naval Appropriations Act. The Advisory Committee acquired the word *National* in its title only at its first meeting, held on 23 April 1915, when rules and regulations were formulated and a temporary organization put into effect.[20] President Wilson opposed national support of aeronautics for fear of compromising the appearance of neutrality.[21]

Walcott's original proposal envisioned 14 members on the committee. The number was reduced to 12 at the suggestion of Franklin Delano Roosevelt, the Acting Secretary of the Navy. He urged Walcott to keep a majority government voice on the Committee to minimize the influence of private, commercial interests on the formulation of national aviation policy.[22] The committee consisted of 12 members appointed by the President—two officers from the Aviation Section in the Army, two from the office in charge of aeronautics in the Navy; a representative each from the Smithsonian Institution, the U.S. Weather Bureau, and the Bureau of Standards; together with not more than "five additional persons who shall be acquainted with the needs of aeronautical science, either civil or military, or skilled in aeronautical engineering or its allied sciences." Members were to serve without compensation. The members appointed were:

Professor Joseph S. Ames, The Johns Hopkins University
Captain Marl L. Bristol, U.S. Navy, Director of Naval Aeronautics
Professor William F. Durand, Leland Stanford Junior University

Professor John F. Hayford, Northwestern University
Professor Charles F. Marvin, Chief, U.S. Weather Bureau
Honorable Byron R. Newton, Assistant Secretary of the Treasury
Professor Michael I. Pupin, Columbia University
Lieutenant Colonel Samuel Reber, U.S. Army, Officer in Charge, Aviation Section
Naval Constructor Holden C. Richardson, U.S. Navy, Navy Department
Brigadier General George P. Scriven, U.S. Army, Chief Signal Officer
Dr. S.W. Stratton, Director, U.S. Bureau of Standards
Dr. Charles D. Walcott, Secretary, Smithsonian Institution

Alfred Zahm, director of the Langley Aerodynamical Laboratory, was appointed Recorder of the Committee. The composition of the committee accorded with the sentiments expressed by Bell, Squier, and Zahm in 1910, concerning a broad-based, interdepartmental advisory body. The creation of the NACA was the practical realization of those earlier recommendations.

Chief Signal Officer Scriven, on the other hand, thought that the committee should have been more oriented to military and naval aviation, containing far fewer members from the governmental and private circles. When the pending legislation was sent to Scriven for comment, he recommended the committee be limited to two representatives each from the Army and the Navy and one from civil life. The Chief of Staff demurred. For him, the object of the establishment of the committee was not to control aeronautics in the Army or the Navy, but to present, work out, and give scientific help to all persons engaged in solving the problems of aerial navigation. General Hugh L. Scott, the Chief of Staff, advised the Secretary of War that a conception of the committee, such as proposed by the Chief, would limit the committee's usefulness.[23] Scriven misread the role of cooperation in tapping the wellsprings of aeronautical research available to the military through a mix of private enterprise, scientific institutions, and government agencies, civil and military. Like Allen on the issue of the balloon plant at Fort Leavenworth, Scriven isolated Army aviation from a ready constituency. In Squier's absence, the Army appeared better informed than the Signal Corps on how to promote aviation.

Scriven should naturally be concerned primarily about military aviation and how the NACA might assist him in promoting support for it in Congress. Soon after its establishment, and even before the first meeting, Scriven tried to turn the NACA to his own purposes. In a lengthy memorandum for consideration of the committee he stressed two points. The most pressing and intelligent application of aeronautics was that relating to military aircraft and the study of aeronautics from the viewpoint of national defense. He argued that, since the NACA's advice carried great authority, it should survey the aviation needs of the Army, Navy, and other bureaus. It should also urge Congress to grant the aeronautical budget request of its various members to achieve sound aeronautical development and adequate defense. Confident of the committee's prestige, Scriven explained that this recommendation "will give Congress a satisfactory ground that shall be standard and beyond cavil."[24] Scriven had failed in his own efforts to obtain a large aviation budget request the preceding months. When Scriven submitted a request of $1 million for aviation to Secretary of War Garrisson it was cut by $600,000. Congress trimmed another $100,000 from the appropriation request.[25] Despite a war in Europe in which aeronautics was daily demonstrating its indispensable value, Scriven received a meager $300,000 for aviation in fiscal year 1916.

Scriven's expectation that by reason of influence and prestige the NACA could stimulate anyone to provide more money for military aeronautics proved unwarranted. The NACA itself had only $5,000 in its annual appropriation. Its only work in 1915 consisted of two meetings, preparation of contracts for seven reports on special topics, and a canvass of national research resources in aeronautics.[26] American universities, engineering schools, engineering societies, and industrial establishments were surveyed to determine what facilities and personnel were available. Most universities regarded flight as more a curiosity than an engineering problem. Only two schools offered courses in aeronautics in 1915. A handful of laboratories were equipped for aeronautical research, very little of which was being conducted anywhere. When fiscal year 1915 ended, the committee had spent $3,300 of its $5,000 appropriation on reports. It returned $1,000 to the Treasury.[27]

As the year 1916 opened, the committee was settled in new quarters located in the Munsey Building in downtown Washington, D.C. The committee agreed to entrust their work to the hands of an Executive Committee, which would meet on a regular monthly schedule. Ten subcommittees were appointed to monitor progress in various aeronautical fields and to maintain the bibliography on aeronautics. Limited funds permitted only one technical investigation, a seven-part report on a gasoline carburetor design.[28] The Secretary of the NACA, Naval Constructor Jerome Hunsaker, vigorously carried out the responsibilities of his office; perhaps too vigorously, in the opinion of Walcott. Walcott wrote Squier in London, in March 1916, that there was "a strong agitation for a great government laboratory to be under the control of the Navy Department." Turning to a theme which he was sure would find strong sympathy in Squier, he expressed the conviction that "the plan should be broadened out so that it would be under the administration of a commission which would not be under the control of any one of the Departments."[29]

Walcott's apprehensions were predictable in light of the momentum developed by the Naval Consulting Board under the prestigious direction of Thomas A. Edison. Created in mid–1915 without congressional authority or appropriation, it forged ahead with such vigor in so many research fields it would have monopolized research not only in the Navy but in the whole government.[30] In March 1916, the same month Walcott wrote Squier, members of the Navy Consulting Board testified before the Naval Affairs Committee on behalf of an authorization for a naval research laboratory. They obtained an initial sum of $1 million. Internal dissension over the location of the laboratory resulted in the entire project being dropped until the year 1923.

Walcott may also have objected to the heavy emphasis upon inventors and engineers represented in the makeup of the Naval Consulting Board. Willis R. Whitney, Frank J. Sprague, Leo H. Baekeland, and Elmer A. Sperry were prominent members. Very few members had links either to government or to university science. The National Academy of Sciences was ignored.[31] As Squier's friend and former accomplice in aeronautics, Walcott might have taken heart when Squier replied that he was "of course, interested in what you say about a government laboratory."[32] At least Squier considered himself a scientist, according to one of the foremost scientists of the day, Robert A. Millikan, of the University of Chicago, and subsequently the Executive Officer of the National Research Council. As a government scientist, Walcott would prefer officials with large sums of money to spend on the expansion of government research in the name of national preparedness and to be sympathetic to university and government science. With an appropriation of $1 million, a broad research warrant, and prominent names in its membership, the Naval Consulting Board must have seemed a Goliath to Walcott and the NACA, which received a niggardly $5,000 annual appropriation.

Squier Takes His Place on the NACA

Soon after his arrival in Washington, President Wilson appointed Squier to the NACA as the Army's second representative.[33] As Chief Signal Officer, Scriven was the principal. The springtime debacle finished Reber's career. Squier's first meeting of the NACA must have seemed like a homecoming. Sitting around the table he would have seen his old friends R.B. Owens, Alfred Zahm, both on the Subcommittee on Publications and Dissemination of Aeronautical Information; S.W. Stratton, Glenn Curtiss, chairman and member, respectively, of the Subcommittee on Applied Aerodynamics and Aeronautical Standards; and Orville Wright, a member of the Subcommittee on Hydromechanic Experiments in Relation to Aeronautics.[34] The sight of Michael Pupin would have brought back memories of his high-speed telegraphy experiments with the high-speed alternator which Pupin gave Crehore and him at Fort Monroe almost two decades earlier. His old mentor, Joseph Ames, completed the circle of friends, colleagues, and associates who had been so much a part of his own professional military and scientific careers.

The Search for Cooperation

If the NACA in its first year of existence had a fault, it was in its heavy representation of government and academic science at the expense of private industry. The Naval Consulting Board was conversely vulnerable with its weakness in representation from institutionalized science. Inability of the Naval Consulting Board to come to an agreement on a location for its laboratory, already generally provided for by Congress, temporarily gave the NACA an opportunity to seize the initiative in aviation research.

The NACA took hold of this new chance by making a public offer of cooperation to the aeronautic industry. On 8 June 1916 the NACA held a public meeting at the Smithsonian Institution. Present were representatives from various manufacturers of aircraft engines and government bureaus. The principal item of discussion was the lack of a good American engine for aviation. Three prepared statements preceded a question-and-answer session, held in the very rooms of the Smithsonian where Langley "first developed his idea of the physics of the air and also of aircraft."[35] Mr. Henry Souther, an engineer of some note and recently hired by the Signal Corps to advise on aeronautic matters, led off with a statement on military aeronautics. Squier followed. The third speaker, Byron R. Newton, presented the interests of the Coast Guard. He talked of their desire to obtain $300,000 from Congress for the purpose of instituting aeronautical activities at ten Coast Guard stations.[36]

When Squier spoke, he praised the British national competition that led to the Green engine and freed Britain from its dependence upon French aviation engines. He said simply that if America wanted to develop superior aeronautical engines she must do it for herself. Europeans learned from each other by capturing their enemies' aircraft and studying them in minute detail. One could not expect the belligerents to share those plans with the Americans.[37] Squier's fascination with a single, standard engine, capable of great power, remained vital for as long as he controlled Army aviation.[38] The meeting ended on a confident note. Committee members considered one of the results of the meeting to be "the inauguration of an important movement for the development of satisfactory aeronautic engines."[39] As part of the movement the committee decided to cooperate with the Society of Automobile Engineers on the subject of motive power, through the NACA Subcommittee on Power Plants.

In the larger perspective of solving the backwardness of American aviation, other subcommittees worked with the Bureau of Standards on the characteristics of materials used in airplane construction and the establishment of standards in design, construction, and testing of materials.[40] These areas formed the crux of successful scientific and industrial cooperation in developing an aircraft industry.

Squier imprinted his own vision of military, industrial, scientific cooperation upon the proceedings by serving as a member of the NACA subcommittees of Standardization and Investigation of Materials, Power Plants, and Nomenclature.[41] He was Chairman of the Design, Construction and Navigation of Aircraft and Airplane Mapping subcommittees as well.[42] John F. Victory, first civilian employee of NACA and its longtime executive director, recalled those early days. He said Squier was a "working member," not just a name on the roster of members.

One of Squier's preoccupations in 1916 was the selection of a site for the experimental work of the Aviation Section of the Signal Corps and of the NACA Committee. Without an experimental field, very little progress could be made in achieving cooperation between government, science, and industry. The acquisition of Langley Field promised the great government research laboratory which Walcott, Squier, and associates had so long urged as essential to the development of American aviation.

The Research Program

Squier saw himself on the ground floor of a great and growing edifice. In an address before the Franklin Institute at Philadelphia, in the fall of 1916, Squier declared: "We are about to launch a conquest of the air which will startle the nations of the world. Tremendous progress will be made in aerial navigation in this country during the next decade." He prophesied that universities would, in time, include instruction on the "physics of the air." City planners, too, would have to provide for municipal airports to contend with "heavy aeroplane traffic."[43] In the following month, Squier addressed a meeting of the National Academy of Sciences at Boston. The areas he identified appear to be the program of research adopted by the NACA. Some of the areas were aerodynamics; engine problems connected with fuel, engine cooling, metal coating, and sound; physiological and psychological effects of high-altitude flying; light alloys for airplane construction; structure of wind gusts; radio apparatus for aircraft; automatic stabilizing devices for aircraft; bullet-proof gasoline tanks; ground-speed indicators; aviator's clothing; fabrics for covering airplanes; and photography. He also asked for a study of the physics of the air as it related to high-frequency vibrations, elasticity, and friction. He concluded his remarks with this observation:

> It appears that we have been passing our lives on the bottom of a deep ocean of comparatively heavy fluid, and that only recently have we learned how to utilize the dynamic reaction of this fluid to construct machines for transportation freely in three dimensions in the interior of this ocean. Who among us is wise enough to foretell what these machines may yet accomplish?[44]

Whatever might be accomplished, Squier was certain that America should be self-sufficient. He told the National Academy of Sciences: "We must become independent in all lines affecting our military aviation."[45]

Despite Squier's confident tone and well-defined program, there is some evidence to suggest that appearances were misleading. In early January 1917, Henry Woodhouse, editor of the magazine *Flying*, asked Walcott for a statement of the coming year's work at NACA to publish in his magazine.[46] Woodhouse had recently written a series of highly critical articles on the work of the Aviation Section.[47] Because of this and an unsavory reputation, some members opposed the project. Walcott wrote Michael Pupin, a member of the executive committee of the NACA, relating the opposition to Woodhouse's request and suggesting that "if anyone has a clear plan for the work of the Committee for the next twelve months I think it would be well to have it considered." The basic weaknesses of the NACA position stood exposed: insufficient funds, lack of experimental grounds, and no available scientific laboratories. They approached problems on an *ad hoc* basis. In the absence of a grand plan, Walcott offered his "general conception ... that the Committee will do all that it can to help along the art of aviation by acting upon such matters as are brought before it from time to time."[48]

The Patent Crisis

But more than the lack of a program, adequate test facilities, and appropriations plagued those who tried to get an American aviation program underway.[49] Patent controversies stemming from the Wright and Curtiss litigations created confusion and division among aircraft designers, manufacturers, and distributors.[50] By 1917, there were approximately 130 airplane patents in force. More were pending. While many were doubtlessly of little value, some were so basic that airplanes could not be made without them. Valuable or worthless, each could cause a lawsuit. Every airplane that left the construction shed laid open its manufacturer to new infringement suits. Manufacturers might pay royalties to more than one claimant and face claims from others.[51] Thus, every airplane was over-priced. Manufacturers operated under a cloud of impending infringement squabbles. The customers who suffered most were the largest potential users, namely, the Army and the Navy.[52] So they encouraged efforts in January 1917 to find peace in the airplane industry.

Numerous meetings between government officials, owners of patents, and aircraft manufacturers ensued in the succeeding months. Solving the patent issue became fundamental to establishing a stable, productive aircraft industry. The Subcommittee on Governmental Relations, Walcott's ingenious creation designed to serve as trouble-shooter for NACA, sat with Squier and Howard Coffin, member of the Council of National Defense, to build agreement on the patent issue. Walcott chaired the governmental relations committee of two; S.W. Stratton was its only other member. Everyone agreed that without Glenn Curtiss's cooperation, failure was certain. According to John P. Tarbox of the Curtiss Aeroplane and Motor Corporation, Curtiss held about two dozen controlling patents. Pressures were intense to clear away the barriers to plentiful production of airplanes on a sound legal and financial basis. Tarbox said:

> Mr. Curtiss has no idea of demanding royalties from other manufacturers under existing conditions, or of using his patents against competitors, that his purpose in obtaining patents was to fortify and defend himself in any patent litigation that might be started by others.

Tarbox and Curtiss liked the idea of organizing airplane manufacturers along the same lines as automobile manufacturers were, and thought it should be easier to accomplish. They both pledged themselves to cooperate in forming an association of aircraft manufacturers.[53]

Discussions, meetings, and negotiations proceeded in the following months, sometimes extending throughout day and night. Squier removed himself from the Motive Power Subcommittee and joined the Subcommittee on Patents, which Walcott chaired. In late March, the Patent Subcommittee composed a draft agreement on patents. It called for cross-licensing in the aircraft industry on a basis similar to that existing in the National Automobile Chamber of Commerce. For every airplane manufactured and sold, with or without an engine, a $200 fee would be paid to the treasury of the manufacturers' association, or more properly, a corporation holding in trust the patents of airplane manufacturers. Of the $200 fee, $135 would go to the Wright-Martin Aircraft Corporation, $40 to the Curtiss Aeroplane and Motor Company (including the Burgess Company), and the remainder of $25 would stay in the treasury to cover the expenses of the association. Representatives of the Wright-Martin and Curtiss companies sat in the waiting room while the proposed agreement was discussed. After some time they were invited to join the patent subcommittee for a discussion of the agreement.[54]

Although a record of their discussion is unavailable, the agreement clearly followed the lines laid down in the cross-licensing arrangement within the automobile industry. W. Benton Crisp's membership on the Patent Subcommittee almost assured it, since he served as Henry Ford's attorney and chief strategist in the Selden Automobile Patent case and representative of the Hudson interests in the Hudson Crank Shaft Case.[55] The outcomes of these famous cases led to the automobile cross-licensing covenants that finally brought industrial peace among automobile companies.

More conferences between the manufacturers and the Patent Committee followed. The NACA finally recommended the organization of an association among aircraft manufacturers for the purpose of cross-licensing aeronautic patents.[56] They offered the draft prepared by Walcott, Squier, and Crisp as the model. Delays in obtaining agreement among the manufacturers, because of jealousies, personal hostilities, and distrust in the industry tormented the negotiators. Negotiations were reviewed and analyzed by the highest officials of all the companies, including Curtiss and Wright themselves.

Then on 6 April 1917, America entered the war. Squier queried the Aircraft Manufacturers Association if their members could supply the estimated requirement of about 3,000 airplanes for the armed services. The association answered officials that they could supply their requirements without further expansion of manufacturing facilities.[57] At a meeting of the Patent Subcommittee on 24 April, 18 days after the declaration of war, it was learned that the Curtiss Company was "favorably inclined" toward formation of a holding company to administer a cross-licensing arrangement.[58]

The key to understanding the exasperatingly slow progress in negotiations on a cross-licensing agreement was the commanding position of the Curtiss interests. A special Committee on Production, formed shortly after the war began, described the American airplane industry as consisting of two parts: "The Curtiss Airplane and Motor Corporation, and others."[59]

The desperate lack of preparation and sound legal conditions for expansion in the aircraft industry were exacerbated by war. Squier took an immediate shortcut to help. Without proceeding through regular diplomatic channels, he personally visited the military attaches of France, England, and Italy in Washington. He asked each "to send forthwith the most experienced and trained flyers, aeronautical engineers, and designers who could be spared to assist us." Simultaneously, a strong technical mission headed by Major Raynal C. Bolling was sent to Europe.[60]

With the arrival, on 24 May 1917, of Premier Alexander Ribot's famous telegram, in which the French leader asked the Americans for the staggering total of 16,500 airplanes and 30,000 engines, the airplane industry realized their sudden inability to supply airplanes in adequate numbers. Benjamin

Foulois, Squier's assistant at the time, remarked in his memoirs that the request "completely shattered all previous American plans."[61] Clearly, more industrial might would be needed to supply all the aeronautical machines and equipment. A survey by the National Council of Defense showed that the automobile industry could handle much of the additional work. But the auto manufacturers looked askance at entering a field in which patent suits and costly litigation might attend the delivery of thousands of aircraft to their purchasers. They vividly recalled the difficulties which the Selden Automobile Patent caused in their own industry. They were understandably cautious about becoming involved.

Once more, the Aircraft Manufacturers Association was called into consultation. The association promptly appointed a committee, including Tarbox and Crisp, to draw up a cross-license agreement which would allow manufacturers for the government unrestrained use of all airplane patents. After two weeks, the plan was submitted to the NACA Subcommittee on Patents.[62] They, in turn, recommended that the new organization, charged with administering the cross-license agreement, be called the Manufacturers Aircraft Association, perhaps in recognition that more than airplane manufacturers were vitally involved in building aircraft for the Allies.[63] On 24 July 1917, the major aircraft companies acceded to the arrangement and joined the Manufacturers Aircraft Association.[64] Two weeks later, the NACA Subcommittee on Patents dissolved; their important work was completed.[65]

The agreement consisted of the same terms first cast the preceding March. Soon thereafter, when all appeared as if aircraft production could proceed unhindered, certain persons sought to establish that the Manufacturers Aircraft Association was a trust, an illegal combination acting in restraint of competition. The adverse publicity hit upon such catch phrases as "trust" and "highbinders" to attack the new organization. Then the Aeronautical Society of America let loose a blast against the Manufacturers Aircraft Association. It was followed by a "virtual hymn of hate" against the new organization.[66] On 17 September 1917, the Secretary of War requested an opinion of the Attorney General regarding the legality of the cross-license agreement and whether it contravened the anti-trust statutes of the United States.[67] Pending the Attorney General's opinion, disbursing officers in the Equipment Division of the Signal Corps refused to make royalty payments to the Manufacturers Aircraft Association.[68] The disbursing officers acted on instructions from the War Department, not from Squier, who was more frustrated than ever in getting aviation off the ground.[69] Later, in October, Attorney General Gregory gave the association a clear bill of health after an investigation. The net result was the loss of a valuable six weeks in preparing American aviation for European service. The Attorney General's investigation was only the first of a half-dozen such inquiries, each accompanied by public recriminations and torrents of hate.

The Struggle for Airplanes

Criticism of Squier's management of military aviation came from nearly every direction. Loening thought him a dupe of the automobile manufacturers.[70] The prominent physicist, head of the Science and Research Division of the Signal Corps, and high official of the National Research Council, Robert Millikan, considered Squier a poor organizer.[71] As actual aircraft production failed to achieve glowing public statements on how clouds of Yankee airplanes would overwhelm the Kaiser, the crescendo of criticism grew louder.[72] While Squier did have a certain penchant for conducting research on the side, that is, across organizational lines, he also had some serious constraints, even hindrances, in fulfilling his responsibilities. Those constraints could have led many, in the absence of full knowledge of circumstances, to conclude that Squier was a poor organizer.

The Appropriation

When the Ribot telegram, asking for tens of thousands of aircraft, arrived in Washington, Squier assigned his deputy, Benjamin Foulois, to the task of computing the cost of such an undertaking. Foulois's figures surprised and shocked everyone. He estimated $640 million would be needed. When Foulois showed his figures to officials responsible for planning in the War College Division, they were flabbergasted. That the air arm should be on a funding par with the rest of the Army left them incredulous. Getting nowhere, Foulois turned to Squier.

Squier knew his proposals for aviation could be staffed to death. The generals commanding the line elements of the army, artillery, cavalry, and infantry, would find Foulois's estimates inconceivable and Squier's vision of the role of aviation in modern warfare unacceptable. So Squier went directly to Secretary of War Newton Baker, an act of considerable courage for a career officer.

He told Baker he was sure his aviation appropriation bill would pass Congress if submitted. It was Baker's turn for unprecedented action. He sent the appropriation request directly to Congress without approval from the General Staff. He even bypassed the Appropriations Committee in sending it to the House Military Affairs Committee. The committee unanimously reported out the bill. The bill spent a total of two weeks in gaining passage by Congress. The President signed the bill—the largest for a single purpose in the history of the American Congress—on 24 July 1917.[73]

A Visit to the Chief of Staff

Squier returned to the Army Chief of Staff for administrative and technical assistance in establishing an aeronautical organization. Considering the way that he obtained the aeronautical appropriation, he was extraordinarily bold, if not audacious. He was, however, dedicated to the idea that aviation could indeed lessen the number of American troops in the trenches. The House Military Affairs Committee, after deliberating over testimony given before it by Squier and representatives from Great Britain, France and Italy, noted:

> The control of the air was at this time at least the most important thing that could be done. In fact, it was boldly stated that the control of the air would naturally result in reducing to a large extent the number of men in the trenches.[74]

Squier's vision of the magnitude and scope of aviation was singular among senior Army officers in Washington. Although Theodore Roosevelt approved of few general officers on duty in Washington, he identified Squier as one he thoroughly accepted.[75] Squier's Chief of Aircraft Production, E.A. Deeds, an industrialist who left the presidency of Delco Corporation to become a dollar-a-year man, later commissioned colonel in the Signal Corps, said "of all the Army officers in Washington he was the only one with the vision to comprehend our aviation requirements."[76] Even Millikan, who thought Squier a strange character and not a man of balanced judgment, said he "had one great quality much needed at that time, namely a willingness to assume responsibility and go ahead."[77] After all, Squier had seen the war firsthand and understood the necessity for extensive preparations. He opposed all halfway measures.[78] So, despite his indiscretion in obtaining the huge $640 million appropriation and in visiting Allied attaches, Squier visited the Chief of Staff for help.

In the Office of the Chief Signal Officer in Washington, there worked 11 officers, ten enlisted men, and 103 civilian employees.[79] In the entire Signal Corps there were only 55 officers, 1,100 enlisted men, and 210 civilians.[80] And of the fifty-plus officers assigned, 42 were detailed on temporary duty to the Signal Corps.[81]

Squier might have tried to explain to Chief of Staff Hugh Scott the importance of aviation to winning the war and the dismal effects created by the practical neglect of aeronautics. There were but a handful of flyers, indeed, about 35 qualified pilots at the time, all of company grade.[82] Squier probably

Main floor of General Squier's office, 14th Street and Park Road, Washington, D.C., c. 1918 (courtesy CECOM Historical Office).

advised Scott what he wrote in his annual report: "There was practically no aviation industry in this country, and the number of professional men trained as aeronautical engineers and designers was so small as to be practically negligible."[83] Having returned only the year before from London and the front lines in France, he perhaps recounted his own observation that wars from then on will be fought in the air, like it or not.

The Chief's response was direct, simple, and unmistakably clear. He told Squier that every man and every dollar he took away from the Army would require one more man and one more dollar to win the war. Then he ordered Squier to get out of his office and "stay out for the duration."[84]

Squier Asks Walcott for Help

On the next day there was a meeting of the Executive Committee of the NACA. Squier tarried after the meeting closed, asking Walcott to speak with him in private for a moment. Squier explained what had happened with the

Chief of Staff. Walcott asked Squier why he did not visit Secretary Baker. Squier replied he would not circumvent the War College Division and the Chief of Staff. Walcott thought for a minute. Then he instructed Squier to come to his home promptly at eight o'clock the next morning with breakfast in his stomach.

When Squier arrived, Walcott was breakfasting with Baker. Squier understandably looked surprised. Squier tried to excuse himself. Walcott asked him to join them for breakfast. Squier, of course, replied he had already eaten. Baker told Squier to come in and have a cup of coffee. When Squier was seated and served, Baker asked how things were going with him. Thus, Walcott created the entree that Squier wanted. Squier could now unburden himself without offense to military decorum.

After Squier told the secretary of his problems in securing Army cooperation in establishing a military aviation program or policy, Baker allegedly suggested holding informal meetings between Squier, David Taylor, his Navy counterpart, Secretary of the Navy Josephus Daniels, and himself.[85]

A day later, Squier announced to the public that the Allied governments intended "to enter Germany by the air route." Robert Millikan said Squier's "method in everything was to go ahead, take in new personnel in a quite extravagant way, and count on each man finding his job, literally *creating* his own place." Squier asserted that the NACA had no full-time personnel, thus, he must take in personnel who could, in turn, attract good people. In this way he would create the necessary organization for aviation. Squier's viewpoint was approved by George Ellery Hale, and not opposed by Walcott and others of the NACA. Thus, Millikan was "well-nigh appropriated" to help staff the Signal Corps with scientists for its great mission.[86]

The Signal Corps Loses Aviation

In the end, production failed to meet public statements on the number of airplanes that would be sent to Europe in the short time expected. By spring 1918, the political pressure was immense. So, in March, the Acting Secretary of War appointed an investigating committee. H. Snowden Marshall headed the committee of three.[87] On the question of exaggerated claims for production, Marshall concluded that the "Signal Corps forecasts are probably more nearly accurate and certainly more conservative than ... others, and also are complete."[88] Still, the press had been filled with optimistic predictions of American air might soon arriving in force at the front. The committee failed to locate the source of the statements which so misled the public into expecting unreasonable, even impossible, achievements in airplane production.[89] When

Marshall submitted his findings in May 1918, he "advised the Secretary of War that the contractors had surmounted their big difficulties and quantity production could be expected within thirty to sixty days."[90]

Marshall's most notable recommendation concerned reorganization of aeronautical activities. On 12 April 1918, Marshall's committee submitted a preliminary report in which they recommended the removal of all aircraft production and aviation operations from the Signal Corps.[91] Three days later, Walcott pleaded with the President to keep aeronautics under Squier and remove from him only the control of aircraft production. He said:

> The Chief Signal Officer, General George O. Squier, by training, technical and military ability, was the best qualified man that America had in 1915 to undertake to organize and develop a new art in warfare. At the present time he is an inspiring leader to the entire army aeronautic organization. The selection of civilians for important places in connection with the development of the air service, has been most marked. Such men as Montgomery, Waldon, Deeds, Potter, Bolling, Millikan and Stettinius (assigned to other duty in the War Department), indicate the character of the civilian aids called to his assistance.
>
> The closest cooperation with Stratton of the Bureau of Standards, Carty of the Telephone System, Whitney of the General Electric, indicates a clear insight into the extent and character of the problems before him.
>
> It would be a material loss to the service and the Allies' cause to take anything more than the actual aircraft production work from the control of the Chief Signal Officer.[92]

However, the die was cast otherwise. On 25 April 1918, Secretary Baker told the press that military aviation would be placed in a new Division of Military Aeronautics under the direction of Brigadier General William Kenley. Squier would, in the words of the announcement, devote "his attention to the Administration of Signals."[93]

Who Gave Squier the Biggest Thing of the War?

A *New York Times* editorial hailed the decision to remove aviation from Squier and judged him "the genius of army aviation [who] has been a lamentable failure."[94] The damning attitude of the *New York Times* doubtlessly represented the feelings of many people. The public was frustrated with shortages in most of munitions and supplies, glowing predictions unmatched by deeds, recurrent talk about sabotage, and heavy losses in the trenches. Squier tried futilely to protect his reputation by asking for a court of inquiry.[95] He was advised that a decision would be made on his request following yet another investigation.[96] The best indication of Secretary Baker's own attitude about Squier's stewardship of American aviation may be reflected in his desire that

Squier deliver an address in January 1919 on aviation in the war before the annual convention of the American Institute of Electrical Engineers.[97] Even nine months after the hateful episode, Baker turned first to Squier to tell the story of American aviation, not to those who won acclaim as aviators or who commanded military aeronautics.

Baker's desire that Squier deliver the address may answer a question raised in the editorial of the *New York Times*:

> It may always remain a mystery why George Owen Squier was selected to do one of the biggest things of the war.... If there was any other reason [than being Chief Signal Officer] for giving him the authority to create an air fleet for an army greater than Wellington commanded at Waterloo it has not been divulged.[98]

The answer to the mystery is no one selected Squier to do one of the biggest things of the war—he, himself, took the initiative and undertook the development of American aviation. He did first, and asked later. He helped bring peace to the aviation industry through cross-licensing; he sought the unprecedented appropriation of $640 million directly from Baker; he requested technical assistance from the Allies without obtaining prior approval from the State Department; and he oriented advisory bodies, such as the NACA, toward policy-making bodies for the direction of aviation. The Chief of Staff did not give Squier the authority—Squier took it upon himself. Squier's personal conviction on the importance of aviation was evident in his remarks to a reporter at the time:

> We want enough [aircraft] to operate in regiments and brigades if necessary, to make all Germany unsafe, to force her to demobilize her air forces at the front and send the men and machines back to protect the cities. This will blind her artillery and render it helpless by depriving it of the rangefinders. Furthermore, our regiments and brigades will be able to destroy all of Germany's interior lines of transportation for the movement of troops and supplies. They will be able to rob the enemy fleet of the security it now enjoys at Kiel. An airplane can now carry and drop a thousand pounds of explosives. If that is not enough to sink a dreadnought, it certainly can disable her for three or four months, and then she can be disabled again in the same way after repairs. We can make the Kiel Canal itself useless.[99]

Squier found dissension and distrust widespread in the Aviation Section upon his return from London. Young flying officers were resentful of "old Army" leadership. The General Staff wanted an older officer to restore discipline and reorganize the Section. The Army Chief of Staff got more from Squier than he expected or perhaps wanted. Skirting Army procedures, forging new working arrangements with outside agencies, and pursuing his own vision of aviation, Squier grasped what, in the popular mind, was the biggest thing of the war.

The failure of the Signal Corps to achieve publicized goals of airplane production cast a pall over Squier's reputation. Measured against what the Allies had achieved in comparable periods of time and against the stated priorities in establishing a modern air service, the American achievement in aviation was prodigious.[100] When the war began, 55 officers and 1,570 enlisted men served in the Signal Corps. When the aviation mission was removed from the Signal Corps in May 1918, just 13 months later, there were 10,336 officers, 150,120 enlisted men, and 7,375 civilians.[101]

Such growth rates testify to an explosive expansion in the responsibilities that Squier carried. On balance he carried them well. During 19 months of war the United States produced 50 percent more airplanes than Great Britain in its first 31 months of war.[102] In April 1917 the Army owned 55 air-planes, 51 of which were obsolete and four were obsolescent.[103] At the signing of the Armistice, American factories had produced almost 4,100 combat airplanes and over 16,000 replacement engines for combat planes. Quantity production of training planes and their engines was achieved in 1918. By the end of the war American manufacturers had turned out 8,000 trainers and nearly 17,000 training engines.[104] The engines were of the renowned Liberty type which Squier ordered when he set American aviation on the path of standardized, interchangeable, mass-produced manufacture.[105] Within 19 months, American production of airplanes exceeded total Allied production by the ratio of 13 to 12. Maintenance and operation of aircraft, the most technically demanding machines in the Army, required long preparation and specialized training. The average soldier underwent six months of training before shipping to France.[106] The aviators needed much more time to complete their training, which consisted of three stages in the United States, and a fourth in France.

When the Armistice was signed, 5,000 pilots and observers were in France, as promised to the French on America's entry into the war. In addition, there were about six thousand flying officers in the United States or en route to France.[107] Virtually all the planning, training, letting of contracts, and constructing of factories that led to the creation of the American Air Service that emerged from the First World War were accomplished under Squier's direction. Measured against the time required to bring other major technological innovations to production-line status since World War I, Squier's achievement looks quite creditable indeed.

Criticism by Charles J. Gross

Dr. Charles J. Gross, who holds a Ph.D. in military history from The Ohio State University, has worked as a U.S. Air Force civilian historian and is currently the chief of the Air National Guard history program. In 1990, Gross published a paper critical of Squier's management of the nation's World War I aviation program, but also praising his leadership in enabling the program to occur in the first place.[108] In the opinion of the authors, Dr. Gross's conclusions are a fair evaluation of George Squier's role in the major expansion of the American aviation necessary for World War I:

> Although Squier's reputation was permanently damaged by the delays and scandals associated with the aircraft production program, he deserved much of the credit for its ultimate success. He had been the only senior army officer in Washington, D.C., who possessed the courage and vision to press for it at the highest levels of government in 1917. Once the program was approved, he supervised nearly all of the planning and contracting as well as the construction of research and development and manufacturing facilities associated with the American air program in World War I. The NACA, its Langley Memorial Aeronautical Laboratory, and McCook Field were monuments to the conviction of Squier and men like him that modern science and engineering could serve the nation well....
>
> Squier was a controversial and enigmatic officer who was made a scapegoat for the World War I aviation scandal. He was a scientist and inventor as well as a dedicated military professional.... Men were drawn to him by his passionate sincerity and strong advocacy of science in the service of the nation. He gave his subordinates unlimited authority inspiring great loyalty but also creating serious problems for himself when his staff performed poorly....
>
> On the negative side, the wartime aviation production program was charged with widespread fraud, waste, corruption, and profiteering. The mistakes that Squier and others made plus the difficulties inherent in that wartime program encouraged public distrust of all arms makers and weakened until the late 1930s political support for government spending on aviation and defense programs.
>
> Yet in final analysis, the assessment of George Owen Squier must be positive. He represented an emerging class of uniformed technocrats that played a pivotal role in securing the future of military and naval aviation.... He and others working with him ardently championed the nation's first wartime aircraft production program that, despite whatever shortcomings it may have had, hastened the rise of a modern American aviation industry based upon professional science and engineering. Squier was a practical realist who recognized the opportunities which the First World War had provided to place American military aviation on a sound technical footing. Ignoring bureaucratic norms, he moved forcefully to implement his ideas. Squier was by no means a cautious bureaucrat in uniform. The significant achievements of the nation's first aircraft production program owed a great deal to his vision, courage, and skill.

The authors offer this final comment on Squier's handling of the aviation program: his decision to entrust the mass production of aircraft to the automobile industry looks poorly thought out. It brought with it a host of problems, the worst of which is the obvious one: building an airplane is not the same as building a car. Engine failure, for example, is inconvenient for a driver, but likely fatal for a pilot. Aircraft performance and safety requirements impose much greater precision and attention to quality control than for automobiles. Why, then, did Squier turn to the automobile manufacturers? When the French asked for 16,500 airplanes and 30,000 engines, the automobile industry was the only one in the U.S. that seemed capable of meeting the French request, let alone what the Americans would need. For example, when Squier asked Glenn Curtiss how many aircraft he could produce in a month, the answer was less than ten. Squier had little choice. In hindsight, some sort of cooperative arrangement that would couple the aircraft designers' expertise to the mass-production expertise of the automobile industry would have been (and later was) the best approach, but Squier had little time to decide. Under the circumstances, he took the best approach available to him.

12
New Weapons

Squier exploited national scientific and engineering resources in a variety of ways. One technique was highly individualistic. It involved the support of several research projects outside established War Department programs. Often they were operated in secrecy, even from War Department authorities. The projects gave the appearance of being run from his hip pocket.

Yet it would be misleading to suggest that they were conducted for scientific curiosity or self-amusement. Squier believed that wars were won with new weapons or devices that gave one side an advantage for a period of time longer than the enemy requires to develop and produce a comparable weapon or countermeasure. The element of technological surprise and advantage counts heavily in modern warfare. He observed in his "granular theory of armies" that the one unchanging factor in warfare is the individual physical strength of a man. In consequence of this fact,

> military supremacy must be looked for primarily in the weapons and agencies provided by scientists and engineers and placed in the hands of these combatant units to multiply their military strength.[1]

It was thus that Squier supported two projects which he believed could affect the course of the war if perfected in time. He provided the first military funding of the rocket research conducted by Robert Goddard. And at his direction, Orville Wright and Charles Kettering readied the world's first flying bomb, that which another generation of people would identify with Germany and the term V-1.

The Liberty Eagle

One of the most imaginative proposals of the First World War concerned automatic carriers, flying bombs, or aerial torpedoes—the Liberty Eagles. The

Liberty Eagle was intended as America's secret weapon to end the war decisively in the expected spring campaign of 1919, but cessation of hostilities in late 1918 led to the abandonment of the project.

Those working on the Liberty Eagle project sought to produce an automatic airplane which could fly by mechanical control, without the aid of a human pilot. Designed to carry a charge of high explosives, fly a given direction for a specified distance, it would then dive suddenly, crash to earth, and explode, destroying itself and causing heavy destruction of the target. The idea of an automatic flying bomb, constructed along aeronautical principles and guided by mechanical or radio means, was not new. Peter Cooper Hewitt proposed the first practical suggestion for achieving a self-controlled aerial torpedo or flying bomb.

As a member of the Naval Consulting Board, Hewitt approached Elmer A. Sperry of the Sperry Gyroscope Company in July 1915 for technical assistance in planning and developing an automatic airplane.[2] Sperry, also a member of the Naval Consulting Board, had studied the application of gyroscopes to the automatic control of flight for some time. With funds personally furnished by Hewitt, Sperry examined the concept of an automatic flying bomb at an aviation field which they acquired near Amityville, Long Island. A control device was perfected and adapted to a Navy hydroplane which flew under automatic control over a seven-mile test course. At a predetermined distance, the automatic device caused the plane carrying Hewitt and Lieutenant Commander T.S. Wilkinson of the Bureau of Navy Ordnance to take a sharp dive. Only Sperry's resumption of manual control prevented a crash.

Progress of the experimental work seemed so promising to Naval officers and members of the Naval Consulting Board that the Board passed a resolution on 14 April 1917, which read:

> RESOLVE: That the Secretary of the Navy be requested to apportion from such as may be at his disposal the sum of $50,000 to carry on experimental work on the subject of aerial torpedoes in the nature of automatically controlled aeroplanes or aerial machines carrying high explosives capable of being initially directed and thereafter automatically managed. It is recommended that this work be carried on under the direction of the proper committee of the Naval Consulting Board.[3]

In early May, Secretary of the Navy Josephus Daniels approved the resolution and quadrupled the requested appropriation. Offered one week following the declaration of war against Germany, this resolution seemed, at last, to promise a solution to the fearful problems created by German submarines. The Secretary of the Navy was evidently infected with the enthusiasm of the Chairman of the Naval Consulting Board for its military prospects. The Chairman, W.L. Saunders, stated publicly that the submarine was doomed because of a new weapon, meaning the flying bomb. Meanwhile, experiments conducted under the supervision of the Bureau of Ordnance progressed so well that the Consulting Board decided to invite the Army's attention to their work. In late November, Squier, who had been kept informally advised of experimental progress by his old friend Hewitt, visited Amityville at the invitation of the Consulting Board.

Whether Squier viewed them as an augmentation to airpower, as the concept would have been understood by men like Billy Mitchell, Hugh Trenchard, or Guiglio Douhet, is an open question. Squier did regard flying bombs, though, as a forward step in the development of aircraft for war. Like modern-day missilemen, he conceived of flying bombs as an extension of artillery. They would have served admirably for tactical bombardment of such large targets as petrol and ammunition dumps located beyond the range of artillery. Since the Liberty Eagles could carry more bomb tonnage than contemporary bombers, their destructive potential made them more lethal than bombers. Strategic bombing would also be possible against large area targets like cities or heavily industrialized regions like the Ruhr valley. That he viewed such a weapon as a signal development is manifest in a memorandum to the Chief of Staff. He said:

> The development of this new weapon marks an epoch in the evolution of artillery for war purposes, of the first magnitude, and comparable, for instance, with the invention of gunpowder in the fourteenth century.[4]

Squier wrote the Chairman of the Aircraft Production Board that the "time has come when serious attention should be given ... to the development of a new type of artillery based primarily upon the principles of the flying machine."[5] With evident pride in Yankee ingenuity and a conviction in the winning effect of new instrumentalities in war, he urged the Aircraft Board in secret session to take to Europe "something new in war rather than contenting ourselves as in the past with following the innovations that have been offered from time to time since the beginning of the war by the enemy."[6] The Aircraft Board concurred in Squier's recommendations. The next day Squier detailed the proposed project to Secretary of War Newton D. Baker, who gave verbal approval to start the experimental work. American leadership overseas was not advised of the project.

Squier promptly began recruiting a research team. Much of the flying bomb development after the Amityville visit now hinged on the Signal Corps' efforts since the Navy suspended further research in that area, reportedly due to lack of funds. After considerable effort, Charles Kettering, the renowned engineer, was persuaded to direct the Army development of an aerial torpedo, dubbed the Liberty Eagle, with the understanding that Orville Wright and C.H. Wills, of the Ford Motor Company, would assist. Kettering undertook specific responsibility for perfection of a suitable control system, in addition to his general supervisory duties. It was agreed that Wright would advise on the flying features of the Liberty Eagle and that Wills would perfect a suitable engine. Squier signed contracts with the Dayton Wright Airplane Company to construct the airplane part of the torpedo; the Dayton Metal Products Company to design and perfect the control features; and the DePalma Manufacturing Company of Detroit to build the engine. After some frustrating and unsuccessful months of trying to develop an electrical control system, Kettering struck upon the use of vacuum. The Aeolian Piano Company, widely experienced in manufacturing vacuum-controlled player pianos, thereby became one of the handful of Liberty Eagle contractors.

Progress was not as rapid as planned, a symptom that plagued the entire aircraft production program. By April 1918, it was necessary to augment military personnel to help Kettering. Squier obtained the assignment of Colonel F.E. Harris, a ballistics expert from the Coast Artillery Corps. Although the Aviation program passed from Signal Corps hands the following month, Liberty Eagle remained under Squier's direction until fall. When it, too, finally came under the auspices of the Director of Military Aeronautics, General W.L. Kenly, a new Army field representative was appointed. Lieutenant Colonel Bion J. Arnold, a prominent electrical engineer in private life and longtime professional acquaintance of Squier, monitored the project until February 1919, when it passed to Colonel Thurman Bane at Wright Field. Bane received about two dozen successfully flight-tested aircraft.

The Liberty Eagle was basically a miniature biplane without ailerons or wheels. It possessed internal controls enabling it to rise "to a predetermined altitude and to hold a fixed compass course. Without a load, it weighed about 350 pounds. It measured approximately 12 feet in length, and the wings spanned 12 feet. Constructed of a wooden frame, it was covered by heavily shellacked wrapping paper. C.H. Wills designed a new, unusual engine of the four cylinder, two-cycle, air-cooled type. The engine block was cast of aluminum. Cylinder sleeves were first made of cast iron; in later designs the cylinder walls were formed from electro-plated gray iron. Weighing a favorable 110 pounds, the engine delivered about 45 horsepower at 2,000 RPM. Reduction of engine weight had reached a point where between 180 and 200 pounds of TNT could be stowed in the fuselage.

The control system to guide the aerial torpedo toward its destination consisted of a gyroscope, an aneroid barometer, and a vacuum system. Kept in continuous motion by an alternating current, the gyroscope maintained a fixed compass heading and a proper climbing angle. Once the torpedo achieved a predetermined operating altitude, an altimeter took over control of the position of the elevator. It was so sensitive that it could regulate changes in altitude of as little as five feet. Changes were affected by a series of pistons actuated by vacuum pressure derived from the engine's manifold system. A series of ratchet wheels or counters similar to those used in cash registers or cyclometers provided the means for recording distance traveled toward the target. Once the predetermined distance was reached, the gasoline supply was shut off and the craft was made to dive. As mentioned above, the aerial torpedo had no wheels. Instead it was mounted on a four-wheel cart which moved along an improvised light railroad track. When the aircraft attained flying speed it lifted itself from the cart and climbed until reaching the operational altitude preset by operators on the ground. Although the control system worked on mechanical and pneumatic principles, one early design envisioned radio control. The first successful flight took place on 2 October 1918. Following this flight, Colonel Harris wired Squier: "Dayton, Ohio, October 2, 1918, Maiden flight of ship successfully accomplished six thirty this afternoon. Congratulations."[7]

Two days later another test established the reliability of the unusual engine in an unexpected way. Although several engineers, including Bion Arnold, were still skeptical about the durability of the Liberty Eagle despite its successful maiden flight, the second flight dispelled all doubts. The flight started in late afternoon and lasted nearly an hour. When the Liberty Eagle ascended from its carriage it flew in a straight line for a short distance. Then it curved slightly to the right, climbing until it almost stalled. Slipping back into a more stable attitude, it continued climbing in great wide arcs, finally disappearing from sight at an altitude of about 5,000 feet. Soon it returned from the direction in which it disappeared, but at a much higher altitude. These wide sweeping horizontal motions reoccurred before it disappeared altogether from the view of observers at Dayton. In a fit of disgust, Kettering muttered that the thing could just stay up there—he was going to bed.

Exhausting its limited fuel supply, the Liberty Eagle fell to earth near Xenia, Ohio. A flurry of reports to the local police accompanied its flight over Dayton. The Chief of Police at Xenia informed searchers dispatched from Dayton that a flying object was reported down in a local farmer's field. Hap Arnold, later Chief of Staff of the Air Force, was among the observers from Dayton in pursuit of the Liberty Eagle. He recounted the scene as the anxious party from Dayton encountered the excited and confused farmers.

> "Did you see an airplane crash around here?" we asked. One farmer said, "Right over there! But strange thing, there's no trace of the pilot!" Colonel B.J. Arnold, the Army officer in charge of the project, remembered quickly that we had a flying officer in a leather coat and goggles in the car. "Here's the pilot," he said. "He jumped out in his parachute back-a-piece. Let's go pick up that wreck." Our secret was secure. The awed farmers didn't know that the United States Air Corps had no parachutes yet.[8]

Impressed by the power and dependability of Will's engine, Bion Arnold became convinced that the Liberty Eagle was a reliable, practical weapon. He then wired his recommendations to the Director of Military Aeronautics.

> Dayton, Ohio, October 5, 1918. From a sustained flight last night of over twenty miles, ending only when gas was exhausted and from information gathered from other tests and investigations made since observing the action of the experimental ammunition carrier here I am convinced the device, although at present lacking in accuracy of control, has great potentialities and that further experimentation with it under suitable conditions will so perfect the control as to enable the device to be handled with reasonable accuracy, I believe this device has military value ... no time should be lost in getting it into production and I recommend as follows:
>
> That suitable preparation be made and authority at once issued for the construction of one hundred of these devices to be used in further development. These when constructed should not cost to exceed one thousand dollars each.
>
> That steps be promptly taken for the placing of an order for ten thousand of them.... In quantities of fifty thousand to one hundred thousand these devices should be built at a cost of from four hundred to five hundred dollars each.
>
> That if practicable the work of the Army and Navy be coordinated and a suitable site in an uninhabited region be promptly selected for further experimentation.[9]

General Kenly, the Director, accepted Bion Arnold's recommendations. Bion Arnold immediately ordered a hundred Liberty Eagles from Dayton Metal Products. In the meantime, Colonel Hap Arnold was ordered to France to organize Liberty Eagle tactical units. Until this time no one in Europe knew of its development. They were not to find out until after the Armistice. On the way over, Hap Arnold contracted influenza, then ravaging Europe and America, and found himself in a hospital in Southampton. By the time he reported to General John Pershing the war had ended.[10] Meanwhile, back in Dayton, several more tests of the aerial torpedoes on hand followed.

On 18 October 1918, the Navy, in conjunction with Dayton researchers, catapulted one of their experimental craft at Amityville. The torpedo flew level on a true course directly out to sea. It was observed for five miles of its estimated 50-mile flight path before it passed from view. No one ever heard or saw it again. The test was considered eminently successful.

Four days later Bion Arnold and his co-workers launched another Liberty Eagle on its first completely successful test flight. The torpedo lifted itself from its carriage upon attaining flying speed and flew a straight course at the predetermined altitude. When the engine shut off at the programmed distance from the launch site, the torpedo crash-dived on its target. According to Hap Arnold, the torpedo had a planned accuracy of 100 yards after a 40-mile run in clear weather. The first phase of experimental trials was satisfactorily concluded. Colonel Harris sent an enthusiastic telegram to Squier:

Final test of experimental work done here under your direction made today. Result obtained wonderfully successful in every way and amply justified the high hopes always entertained by you for its final success. Idea has now passed into practical reality. Production in vast quantities should no longer be delayed. Heartiest congratulations from myself and force associated with me. Will see you in Washington Thursday.[11]

In the meantime, responsibility for the Liberty Eagle project was transferred once again to the Bureau of Aircraft Production. Research funds were more available from the Bureau than from the Director of Military Aeronautics.

Because of the rapidly diminishing prospects for secrecy at Dayton, Bion Arnold decided to look elsewhere for a test site where long-distance trials could be conducted. He left on 28 October on a grand tour of California, Texas, Florida, and the Washington, D.C., vicinity in search of suitable testing grounds. While inspecting sites near Houston, Texas, Bion Arnold learned of the Armistice in Europe. The war had ended.

Bion Arnold conferred in Washington with the Aircraft Production Board. They decided to proceed with further tests to establish the torpedo's value as a long-distance weapon. When the government canceled aircraft orders and contracts, Kettering and his associates requested to be relieved of further responsibilities for the Liberty Eagle. They wished to be free to attend to changing business affairs. As a consequence, further Liberty Eagle tests were postponed. Bion Arnold himself now expressed a desire to leave the service in view of the postponement. When he left the Army in mid–December 1918, the project was once again transferred. This time the Airplane Engineering Division at Wright Field assumed responsibility. By World War II the Liberty Eagle project was only a memory to some and totally unknown to most.[12]

First Military Funding to Robert Goddard

In the fall of 1916, a young professor of physics from Clark University came down to Washington, D.C. He gave a presentation on some of his experiments to the Smithsonian Institution in the hopes of procuring a grant. The experiments held such scientific promise that the young man, Robert Goddard, received $5,000 from the Smithsonian to support his rocket work in the coming year.

At the beginning of the year 1918, Walcott approached Squier with the Goddard project in hand. Walcott's only son, Stuart, had been killed in aerial combat over France as a member of the Lafayette Escadrille in December 1917.[13] He suggested that more money invested in rocket research would hasten development of a militarily useful weapon.[14]

Squier immediately transferred $10,000 from Signal Corps' appropriations to the Smithsonian and the Bureau of Standards for support of the rocket project. Squier, S.W. Stratton, Chief of the Bureau of Standards, and Walcott constituted themselves as a committee of three to supervise Goddard's work.[15] Goddard then returned to Clark University to conduct his research. Work progressed well, but by April it was apparent to the supervisory committee that more money would be needed. As the original sponsor of Goddard's work, Walcott made the formal request to Squier for another $10,000. Walcott optimistically reported that the new money would bring them to actual rocket trials.[16] Three weeks after approving Walcott's request, Squier doubled the appropriation.[17] Trouble attended the otherwise-happy news of a renewed grant. For some time Goddard had been displeased with the work of his foreman, C.D. Haigis. In mid–May, the troubled relationship ruptured and Goddard terminated Haigis.[18] Haigis promptly accepted employment under G.I. Rockwood, owner of the Rockwood Sprinkler Company of Massachusetts. At one time Goddard had considered forming a joint enterprise with Rockwood to develop rockets, but later abandoned the idea.

Just three days after Haigis began his new job, Rockwood wrote a Colonel E.M. Shinkle in the Ordnance Bureau. Rockwood asked for a letter from Shinkle requesting the Rockwood firm to produce for the Ordnance Bureau an efficient rocket, containing the features developed by Goddard.[19] Goddard was already aware that his erstwhile foreman and Rockwood were developing an infringement on the rocket apparatus. He anxiously informed Walcott that Colonel Shinkle had reportedly requested Rockwood to proceed with manufacturing. Goddard asked in disbelief if the request to manufacture was authentic.[20]

In reality, Colonel Shinkle wrote Rockwood that interest expressed by other government departments in Goddard's work prevented him from requesting anything from Rockwood.[21] Meanwhile, Brigadier General Charles Saltzman, Acting Chief Signal Officer, recommended that Shinkle pay a visit to Goddard in Worcester, Massachusetts, to observe the rocket experiments.[22]

Two days later Colonel Shinkle arrived in Worcester, representing himself to Goddard's workers as one of Squier's men. Goddard reacted with suspicion when Shinkle asked for an appointment. Putting Shinkle off until afternoon, Goddard immediately called the Smithsonian for instructions. He was informed that Shinkle had indeed conferred with Squier "before General Squier understood all the circumstances of the case."[23] Goddard was to say nothing.

When the Colonel received no further word from Goddard he became so angry he warned Goddard's father that if Goddard "did not show up at the hotel [for their meeting] he would close the laboratory before he left and Dr. Goddard could put that in his pipe and smoke it."[24]

12. New Weapons

Uncertain just what authority Shinkle possessed, Goddard instructed his workers to begin removing certain pieces of equipment. Then he called the Smithsonian again. Dr. Charles G. Abbot, a world-renowned astrophysicist and one of the Smithsonian scientists who consulted on the rocket work, told him to tell the Colonel that new instructions were forthcoming from General Squier. He should say nothing more except a council was awaiting the Colonel's return to Washington, and he could shut up the laboratory if he wished.

Walcott asked Squier if the project might now pass to the Ordnance Bureau. A definition of responsibilities was necessary since two weeks had been lost as a result of the Shinkle intervention. The secrecy that was so important to the project was lost. Walcott offered to accept new project leadership "in view of the importance of this invention both as a signaling rocket, and also of its possibility as a carrier of high explosives."[25]

Squier would not relinquish his leadership in the project. He and Walcott decided to look for a more secret location to continue the work. After making arrangements with the Carnegie Institution of Washington, Walcott instructed Goddard to remove his laboratory to the Solar Observatory at Pasadena, California. Squier considered the location ideal.[26] Less than a week after Shinkle's stormy visit, Goddard had his suitcases packed with propellant and ready to depart for California.[27]

By mid–July, Goddard's work on single- and multiple-charge rockets had progressed to the point where Squier decided to invite the Chief of Ordnance to send an inspector to California. He suggested the inspector determine the usefulness of the rocket principle for Ordnance purposes.[28] A month passed and no representative from Ordnance appeared at Pasadena. In the meantime, Squier allotted another $5,000 to the project.[29] Squier gently nudged the Chief of Ordnance with a request for the results of his inspection.[30]

On 13 September, two captains finally arrived at Goddard's laboratory to inspect his work. They asked in wonder if Goddard realized "that that stuff you've got there is going to revolutionize things?... We thought it was going to be some kind of toy."[31]

Squier's interest in rocketry derived from several considerations. First, his old friend and colleague, Walcott, had asked for the support. Squier knew that, on the face of it, it was more properly an ordnance problem than a Signal Corps issue, although the rocket might have been a useful signaling device. George Ellery Hale, who helped arrange the move to Pasadena and who tried to command American science resources for the war effort through the National Research Council, wrote the Chief of Ordnance: "The appropriations for this work were made by the Chief Signal Officer, but it is strictly an ordnance problem and sooner or later would be transferred to your field."[32] Another Ordnance Corps inspector, Major W.A. Borden, subsequently recommended a thorough testing of Goddard's work at the Aberdeen Proving Ground.[33] Goddard began his demonstration of rockets as recoilless guns on 6 November at Aberdeen. On the second day of rocket trials, he received the news of Germany's surrender.[34] By spring 1919, it was apparent that continued research on rockets under military sponsorship would suffer an "indefinite postponement."[35]

Another consideration motivated Squier's sponsorship of rockets as carriers of high explosives. The relentless loss of life in France had appalled everyone. Squier saw it firsthand. There is evidence that his research interests were guided by the desirability of throwing machines rather than men against the enemy. In this context, his research in rockets, which would promise to remove American soldiers a greater distance from enemy fire, was predictable. Goddard's rockets would have accomplished on the ground what he was simultaneously trying to accomplish in the air; namely, develop automatic-weapon carriers to deliver ordnance against the enemy.

Those who worked for the pilotless airplane concept emphasized two key thoughts. Squier favored its revolutionary character. The element of surprise and the research time necessary for the Germans to catch up could assure the Americans a competitive edge in battle for some time. Especially so, since the aerial torpedo destroyed itself upon impact. Goddard's research held out the same promise as the Liberty Eagle. By throwing machines against the enemy instead of men and by increasing the technological terror of the conflict to a new level, the Germans would be forced to withdraw from the battle.

Squier's sponsorship of the Liberty Eagle and rocket projects revealed his conviction that the responsibility of scientists and engineers in wartime lay in magnifying the individual physical strength of every soldier through new weapons and technological surprise. This stress upon efficiency stood in strong contrast to traditional military emphasis upon morale and leadership as the principal elements of victory. In this regard, Squier presaged much of the tone of future Air Force doctrine which likewise stressed technological superiority.

13
Science Joins the Army

The First World War was widely viewed as a struggle for survival, waged both on the battlefield and in the laboratory. The British, French, and Italians acted quickly at the outset of hostilities to organize scientific talent in hopes of overcoming their industrial backwardness. In Berlin, the popular slogan was "chemistry will win the war." Soon after the United States entered the war, Professor Joseph Ames of the Johns Hopkins University and a member of the NACA, headed a special scientific commission to Europe. Ames remarked that even the geologist thought his scientific knowledge could benefit a front-line general. Early in the European phase of the war Chief Signal Officer George Scriven observed that the application of science had

> enormously augmented ... killing powers. At no period of the world's history has this truth been better illustrated than in the tremendous application of science to war made by the fighting nations of Europe.[1]

The National Research Council noted that "one of the most striking results of the war is its demonstration of the importance of scientific research in strengthening the national defense."[2]

Squier brought professional science and engineering into the Army by commissioning established scientists and engineers. He assigned them to high positions of authority from which they directed the scientific side of the war. By placing the scientists and engineers in uniform Squier was able to obtain substantial control over their creative and productive energies. He also deterred any developing expectations that scientists might deliver their services through an autonomous agency. Near the beginning of the American phase of the war, scientists and engineers were encouraged to work for the Signal Corps at their normal places of business. In time, Squier saw to the establishment of military research laboratories under his control, but available to other government agencies or bodies.

Squier adopted a bipartite approach to exploiting the scientific and engi-

Colonel George O. Squier at Fort Monmouth, New Jersey (courtesy CECOM Historical Office).

neering resources of the country. The first part involved the large-scale absorption of scientific, engineering, and technical personnel into the Signal Corps. It was an expedient, but short-lived, response styled for the urgent need for scientifically trained personnel. The second part of the approach consisted of building government laboratories for scientific research on military problems. It has endured and proven successful. It is exemplified in two of the most important research facilities founded during World War I, namely Fort Monmouth and Langley Air Force Base.

Recruitment of Personnel

The Signal Corps was charged with two large, very technical responsibilities in the war. When Congress declared war in April 1917, there were 55 officers and 1,570 enlisted men in the entire Signal Corps. The need for personnel to support the wartime responsibilities of the Signal Corps far exceeded the number initially assigned. Moreover, most of the commissioned officers in the Signal Corps were on detail from other corps. The detail system further handicapped the Signal Corps in performing its mission due to regular loss of skill and experience. A detailed officer generally remained only four years, then returned to his permanent service arm.

Squier's own observations on the importance of an effective signal service in combat stimulated him to prepare for auxiliary personnel in advance of the war. While the Signal Corps normally constituted between 2.5 and 4.0 percent of the manpower of the army, its importance was out of proportion to its numbers. Squier noted that no other 2 to 4 percent of army personnel in any other arm of the Army were so important or vital. The loss of that 4 percent or less represented by the Signal Corps would cause the utter collapse of the whole military machine.[3]

Thus, before the war began, Squier approached the leaders of the great telephone, telegraph, and electrical industries to obtain their help in the event the United States should become militarily involved in Europe.[4] An arrangement to that effect was worked out in an informal conference in New York, in January 1917. In attendance were Theodore N. Vail, president of the American Telephone and Telegraph Company; Newcomb Carlton, President of the Western Union Telegraph Company; John J. Carty, Chief Engineer of the American Telephone and Telegraph Company; and Squier.

In case of war, heavy demands would be imposed upon these companies. To minimize interruptions in maintaining domestic service and to allow maximum freedom to company officials in reconciling all the demands on their services, they recommended commissioning four or five leading engineers and executives in the commercial telephone and telegraph companies.[5] Under this so-called Affiliated Plan, the commissioned engineers and executives would organize and train the first units of reserve signalmen from their own ranks to go overseas. When war did come, the Bell Telephone System alone offered 12 battalions of highly trained technical men ready for immediate overseas service.[6]

As a result of the Affiliated Plan, Squier obtained the services as commissioned officers in the Signal Corps of John J. Carty, Chief Engineer of the American Telephone and Telegraph Company; Frank B. Jewett, Chief Engineer of the Western Electric Company; Charles P. Bruch, Vice-President of the Postal Telegraph Company; and, G.M. Yorke, Vice-President of the Western Union Telegraph Company. Carty and Jewett then selected such engineers from their own company laboratories as were necessary to balance the needs of their civilian and military organizations. What the Signal Corps lacked in laboratory facilities, equipment, and technicians could be supplemented by private companies until the Signal Corps could secure space, equipment, and people for itself.

From other industries, Squier commissioned men experienced in industrial production. Those equivalent in experience and executive responsibility to the men he drew from the electrical industry were E.A. Deeds, Vice-President of the National Cash Register Company; C.G. Edgar, Detroit financier, Managing Partner of W.H. Edgar and Son, and President of Edgar Sugar House; William C. Potter, mining engineer, member of Guggenheim Brothers, and Director of Guaranty Trust of New York City; Robert Montgomery, Assistant Professor of Economics at Columbia University and a founder of the accounting firm of Lybrand, Ross Brothers, and Montgomery, in New York City; Robert H. Morse, Chief Executive of Fairbanks, Morse Manufacturing Company; and Sidney D. Waldon, Vice-President of the Packard Motor Car Company.

Squier procured trained scientists from government bureaus as well. As early as June 1917, soon after the first American units arrived in France, a request for meteorologists was received. While the National Research Council sought qualified persons, six employees of the United States Weather Bureau were commissioned as officers and dispatched to Europe.[7] W.R. Blair, Commissioned Major in the Signal Corps, left his position as Chief of the Aerology Division in the Weather Bureau to head the first contingent abroad. Other scientists came from the Bureau of Mines and the Bureau of Standards.

Squier also tapped academic science for assistance. One of the most unusual sciences of the time was psychology, especially among Army officers, who considered "its methods as akin to those of the spiritualist, the devotee of psychical research or those of the 'medium.'" While some psychologists were promoting placement testing and intelligence testing with the Surgeon General's and the Adjutant General's offices with mixed results, the Signal Corps received the proposals of psychologists with an enthusiasm unmatched in the rest of the military establishment.[8] Squier recognized early the value of psychological studies in aviation. In an address before the National Academy of Sciences in November 1916, he talked of the need "to study the physiological and psychological effect of low density air at high altitudes on the performance of pilots."[9] According to one student of the history of psychology in the First World War, Squier's leadership accounted for the psychologists' successful entry into Army aviation activities. This event was unique in the development of World War I psychological services since the psychologists won instant acceptance without first demonstrating applicable services.[10]

One reason for their instant acceptance was the Chief Signal Officer's conviction that psychologists could establish the qualities necessary in competent aviators and could devise tests to identify such qualities. The goal of all such studies and tests was to guard against acceptance of men constitutionally unfit to fly.[11] Large numbers of men failed to pass the training program of ground schools. Thus, ways were sought to select those men most likely to succeed in the aviation training program, through the application of the science of psychology.

The first of the psychologists commissioned in the Signal Corps was John B. Watson, promoted to the rank of Major in August 1917. By the end of 1917, 23 psychologists were in the Signal Corps. Five of them supervised the nationwide system of Aviation Examining Boards; the balance worked on various research projects in the development of tests to be used in selecting candidates for the aviation schools.[12] By January 1918, John Watson organized 70 examining boards.[13] They functioned for the duration of the war, using procedures and tests supplied to them from the Adjutant General's Committee on Classification of Personnel and the Signal Corps' psychologists.

By far the most significant absorption of scientists into the Signal Corps occurred under the aegis of Robert A. Millikan. Squier "well-nigh appropriated"[14] him while he was serving as Third Vice-Chairman, Director of Research, and Executive Officer of the National Research Council (NRC).[15] Formed in 1916, the NRC was largely the product of the efforts of George Ellery Hale and Robert Millikan. They organized it during the pre-war period of national preparedness as a parallel response of some scientists in mobilizing American resources for war. The outbreak of war caused the NRC to be inundated with military problems. Those problems related to airplanes and new methods of signaling were most urgent. It soon became evident that the funds and personnel of the NRC were unequal to the military research load created by the war.

Squier stood ready with financial support if the NRC would assist in the formation of a Science and Research Division in the Signal Corps.[16] Squier asked Hale to allow the NRC "to act as the advisory agent of the Signal Corps in the organization of its various scientific services and the solution of research problems." To this end, Squier suggested that Robert Millikan apply for a major's commission in the Officers' Reserve Corps.[17] Hale obliged Squier by arranging the applications of Millikan and C.E. Mendenhall, the Chairman and Vice-Chairman, respectively, of the Physics Committee of the NRC. When Millikan took charge of the newly created division, Squier virtually captured the Physics Committee of the NRC.[18] It took some time for Millikan to acquire all the personnel he desired so the formal establishment of the Science and Research Division occurred on 22 October 1917.[19] According to Office Memorandum No. 162 of that date, it was to "have charge of all development and research work assigned to it by the Chief Signal Officer."[20] No definite work assignments were given. As problems arose, designated members of the division undertook their solution. In February 1918, the division was fully constituted. Its charter in Office Memorandum No. 35 granted a larger measure of independent responsibilities.[21] In the fields of meteorology, photography, sound ranging, chemistry, signaling, and instrumentation Millikan brought in or recommended to the Signal Corps a select group of scientific personnel to work on specialized projects in aviation and signaling. W.R. Blair, already mentioned in connection with meteorology, was one of Millikan's own Ph.D.s. Henry Gale, Millikan's colleague at the University of Chicago, accompanied Blair to France with a considerable group of other young physicists enlisted from the universities and the Weather Bureau.[22] C.E. Mendenhall headed the Aeronautics Instrument Section of the Science and Research Division. He and his subordinates worked with General Electric Company in perfecting trench-signaling lamps, air-speed indicators, improved venturi pitot tubes, gyroscopic-turn indicators, and compasses. Ten thousand of the aeronautical compasses were ordered, since they were regarded as superior to all others in use.[23] A host of other devices and scientific procedures were tested, including secret ultraviolet signaling lamps, the production of helium, and special photographic-sensitizing dyes. Even special color filters for use on aerial cameras were produced for detecting camouflage.[24]

The work eventually grew so heavy much of it was being sent to existing laboratories located in government organizations, educational and research institutions, and commercial and industrial firms. In many cases, the Science and Research Division augmented the staffs of outside laboratories with its own personnel.[25] One of the most notable augmentations took place in the laboratory of the Bureau of Standards in December 1917. A conference composed of the Secretary of War, the Secretary of Commerce, and War Department officials responsible for research agreed in December "that drafted scientific men might be detailed to the various bureaus of the War Department and reassigned to the Bureau of Standards or other civil bureaus for military work." Millikan represented the Science and Research Division.[26] Since the Bureau of Standards carried on a large amount of testing and research for the Signal Corps, Stratton invited Squier to participate by supplementing the scientific manpower of the Bureau.[27] At Squier's direction, Millikan soon arranged for the transfer of 50 scientists and technicians in uniform to the Bureau of Standards.[28]

Squier's personal style of leadership in a scientific setting minimized latent conflicts between professional soldiers and professional scientists and engineers in uniform. His newly commissioned scientists and engineers recognized Squier's capacity for technical depth.[29] Squier, in turn, encouraged individual initiative. His general principles for guiding people engaged in research were best expressed in his Victory Creed:

> To foster individual talent, imagination and initiative, to couple with this a high degree of cooperation, and to subject these to a not too minute direction; the whole vitalized by a supreme purpose which serves as the magic key to unlock the upper strata of the energies of man.[30]

While Squier might minimize the occasions for conflict he could not realistically eliminate them. The most serious challenge to the use of scientists in uniform involved Millikan and the Inspector General of the Army. A dispute between an inventor of a special gun and Millikan resulted in an investigation by a representative from the office of the Inspector General of the Army.[31] The inventor, E.L. Rice, claimed that Millikan had conspired to steal his invention for personal considerations. The inspector on the case agreed with Rice. He charged that Millikan appeared to have "tried to coerce Rice into silence regarding certain misfeasance and malfeasance."[32] The inspector recommended a reprimand for Millikan. He also thought that Millikan should be "informed that he has shown himself to be temperamentally unqualified for the important office of Vice-Chairman of the National Research Council."[33] Inspector General W.T. Wood recommended harsher consequences. He believed that Millikan, by his attitude and conduct, has "conclusively demonstrated his unfitness for the responsible duties of his office, and recommends that his commission as an officer of the United States Army be terminated."[34] This action would have necessitated court-martial proceedings. The tide of impending adversity for Millikan reached its high level in Wood's recommendations; then quickly receded. Army Chief of Staff Peyton C. March appointed his executive assistant, General Frank McIntyre, to review the entire case. McIntyre's conclusions contrasted the difficulties of scientist and inventor trying to work together. In exonerating Millikan, he wrote that "this whole controversy all hinges on the trouble which is usually experienced when an inventor, who becomes obsessed with his own idea, comes in contact, or tries to work with men of scientific instruction."[35]

He might have also contrasted the difficulty experienced by professional military men and professional scientists in working together. Had the Inspector General succeeded in humiliating Millikan, the future of relations between the military and scientific communities might have been sour indeed. Squier's success in obtaining a measure of military control over a weak scientific body in need of money to conduct its program was a victory for him. The technique of commissioning scientific, engineering, and industrial men of advanced training and established reputation failed to survive the war as a suitable vehicle for promoting military and scientific cooperation. Squier's efforts to incorporate professional science and engineering within the Army proved more successful within the two great research laboratories he founded.

The Founding of Langley Field

The need for an aviation proving ground was recognized even before the doors of the Smithsonian's Langley Aerodynamical Laboratory opened in 1913.

The Smithsonian had insufficient funds to purchase such a site. The Navy experimented at the Naval Basin, and the Army did what it could at its flying field at San Diego. Of the three organizations, the Navy became the most aggressive in developing research facilities and gaining support for aeronautical research. Walcott told Squier in early 1916 how pleased he was to hear of Squier's imminent return because the Navy was on the verge of taking over aeronautical research.

Shortly after Squier's return, the Technical Aeronautical Advisory and Inspection Board of the Signal Corps formally asked him to acquire a permanent experimental and inspection station. Their recommendation outlined a tripartite mission for such a station. First, materials, motors, radio sets, machine-gun mounts, instruments, and accessories should be tested. Second, aeronautical experiments should be conducted in wind tunnels and laboratories. Living and administrative facilities would be built to accommodate a military part in residence. Third, the Signal Corps should train inspectors for assignment to the factories of aircraft manufacturers under contract to the government.[36] Their task was to assure high engineering and quality standards in aeronautical equipment supplied to the Army.

Despite current attention to the project, a request for an allotment in the Army Appropriation Bill for Aeronautics was unaccountably omitted. Secretary of War Newton Baker caught the omission and asked Chief Signal Officer Scriven for his desires in the matter.[37] Squire replied to the Secretary that the Aviation Section needed two kinds of facilities—aviation-training sites and experimental and testing stations—on government-owned land. The facilities should consist of permanent buildings, unlike so many Army structures which were often built for temporary use. Squier emphasized the "pressing importance" of an Army experimental station in view of the limited development in the American airplane industry. To this end, the station should be permanent and situated in a place convenient to all interests.[38]

Confident that the Army Appropriation Act would be carried forward on the wave of public support for national preparedness, Squier informed the Executive Committee of the NACA that Secretary Baker would probably request assistance in selecting a site for the aviation proving ground.[39] As a member of the Executive Committee, Squier was in an ideal position to serve aviation in both its military and civilian aspects.

The provisions of the Appropriation Act required the Army to investigate the suitability of existing military installations for aviation purposes.[40] If none could be found or released for aviation, the Secretary of War was authorized to purchase, by condemnation or otherwise, necessary land for as much as $300,000.

Early in the site-selection process it became apparent that existing military installations were unsuitable. Secretary Baker appointed a board of officers to select a new site for an aviation proving ground. He chose Squier, as Chairman, Captain Virginius Clark, Captain Thomas DeW. Milling, and Captain Richard C. Marshall of the Quartermaster Corps. Clark and Milling were both trained in aerodynamics at the Massachusetts Institute of Technology.[41] They investigated locations at Aberdeen, Maryland; Brooke, Virginia; Bush River, Maryland; and Hampton, Virginia. Their visits were often shrouded in mystery for fear that landowners would drive up prices if they discovered the Army's interest in their land. Nevertheless, owners of large tracts of land learned from newspapers that the War Department would be surveying sites for a proving ground. The *New York Times*, for example, announced plans for acquiring land to establish an Army Aircraft Experimental Plant and Proving Ground.

The *Times* took note of the innovative relationships between the armed services, industry, and independent inventors that such an installation introduced. For scientists and inventors, Squier announced an open door policy. For manufacturers, the proving ground personnel would conduct static and dynamic tests free of charge on motors, flight instruments, exact models in wind tunnels, air frames, airborne radios, airborne armament, and dirigibles. Of more import than free testing or research conducted in governmental laboratories was the creation of an inspection department, including an intelligence bureau to oversee aviation intelligence collection from foreign attaches and foreign journals.[42]

Squier subtly involved the NACA in the new project of the Aviation Section. He began by asking the Executive Committee how the new facility should be designated. After some discussion, the Committee recommended the term "Aircraft Proving Ground." Appreciative of this assistance, Squier told the committee that he also desired their help in locating a suitable site for the proving ground. He must, however, first obtain Secretary Baker's approval. It soon followed.

After two weeks of extensive searches for suitable locations, Squier informed the Executive Committee of his travels and difficulties. He suggested that an advisory subcommittee of the Executive Committee be appointed to judge the various sites selected by the War Department site-selection committee, namely, the one he himself chaired.[43]

Squier's quiet campaign to persuade the NACA to collocate their facilities and operations at the proving ground gained favor within the week.[44] On 9 October, the Executive Committee appointed a subcommittee to consider the needs of the committee for its experimental research program at an Army proving ground.[45] A month later, Walcott reported to the committee that arrangements had been made with Secretary of War Baker and Secretary of

Navy Josephus Daniels for joint consultation in considering the requirements of a field to serve as an aeronautical proving ground.[46] Military and naval representatives told the committee at this key meeting of their visits to possible sites at Wheeling, Parkersburg, Cincinnati, St. Louis, Cairo, Memphis, and Aberdeen Proving Ground.[47] They rejected all sites, in part for their unsuitability for naval aviation purposes.

Evidently the site near Hampton, Virginia, originally selected by the Squier committee, gained renewed interest. The Executive Committee voted to query the President of the First National Bank of Hampton and the Surgeon General of the Army (Rupert Blue) about health conditions in the vicinity of Hampton. The committee then voted to visit the proposed site on 18 November.[48]

On the morning following this decisive meeting, the Squier committee submitted a memorandum to Walcott in which they expressed their preference for the Hampton location. Squier proposed recommending purchase of the site to Secretary Baker.[49] After examining some 15 sites, the Army Board (Squier's committee) concluded that a suitable site should meet the following criteria:

> The site ... must comprise at least 1,600 to 1,800 acres of very flat land with a frontage on a large body of water, in the following general locations: East of the Mississippi River, South of the Mason & Dixon Line, not farther than 15 hours travel from New York City, not so close to the unprotected coast as to be extremely vulnerable to attack or capture in the event of war.

At its first meeting after visiting Hampton in the company of Army and Navy representatives, the Executive Committee carefully discussed their impressions and observations concerning the Hampton site.[50] The meeting ended with a unanimous decision to accept the Army Board's recommendation to purchase the Hampton site. Walcott thought it the best of those available.[51]

A week later, Squier recommended immediate purchase of the Hampton site to Secretary Baker. He informed the Secretary that such an experimental station was "an absolute and immediate necessity for the proper and rapid development of Army aviation."[52] Two days later, Baker approved Squier's request and authorized the Quartermaster Corps to purchase the land for $290,000.[53] Squier informed the Executive Committee that Secretary Baker approved the site.

The committee passed a resolution requesting the Secretary of War to use certain portions of the grounds.[54] This resolution culminated repeated invitations extended by Squier to NACA to locate its facilities at the Army proving ground.[55] It also realized fulfillment of Squier's ideal of interdepartmental cooperation in the governmental development of aviation presented to the Smithsonian in 1910.[56] Acting Secretary of War William M. Ingram approved the committee's request in the closing days of 1916.[57] A section of the site known as Plot 16 was subsequently assigned to the committee.[58] Criticism of the Aviation Section welled up from the General Staff. Major General Tasker Bliss, the Assistant Chief of Staff, complained in early January 1917 that no general plan, based on sound military principles, had been worked out.[59] Scriven appeared embarrassed by the tender of many sites for aviation purposes and the absence of criteria or doctrine against which their usefulness or suitability could be judged. The dissatisfaction concerned flight-training sites. The whole issue came to a head when the Commanding General of the Western Division wrote a personal, confidential letter of protest about an impending occupation of a site near San Diego by aviation personnel. The War College Division of the General Staff began a military study intended to correlate available funds to probable future aviation demands. Squier represented the Aviation Section.[60] This conference session was overdue, and much needed. Coordination between the Signal Corps and the General Staff on aviation matters was in disrepair. Correspondence in this study revealed a Scriven (Chief Signal Officer) awash in controversy and unable to cope with the burgeoning technical and industrial problems of aviation. Squier emphasized the necessity of building an air arm upon a strong base of research. Out of these meetings came formal General Staff approval for an experimental station.[61] Squier was fortunate. He had presented a *fait accompli* to the General Staff and come away without a reprimand.

With General Staff approval of a proving ground assured, Squier began publicizing the project. In an address to a meeting of the Aeronautical Society of America, he prophesied that airports would one day be as common as seaports. City agencies would find it as necessary to incorporate airports in their plans as they would other municipal features. He compared their impact upon cities to a Lincoln Highway. He also cited the promise of aviation in carrying the nation's mail, just as General John J. Pershing's headquarters received all its mail.

Squier devoted the greatest part of his speech to the aeronautic research needs that a proving ground would serve. In armament, for example, he wanted "a multi-barrel machine gun" developed. He called attention to mapping upper-atmospheric currents, developing instruments of aerial navigation, eliminating the noise involved in operating aircraft, fabricating bullet-proof fuel tanks, creating aircraft fabrics superior to Irish linen, and developing an airplane-to-airplane communication system.[62] In conclusion, Squier envisioned the role of the proving ground as important to aviation as Sandy Hook was to the Army, and Indian Head to the Navy. This aerial proving ground culminated a half-decade of hopes for American aviation shared with Alfred Zahm, Charles Walcott, Paul Brockett, and Alexander Graham Bell. In the tradition of these "friends of Langley," Squier informed his audience: "If I have any influence in the matter we are going to call that proving ground on the Atlantic 'Langley Field,' and I cannot conceive of any better monument to the memory of Professor Langley."[63] And, of course, Squier had everything to say about the name of the proving ground, for in less than one month he succeeded George Scriven as Chief Signal Officer of the Army and rose overnight from Lieutenant Colonel to Brigadier General.[64]

In the meantime, Squier recommended the expenditure of $75,000 to construct hangars, buildings, and site improvements. After a maddening delay, the Adjutant General approved on 16 March the transfer of Signal Corps appropriations to the Quartermaster General to begin work. Squier made it clear that Langley Field would not be like any ordinary Army post. He wanted it to endure and to become "one of the foremost stations for ... [aeronautical] research and experimental work."[65] To this end, he hired Albert Kahn of Detroit, a nationally noted architect, to design the buildings.[66] The special brick work on the original buildings at Langley is mute testimony to Squier's early ambitions for Langley, which have been amply fulfilled.

Work began at Langley Field. Burdened by war demands, the Quartermaster Corps transferred the construction project back to the Signal Corps, in which a new construction division was created.[67] The first Commanding Officer, Captain C.P. Eartholf, assumed his duties on 2 July 1917.[68] Two weeks later, Squier issued a memorandum officially designating the proving ground as Langley Field. If ever there were doubts about Squier's sentiments about the "friends of Langley," he completely dispelled them. He wrote that the field was named in honor of Professor Samuel P. Langley, who, after years "of patient study, first achieved actual flight by a heavier-than-air machine and must always be regarded as the discoverer of the principles which made the spectacular triumph of later years possible."[69] On 7 August 1917 the site was announced as a permanent military station and designated as "Langley Field, Virginia."[70] Thus began one of the most significant institutionalizations of aeronautical science within the armed forces and the federal government. In terms of scientific institutions for aeronautical research under governmental control, the Americans finally reached the point where the British started in 1910. The mandate of the NACA, contained in their organic act of 1915, could now be fulfilled with an experimental facility and generous financial support from Army appropriations. The establishment of another great laboratory for radio research paralleled the creation of Langley Field.

The Founding of Fort Monmouth

The Signal Corps performed its radio research before the war in two laboratories in Washington, D.C. One was located in the Mills Building (at 1710

Pennsylvania Ave. N.W.) and the other was in the Bureau of Standards Laboratory. The war created a burden of research unsupportable by the two Washington laboratories. Much of the new research was ultimately transferred to two new Signal Corps laboratories. As the work of the Science and Research Division got underway it soon became apparent that "a special laboratory devoted exclusively to development work, and entirely independent of the commercial laboratories ... would be needed."[71] On the first of June 1917, a lieutenant visited the offices of the *Red Bank Register*, the newspaper of Red Bank, New Jersey. He announced the imminent arrival of thousands of men. They were to construct an Army camp on the grounds of a nearby abandoned racetrack. He advertised for farm produce to supply the troops. Three days later two "Model T" Ford trucks and 32 men arrived to prepare the site for construction of buildings.[72]

On 17 June 1917, Lieutenant Colonel Carl F. Hartmann became the first Commanding Officer of Camp Little Silver, as the new camp was first called.[73]

Major General George O. Squier, C.S.O., and Dr. Louis Cohen, radio engineer, at Fort Monmouth (courtesy CECOM Historical Office).

By the end of the month, almost 500 military men occupied the camp. The First and Second Reserve Telegraph Battalions were stationed at the camp.

On 15 September 1917, the camp was officially named Camp Alfred Vail, in memory of a New Jersey inventor and financier in the early days of telegraphy. The camp was used primarily in its first months to train telegraph troops for service in France. By the time the camp acquired its new name, the First and Second Reserve Telegraph Battalions, comprised of commercial telegraph and telephone men, had completed their training and embarked for France. Replacements soon arrived for training and immediate transfer to France. Among the new arrivals were pigeon handlers. They organized the first pigeon service in the United States Army.[74]

Meanwhile, necessary authorization was obtained in October to proceed with the laboratories and flying fields at Camp Alfred Vail. Adverse weather conditions and chaotic transportation conditions delayed greatly the delivery of materials and construction of facilities until March 1918.[75] In February, the first men associated with the laboratories arrived. On 23 February 1918, Major L.B. Chambers was detailed to command the Radio Laboratories.[76] Effective use of the laboratories and flying fields began in April 1918. By 1 May, 13 aircraft were stationed at Camp Alfred Vail for the purpose of testing new airplane radio equipment. The radios perfected at Camp Vail and in other cooperating laboratories helped transform the aircraft from the mounts of lone warriors into integrated regiments of fighters under the control of one mind.

In August 1925, Camp Alfred Vail was redesignated as Fort Monmouth, New Jersey, in honor of those who lost their lives in the Revolutionary War Battle of Monmouth. In 1935, the radio laboratories were named "Squier Signal Laboratory." In 1954, Squier's name was removed from the title of the laboratories. The Academic Building Officers' Department was renamed Squier Hall.[77]

Squier created the largest scientific research organization ever witnessed in the American Army. His eclectic interests stimulated research across a wide spectrum of sciences. He was convinced that the application of science to military problems would give victory to the Allies. To this end, he acquired control over the physical sciences division of the NRC by commissioning its members and funding its work. To provide a scientific focus for future military research, he founded two research facilities, Fort Monmouth and Langley Air Force Base, which grew in importance in radio and aeronautical sciences since their establishment. It should be noted that Fort Monmouth was affected by various Base Realignment and Closure (BRAC) rounds instituted by the Department of Defense in the 1990s and 2000s, and many of its personnel were transferred elsewhere. The Fort was finally shuttered in the BRAC round of 2005, and the remaining military and civilian personnel were moved to Aberdeen proving ground in Maryland.[78] Langley Air Force Base is the home of Langley Research Center, which continues to perform research for the United States Air Force.

General John J. Pershing regarded Squier's technical attainments as without equal in the Army, but rated his administrative and executive abilities toward the bottom of all contemporary major generals.[79] The last three chapters testified to Squier's unorthodox approaches to obtaining scientific and engineering assistance. Whatever may be said of his administrative abilities, he did obtain the services of some of the finest, most able men in the scientific, engineering, and industrial communities in a time of great national need, while maintaining a strong military direction over their efforts. He set them to working in an unprecedented number of new scientific and technological fields within a military structure. The innovation and creativity that flourished under his leadership was illustrated in the union he forged among industrial, military, and scientific men to develop airborne voice radios for combat pilots, which is the subject of the next chapter.

14
Voice-Commanded Squadrons

Airborne radiotelephony was one of the most important, yet least discussed, developments of World War I. To later airmen and scholars it represented an evolution in communications so natural and so obvious that mentions of its origin are rare in the literature of military history. Nevertheless, this was an achievement accomplished by only one of the belligerents, the United States. Despite the ready availability of vacuum tubes, the essential element of airborne voice communications, to all belligerents, none of them developed an operational system.[1] The British came closest, as discussed elsewhere,[2] but were unable to equip more than a few squadrons. Moreover, only the United States recognized the necessity of equipping all of its military arms—infantry, artillery, tanks and aircraft—with radiotelephones that could intercommunicate with each other, an essential element of modern warfare. It appears that the apparent naturalness of its introduction to combat aircraft and its extension to all arms has obscured the origins of this important innovation, which, in particular, led to the "voice-commanded squadron."

The organization Squier created to develop voice radios capable of withstanding the rigors of combat service illustrated one scheme for obtaining the services of the scientific and engineering communities for the military services. It consisted of placing scientists and engineers in uniform and installing them in research facilities built and directed by the military. The first of two parts of the scheme as applied to the development of radio was realized under the Affiliated Plan, which committed the majority of radio scientists and engineers to uniformed status in the event of war. The second part was achieved in the establishment of Fort Monmouth, for almost 90 years one of the world's foremost radio-research facilities. (The institutional phase of the scheme was surveyed in the preceding chapter.) Although it failed to survive the war in its entirety, the scheme performed adequately during the time when the Army had virtually no scientific experience or facilities of its own. The high spirit of patriotism which infused the scientific community, indeed the entire country, provided the necessary adhesive to bind together two disparate groups of men, scientists and military men.

Squier prepared for the war in his own way. While the rest of the country talked of armaments, he quietly began proffering reserve commissions in the Signal Corps to his engineering acquaintances in the great electrical communications companies. He believed that the decision of victory or defeat hung upon the successful organization of science and engineering. After the war, he expressed his conviction that "military supremacy must be looked for primarily in the weapons and agencies provided by scientists and engineers and placed in the hands of ... combatant units to multiply their military strength."[3] Out of this organization, composed of such notable electrical engineers as John J. Carty, Frank B. Jewett, and Nugent H. Slaughter, came a device, which transformed the airplane from a weapon of individual opportunity to a weapon capable of centrally commanded operation. The airborne radiotelephone made possible the application of the military principle of concentration of mass to aerial combat.

Origin of Airborne Radio

The aspiration of talking from airplane to ground and vice versa was not new. The first crude radio transmissions from an aircraft to the ground took place over Belmont Park, Long Island, in August 1910.[4] During the International Aviation Tournament held there two months later, the concept of an aerial fleet commanded by radiotelephone was discussed by a group of Army officers who attended as official observers.[5] In 1911, an airplane pilot radioed a message over a distance of two miles. The distance was extended to 50 miles in the following year.

While the work on increasing transmission distances progressed, other officers experimented on tactical uses of radios in airplanes—still radiotelegraphs; voice transmissions were impossible. In early November 1912, Joseph O. Mauborgne, Jr., traveled from the Signal School at Fort Leavenworth to Fort Riley, Kansas, where he installed a radio set in an airplane in order to test airborne reporting of fire from field artillery. The test consisted of a competition between two airplanes, one using visual signaling methods, and the other sending messages by radio alone. Mauborgne acknowledged receipt of messages and sent new information by means of panels laid out on the ground, since at that time two-way radio communications aloft had not been accomplished in the Army.[6] The aircrew, consisting of a pilot and an observer of the radio-equipped airplane, reeled out a wire antenna over 250 feet long. The observer adjusted spark gaps and tuning inductances of his transmitter. He also held a telegraph key and a map in his lap. A chain connected to the engine drove the electrical generator. On the day of the test, 2 November 1912, the planes were dispatched to observe and direct artillery fire on targets located behind a hill about 7,000 yards from the firing battery. The target consisted of a silhouette wagon train, among which were interspersed vehicles and a large number of soldier silhouettes arranged as if they were columns on the march. The test was entirely satisfactory. After firing ceased, members of the firing battery found shrapnel holes through every target. According to Mauborgne, this was the first use of radio in the control of artillery in the armed forces of the United States.[7]

Two years later, in the Philippine Islands, the first radio messages were received in an Army airplane.[8] Mauborgne was the individual in the Army who first accomplished two-way radiotelegraph communications, repeating the feat on seven occasions. Experiments ended when the pilot, Lieutenant Herbert Dargue, cracked up the old Burgess-Wright plane.[9]

In 1915, the Aviation Section formulated definite plans to realize in practice the Belmont idea. Besides the numerous demonstrations of its utility in artillery spotting, both in Signal Corps exercises and in the French war zone, rapid technical advances in radio promised a solution to the most vexing problems associated with reception on board an aircraft. Long-range radio transmissions from Arlington, Virginia, were received simultaneously in Paris, the Canal Zone, the West Coast, and Hawaii.[10] This accomplishment resulted from advancements in power tubes, oscillators, circuit designs of vacuum tubes connected in parallel, and receiver designs employing amplifiers of both the radio frequency and audio-frequency signals. All these advances were necessary to successful airborne radio communication. Even the change in airplane design from pusher to tractor types, which greatly increased noise in the cockpit, failed to deter experimenters at San Diego.[11] Following a series of Dictaphone recording tests aloft, aviation officers became convinced of the practicability of radiotelephones in aircraft. These developments continued through 1916. Successful transmission from an airplane up to 140 miles, methods for receiving in the noise of wind and motor, and successful telegraphic communication between airplanes in flight were all established.[12] A radio-telephone was also under construction. In February 1917, voice was first transmitted from an airplane to a ground station. Admittedly a crude device, it concretely revealed new possibilities. (The Appendix provides a technical description of the difference between spark and continuous wave transmission and the importance of the vacuum tube to the latter type, the only way to transmit voice.)

Organizing for War

Upon American entry into the war, Squier commanded a Signal Corps without a modern engineering organization. It had fallen far behind European powers in the development of signaling methods suited to European warfare, particularly in radio. The improvement of European radio apparatus under the stimulus of war had left the American Army with inadequate equipment. With this outdated equipment, Squier had the responsibility not only of establishing communications between organized units of the Army but the Air Service and the newly formed Tank Corps.[13] It was a problem of enormous proportions, having principally a scientific and engineering solution.[14]

According to Nugent Slaughter, head of the Signal Corps Radio Development Section and a design engineer for Western Electric, Squier "realized the situation more fully than anyone else."[15] Squier took immediate steps to expand his engineering organization at the same time he invited the French and the British to send scientific missions. Within a short time, the French mission arrived with examples of their latest signaling devices.[16] Many prominent French electrical scientists comprised the mission. Meanwhile, Squier commissioned enough radio engineers directly from the radio industry to organize a Radio Division within his Office of the Chief Signal Officer.[17] In this action he continued in practice an apparently strong sentiment toward self-reliance in the Signal Corps.[18]

Squier held an important conference in his office shortly after the arrival of the European scientific missions. On 22 May 1917, Squier invited Colonel Reese of the Royal Air Force of Great Britain, Captain C.C. Culver, U.S. Army (the officer responsible for airplane radio experiments at San Diego and earlier), and Frank B. Jewett and E.B. Craft, respectively the chief engineer and the assistant chief engineer of the Western Electric Company, to discuss the uses of radiotelephones in aviation. In concluding their meeting, he called upon them to develop at once what the military officers and scientists of Europe had been unable to do during three years of war and the most intense needs of airplane fighting.[19] As with the Liberty engine development, which he also ordered, Squier thought of this work as one of the "major creative efforts in the development of American Air Service."[20]

The Research Problem

Success, it was generally agreed, lay in the successful development of an antenna system which would efficiently radiate sufficient power, but not interfere with tactical maneuvering; a transmitter which would respond to the human voice and be deaf to engine and wind noise; a soundproof helmet and receiver combination; transmitting and receiving circuits in which minimum adjustments were necessary for simple operation, apparatus combining lightness and ruggedness; a power supply suitable for vacuum tube designs which would neither add materially to weight nor increase aerodynamic resistance; and vacuum tubes which were small, rugged, within fixed electrical and physical parameters, and suitable for mass production.[21]

Of all these problems the vacuum tube presented the greatest challenge. It was recognized from the outset that vacuum tubes would be vital to the creation of effective radiotelephone sets for aircraft as well as practically every other piece of Signal Corps communication equipment to be developed. Moreover, they would be required in quantities far larger than ever before attempted. Total vacuum tube production in the United States at the outbreak of war was estimated at approximately three to four hundred tubes per week.[22] To meet Army needs effectively, this figure would have to be increased two hundred fold.

The research problem was divided into three parts. First, it was necessary to decide what kinds of tubes were required and what electrical and physical characteristics they must exhibit. Secondly, sample tubes must be produced which possessed the requisite characteristics and which were suitable for mass production. The third and most unusual problem concerned writing specifications and developing testing methods to ensure standardization of the tube production.[23]

In setting vacuum tube requirements, it was agreed to develop two types of tubes, receiving and transmitting.[24] They were intended to be standard in a variety of circuit designs, balanced with each other, interchangeable, and suited for performing several electronic functions. Never before were so many people and firms engaged in "electronic" research.[25] Simplicity, interchangeability, high vacuum and uniform operation over a 20 percent variation of filament and plate voltages represented radical departures from established vacuum tube technology and use.

Work on a radiotelephone for airplanes pressed forward concurrently. Western Electric undertook the main responsibility for developing a prototype of the aircraft radiotelephone. Investigation began immediately after the May meeting at the laboratories of Western Electric in New York City. Men were drafted from every department to accomplish the project.[26] Less than six weeks later, the first tests of workable airborne radiotelephone were conducted at Langley Field, Virginia. This square, awkward radio was affectionately called the "soap box model."[27] While communication over sufficient distances could not be maintained, the utility of short radio waves was established. Between 30 June and 5 July, successful voice communications took place between two planes transmitting over a distance of about three miles, ground to plane, and plane to ground.[28]

Demonstrations

In August, various officials appeared at Langley Field to witness the novelty of hearing voices from distant airplanes. Many were skeptical. Among those present to hear conversations carried over the "air waves" were Secretary of War Baker and Army Chief of Staff Scott.[29] Several cables were sent to the forces in France, informing them of progress made on the new apparatus. They received the news disbelievingly. By October, a set was produced which was considered to be rugged enough for combat use. Colonel C.C. Culver personally took several suitcases of the new equipment to the American Expeditionary Forces in France to demonstrate them to skeptical officers and to test them under field conditions.[30]

Culver tested his novel equipment at the overseas research and inspection laboratory maintained by the Signal Corps in Paris. Reminiscent of Squier's encounter with Rowland and his laboratory in a lorry on the front in 1914, the forward research laboratory was charged with carrying out "research along original lines and development of new apparatus."[31] While Rowland's example had spread to the establishment of front-line research organizations in the British army, the Paris laboratory represented the first such effort in the American Army. Squier demonstrated the importance he attached to the Paris laboratory by the high-caliber people he assigned to it. First in his selections was Major (later Brigadier General) J.J. Carty, who, as chief engineer of the American Telephone and Telegraph Company, had earned an international reputation for his supervision of the American transcontinental tele-

Major Edwin H. Armstrong, Signal Corps, Paris, France, 1918 (see note 32) (courtesy CECOM Historical Office).

phone system. Carty obtained leading men from industry and universities to staff the Paris laboratory. More than half the enlisted men in the division were university graduates. The officer in charge of the division was H.E. Shreeve, an engineer with 22 years experience in the development of telephone, telegraph, and radio equipment. For two years previous to his commissioning in the Signal Corps he was an executive officer of the research division of Western Electric Company. His deputy in charge of research was O.E. Buckley, a physicist trained at Cornell University and engaged in radiotelephony research at Western Electric when America entered the war.[32]

Meanwhile, another test of the airborne radiotelephone was arranged at Dayton, Ohio, for a larger number of military and civilian officials. Representatives from the Allies were invited, too. Most were dubious about the possibility of using a radiotelephone in aerial combat. Aircraft designers balked at multiplying devices and pilots resisted new equipment on their craft.[33] Visiting officials viewed the equipment in the airplanes before they were led to a nearby hill where a receiving set, fitted with a megaphone, was set up. Attitudes ran from fascination to skepticism to boredom. Officers with combat experience from foreign countries were the least impressed. Soon, crackles issued from the loudspeaker, then a clear voice: "Hello, ground station. This is plane number one speaking. Do you get me all right?"[34] Everyone looked up in amazement. The airplanes were barely visible as tiny specks against the dark blue of space. All attitudes became as one, full of enthusiasm.

Two days later, the airplane radiotelephone was used for artillery-control experiments at Fort Monroe. The value of voice communications was clearly demonstrated to visiting American and Allied military observers. The battery commander could often inform the airborne artillery observer that he had fired before the shell landed near its target.

Production of Radiotelephones

Now the difficult work of transforming these few experimental apparatuses into mass-produced equipment confronted the engineers. The equipment had to be standardized and a multitude of drawings made for the factory. The whole process was complicated by widely separated working locations. Western Electric's factory was located in Chicago, but its drafting facilities and laboratories were in New York. Men carried equipment and drawings between Chicago and New York by only the fastest passenger express trains available. By December, the receiving type of vacuum tube, dubbed the VT-1, had been sufficiently proven, so it was placed into mass production. Despite clearances of $3/100$ of an inch in the internal structures and a highly evacuated glass envelope,

14. Voice-Commanded Squadrons

it combined such ruggedness with well-balanced electrical characteristics that it became known as the "battleship type."[35] The development of a sending tube proceeded more slowly because of greater difficulties associated with the requirements for high-operating power. The first acceptable sending tubes made by Western Electric, called the VT-2, were capable of emitting high-frequency signals at 3 watts, and as high as 5 watts without damage. General Electric improved upon the Western Electric VT-2 in its VT-18, which could produce an output of 30 watts of high-frequency power.[36] Essential to mass-producing the Signal Corps VT series was the standardization of physical sizes, shapes, and testing procedures. Standard designs of bases and sockets (the four-contact bayonet type was selected), socket differences between receiving and sending tube types, and various auxiliary apparatus (such as step-up input transformers and telephone headsets) were established throughout the industry and the Signal Corps. One other problem of significance remained. How the radiotelephone should be powered perplexed designers because of limitations imposed upon them by aircraft designers. The simplest way to obtain the needed current would have been from a dynamo connected to the driving shaft of the airplane engine. The airplane constructors would not allow any such connection. Storage batteries might also have supplied the needed current, but their weight ruled them out, too. Therefore, it became necessary to design a dynamo, which could be attached to the fuselage of the airplane. It was to be driven by a propeller placed in the airstream. Since the airplane might travel anywhere between 90 and 160 miles per hour in a dive, the propeller shaft, which served as the armature of the dynamo, would revolve at speeds of between 4,000 and 14,000 revolutions per minute. This variation in speed seemed to preclude obtaining a constant level of current from even the most poorly designed propeller-driven dynamo.[37] The solution to this problem came from research on thermionic phenomena connected with the development of vacuum tubes. Out of that research came the regulator tube and the ballast lamp. These two tubes were added to the radiotelephone to regulate the voltage delivered from the fan-driven dynamo.[38]

By mid–February 1918, preliminary production samples of the radiotelephone receivers and transmitters (SCR-67 and SCR-68) were delivered. Vacuum tube production was successfully organized on a mass-production basis. Manufacturers achieved a production rate in excess of 1,000,000 vacuum tubes per year in less than six months after Squier's May meeting. For two weeks, daily test flights were made at Langley Field under simulated combat flying conditions. The wind-driven dynamo provided undisturbed speech with special electrical filters to suppress noises created by commutator sparking from the engine.

Successful trials at Langley led immediately to full-scale demonstrations of the concept of voice-commanded squadrons at Gerstner Field, Lake Charles, Louisiana.[39] In May, training of an air squadron maneuvered by the voice of its commander began. On 1 June, an aerial review was conducted for the Director of Military Aeronautics, who succeeded Squier as Chief of Army Aviation when aeronautical activities were removed from the Signal Corps on 20 May 1918. An aerial close-order drill by six airplanes and a tactical problem among the clouds followed the review. Throughout the review and drill, command was exercised by the voice of an airborne leader.[40]

Doctrinal Development: A Signal Corps Tradition

Squier regarded the radiotelephone development as one of his most creative achievements of the war. Nugent Slaughter, wartime director of the Radio Development Section, remarked that Squier was more alive to the uses of military radio than anyone else. Lloyd Espenscheid, an engineer of note with the Bell System, observed that, despite the employment of vacuum tubes for radiotelegraphy in Europe, radiotelephony was not known to have played a part on the Continent.[41] For three years, the Europeans had had the technology—principally the vacuum tube—the immediate lessons of war and an independent air service: yet no radiotelephone development followed. William A. MacDonald, of the Hazeltine Corporation and a lieutenant in the Paris laboratory, insisted that while the French had tube circuits in abundance they had no radiotelephone. They even had no voice communications in their radio plans. To imitate the American radiotelephone feat, the French had only to insert a microphone into the input circuit.[42] So close and yet so far.

In contrast, interest in airborne radiotelephony in the American Army began early and centered on a small group of technical officers. Starting in 1910 at Belmont Park, continuing in Culver's experiments at San Diego, and culminating in the mass production of SCR-67s and SCR-68s, technical officers played the principal role of innovator. Squier's resources, authority as Chief Signal Officer, penchant for research, invention and innovation, and personal and professional bridges to the engineering community laid the essential foundations for success of the project. Too many people and institutions were involved for any one person to take credit; still, the outstanding feature in the entire development is the vision of voice-command held largely by technical officers in the lace of skepticism.

The practical realization of this vision of voice-commanded airplane units held as much portent for the future of combat operations as did tanks and other weapons. Without communications the airplane was as a single steed mounted by a knight, who was left to his own devices. Opportunities for concerted action were limited.[43] Imagine leading a squadron of 70 airplanes and wanting to turn around. Wiggle one's wings and hope the wingmen see the maneuver? How, too, are orders changed if more opportune targets should present themselves, or if aerial support is needed at the home station. General Hugh Drum neither forgot nor forgave Billy Mitchell's mass-bombing raids against German airfields at a time when unopposed Germans bombed American troops in their trenches.[44] The voice-commanded squadron became a new type of fighting unit, bringing into reality its potential as an aerial weapon of concentrated mass. It was this transformation, which realigned aerial combat practice of individual exploit with the traditional military principle of concentrating military power for greatest effectiveness. In modern parlance, radiotelephone-equipped airplanes closed the control loop and subjected airplane pilots to the same principles of command and control incumbent upon the rest of the Army.

Squier's interest in airborne radiotelephone derived from carefully considered thoughts on centralized command. As an officer of the Signal Corps, one recruited by Greely, Squier also fitted well into a Signal Corps tradition, which stood at odds with traditional 19th century command and control doctrine. The situation Squier encountered in 1917 in aviation was nearly the same situation Greely faced in the Spanish-American War. At that time, the leading doctrine of command and control was the Clausewitzian formula of "pointing out the enemy as the objective, and victory as the goal."

The doctrinal conflict was joined on several levels. The theoretical level will be discussed below. The other level might be called the cultural level. The main cultural cleavage in the Army, then as now, existed between staff officers and combat, or line, officers. Technical officers fell in the category of staff officers. The milieu, circumstances, and expectations in which they pursued their careers tended more to separate them than to unify them, even though they wore the same uniform. The average line officer saw precious little value in what services the technical arms provided, especially the Signal Corps. When the Spanish-American War erupted, it was discovered that no provision had been made to expand the Signal Corps with volunteers, so Congress was obliged to pass special enabling legislation to recruit signalmen.[45] Typical of attitudes held by many line officers were those expressed by General William R. Shafter, Commanding General at Tampa, the principal port of embarkation for Cuba. He soon became the Commander of United States forces in Cuba. When he discovered that signalmen and their equipment were present in his jurisdiction awaiting transport to Cuba, he exploded: "I don't want men with flags. Give me men with guns."[46] He absolutely refused to accept a field telegraph system in his command, the important Fifth Army Corps.[47] Only two officers commanding corps or armies, Generals Nelson Miles and Wesley Merritt, asked for signal officers.[48] Naval line officers, too, regarded with incredulity and hostility Allen's interception of the critical information fixing Admiral Cervera's precise location.

The President directed that the Santiago campaign, aimed at bottling Cervera's fleet in Santiago harbor, continue despite the vigorous objections of the Naval War Board, basing his decision solely on the integrity of Greely's word. Greely staked his personal reputation with the President. Naval sources first confirmed Greely's report of Cervera's presence in the harbor, ten days later.[49]

Greely's persistence, confidence, and personal influence with the President created a fundamental transformation of military communications. The Santiago de Cuba campaign was launched on the basis of an electrical message intercepted by a technical officer located nowhere near a line organization or zone of combat. This was a revolutionary occurrence. The traditional line organization was ordered into action as a consequence of information derived from sources without the command structure.[50] Greely recognized the pivotal importance of his information. He said, "I was held morally responsible by the President for the change in the campaign."[51]

There is another dimension. Military communities, regardless of nationality, are bound together by an ancient code of fidelity built upon a foundation of precedence by rank. It is a highly personal relationship, which is strengthened and reinforced by customs of the service and discipline. Commanders know their subordinate commanders in much the same way as a father knows his sons. He is sensitive to their strengths and their weaknesses. When he gives orders, makes suggestions, offers advice, he knows their response in advance. When he wishes to communicate with them, he does so personally. If unable to do so, he personally dispatches a member of his own retinue, a trusted soldier or officer who lives the common life of a man in the field, one exposed to the same dangers and privations, to bear his message and deliver it in person. For these officers and men, the electrical age represented perhaps a depersonalization of a highly intimate and formalized social relationship. It even threatened the very integrity of command authority. Impersonal electrical communications were intruders, for they depended upon outsiders for their transmission and delivery upon uninitiated persons, people without the shared ideals of sword, horse and mess kit. Howeth, the historian of electronics in the United States Navy, relates how naval commanders resisted the installation of radio sets on board their vessels for fear their authority would be diluted by instructions from others who, because of their absence from the scene, could not understand the local situation. The local senior naval officer was in command of all he could survey unless a superior officer's signal flag was observed.[52] Morison tells of the hostile reception that steam-propulsion systems met among naval commanders who feared that the presence of engineers as officers would deteriorate the quality of command schooling that took place on the deck because steam engineers were subject to conflicting professional loyalties. Their ultimate professional ambition in the Navy would not necessarily be that of a commander. Hence, Naval leadership actually rejected steam vessels for a period of time, opting for sailing ships, where the old values of personal leadership and loyalty could be perpetuated.[53] The traditions of military leadership presupposed close and immediate relations between commanding officer and his subordinate commanders. It was a respectable tradition, tested and recommended by the greatest military minds of Europe. Frederick the Great emphasized personal leadership. He advised his generals to exercise a *coup d'oeil* (glance) upon the battlefield, organize their own campgrounds, and appear often before their troops.[54] Napoleon adopted Frederick's principles of leadership and formulated two new ones: corps separated from the main body should not be controlled by the Commander in Chief, but subject to independent leadership; and military commanders should exercise their own judgment, independent of the sovereign or his ministers unless they are present upon the field of battle.[55] Perhaps the most influential 19th-century military thinker, Karl von Clausewitz, adopted the injunctions of Napoleon and Bismarck in his Principles of War. He stated:

> The concerted attacks of the divisions and army corps should not be obtained by trying to direct them from a central point, so that they maintain contact and even align themselves on each other, though they may be far apart or even separated by the enemy. This is a faulty method of bringing about cooperation, open to a thousand mischances. Nothing great can be achieved with it and we are certain to be thoroughly beaten by a strong opponent.
>
> The true method consists in giving each commander ... the main direction of his march, and pointing out the enemy as the objective and victory as the goal.[56]

Marshal Helmuth von Moltke, Prussian Chief of Staff from 1858 to 1888, seemed to depart from traditional reliance upon independent leadership of detached army units in his conception of how electrical communications should be utilized by the Commander in Chief:

> The possibilities of the rapid transmission of information over the greatest distances, and also in long, round about ways, offers the means of leading separate army detachments according to a simple will to obtain common objectives. It is therefore unconditionally submitted that telegraphic connections should be established wherever possible. All independently operating units of the army must always see to it ... that it is possible at all times to transmit information and receive orders.[57]

The promise of unitary command and control implicit in his doctrinal statement was unrealized in practice, however. According to Marshal Joffre, Moltke prepared his troops very well before battle, gave them their initial orders, sent them into battle, and awaited "the results of ... [the] contest whose development he ... [was] powerless to control."[58] Thus, the potential of the telegraph had been recognized but unexploited. In the pre–World War I German military tradition, telegraphic communications were intended to link commands for the purpose of maintaining coordination, but initiative belonged to subordinate commanders.

French teaching under Marshal Foch at the French War College, c. 1900, took as its organizing principle another strain of Napoleonic thought which emphasized strict centralization of command. The advent of electrical communications in military field operations made possible the handling of modern armies with their great mass. British practice tended to favor the French position over the German.[59] American practice following the Spanish-American War pursued the dominant European doctrine of centralization. By 1910, centralization was the official teaching at the Army War College. The day of a General Shafter or a Navy Admiral refusing electrical communications had passed.

The change in the United States Army occurred noticeably during the Spanish-American War, certainly with the Santiago de Cuba campaign. Thereafter, the line of the Army placed increasingly higher value upon electrical communications as a means of achieving centralized control over mass armies distributed over large areas, separated by the most difficult of physical barriers. One of the notable champions of these Signal Corps endeavors was General Arthur MacArthur, commanding officer of forces in the Philippines. As he told the Secretary of State in 1908:

> This service has become so intimately associated with strategic marches and tactical movements on the battlefield that it must be in the future regarded as an integral part of the combat force—a fourth arm so to speak.[60]

While it has been said that Grant brought unity of command to the Army during the Civil War, commanders displayed a greater sympathy during the Spanish-American War for the doctrines of Frederick, Napoleon, Clausewitz, and von Moltke than a thoroughly developed unity of command doctrine.[61] Greely himself forced a communications system upon an unwilling Shafter, who sailed from Tampa without telegraph, telephone, or "even a call bell."[62] His program consisted of two parts: link the United States and Cuba by submarine cables and interlace the entire front of the advancing Fifth Army Corps by telephone lines "*so that its movements in action could be under one controlling mind.*"[63]

Although written much later, the italicized phrase represents an accurate expression of a contending tradition rooted within the Signal Corps, a tradition with respectable authority in Napoleonic maxims, but incapable of being realized without electrical communications. The contending tradition, imposed upon the Army by Greely with the assistance of the President and the electrical companies of America, belongs to a larger realization among armies of the world that centralized command, made viable through electrical communications, was the wave of the future.[64] The mechanism by which this realization was made manifest has not been well investigated. For the American Army, General Greely's actions would be central to any account.[65]

And so would Squier's contribution have to be considered. Squier spent the summer of 1899 on a leave of absence, studying and experimenting in the laboratory of the British Post Office, and testing his alternating current telegraphic transmitter over the Commercial Cable Company's "No. 3" Atlantic Cable from Canso, Nova Scotia, to Waterville, Ireland.[66] Nevertheless, considering the several extensions urged by Greely, Squier's work in London should be viewed as primarily official, albeit at no expense to the government. It has already been stated that he studied with Marconi to learn more of radio science, but it appears there was more to his mission from a hint in one of his subsequent papers on the importance of an American Pacific cable. He stated that it had become his duty to investigate the subject of deep-sea cables, not only from a technical point of view, but also "in relation to their uses for colonial service."[67] Additional evidence of his study of cables for their impact upon strategy, as well as colonial administration, comes from an article he wrote in the Artillery *Journal*.

> A leading English authority has recently stated that the value of the English Navy is increased fully one-half over what it would be considering the number and power of her ships, by the complete system of submarine cable communications connecting her colonies, coaling stations and fortified ports and wholly under British control.

Extrapolating, he argued:

> Control of the sea requires not only ships, naval bases, and coaling stations, but the most efficient means of communication obtainable by which only can each and all be effectively employed.[68]

No sense of a *coup d' oeil* upon the battlefield here. Instead of independent judgment, he implied unitary control and centralized organization. In strategy and tactics he showed himself to be a disciple of Greely and very much in the Signal Corps tradition. It was they who, by their efficiency and effectiveness, convinced MacArthur that communications should be regarded as a fourth arm of the army. Squier later developed the theme of counting communications as important as possessing ships before the Naval War College: "The very foundation of successful naval strategy is efficient and exclusively controlled communications, and lack of them more serious than inferiority in ships."[69] Squier also drew attention to the alarm spread along the entire eastern seaboard of the United Stated arising from the ignorance of Admiral Cervera's exact location, due to a lack of effective communications under the exclusive control of the American government.[70] In a later writing concerning colonial considerations, he pointed to the difficulty Spain experienced in pacifying Cuba. An inefficient, incomplete, and unreliable telegraph system made it "possible for bands of insurgents to move about much at their pleasure, appearing here and there, with no means of locating or concentrating for their destruction." It was not a lack of sufficient troops, so much "as that there were no efficient means of directing the troops in such a way as to make results decisive."[71] This doctrinal lesson and the import of a new environment of centralized communications and smaller detachments of troops were not lost on the leadership of the Signal Corps. Britain's naval and colonial uses of electrical communications served as a model for the subsequent development of military command and control systems. Before Allen departed for the Philippines on the *Hooker*, plans had been carefully drawn for interconnecting army garrisons on the various islands by submarine cables and telegraph lines. Squier said that, from the outset, it had been assumed that the "quickest means of pacifying and civilizing the Philippine Archipelago is to cover it with a network of telegraph wires."[72] Greely observed more cogently that "a system of public order can be maintained or restored with a military force which would be totally inadequate without the advantages of instant communication."[73] The relationship between the needs of colonial administration and maintenance of public order with a scarcity of armed troops was an important consideration in the development of centralized command doctrines ultimately adopted by the French, British, and Americans. The American acquisition of colonies and the need to maintain effective control hastened their evolution in the United States Army. And the successful experience of electrical communications in the Spanish-American War provided Greely with new influence in political and military circles to secure adequate appropriations from Congress, attentive hearings in Army headquarters, and support from the President for his programs.

Ten years later, the relationship between the Signal Corps and the Army was improved. The value of electrical communication to the mobile army was increasingly accepted by officers of the combat arms.

One of the means by which the effectiveness of Signal Corps service became known was through the reforms of Elihu Root. Begun in part as a response to the criticism of military performance during the Spanish-American War, the reforms were instituted in 1903. They marked the beginning of a professionalism in the military built upon full-time service and advanced education in the military arts. Devotion to duty, senior service schools, and a general staff became the hallmarks of the reforms.

Implicit in these reforms was a formal recognition of the importance of

14. Voice-Commanded Squadrons

the Signal Corps to the conduct of modern armies in the field. General MacArthur reported to the Secretary of War in 1908:

> The most recent experience has demonstrated that efficient service of military lines of information is indispensible to successful strategic and tactical operations. The organization charged with this service has become so intimately associated with strategic marches and tactical movements on the battlefield that it must be in the future regarded as an integral part of the combat force—a fourth arm, so to speak.[74]

By 1910, this viewpoint had become the official position of the War College. It was recognized that, without effective communications between all levels of command, a large army in the field would appear as an armed mob, directed by uncertain chance.[75]

Squier had expressed the same thought two years earlier when he recommended adding a new dimension to the military use of electrical communications. He proposed a bold expansion of the war game to large geographical areas and integration of its use in the training of officers destined for general rank in the War College.[76] Previous maneuvers had made use of deployed signalmen operating wire-telegraphy instruments, radio telegraphs, and, in some experimental situations, maintaining control between scattered large bodies of troops. No longer did serious professional opinion doubt the usefulness and necessity of controlling troops by telegraph. Greely's battle had succeeded. French and British armies also accepted the importance of controlling their troops by one mind. The effective employment of the telegraph by the Japanese Generals Oyama and Kuroki, in their 1904 campaigns against the Russians, demonstrated this lesson for all armies. The Japanese use of electrical communications, in one of several ways a preview of the First World War, made a profound impression upon Squier.

In a memorandum sent to the Chief of Staff during Allen's absence, in the fall of 1908, he advised that the Russo-Japanese War revealed the need for generals to practice handling large numbers of troops in the field.[77] Observing the general progress that Root's reforms had made, he called attention to the deficient training senior military officers received and the total absence of training that general officers received at any time. Squier recommended a way for generals to obtain vital experience in commanding large bodies of men in the field, without "the presence of the men at all, namely, skeletonized maneuvers conducted by telegraph."[78]

This recommendation added to the list of fertile uses of electrical communications for military purposes. Greely imposed electrical communications upon the Army and thereby provided the means of central control over all echelons. Allen dramatically demonstrated another aspect of military electrical communications at Tampa. There he acquired vital information about the adversary by obtaining access to the enemy's lines of information.[79] Squier's suggestion introduced another characteristic of reliable electrical communications.[80]

As long as military leaders were assured of receiving information they could prepare for a wide variety of contingencies. It was only when receipt of information was uncertain that electrical communications were valueless for training personnel and preparing their leaders. Effective and reliable communications provided the nerve pathways for alerting and informing an army. Generals could plan field operations removed from the field of battle, never personally viewing the enemy, ensconced in their headquarters at the rear. Advised by their staffs, who were informed by timely messages from a collection of distant points, they might manipulate statistical interpretations of combat progress upon intellectual projections of space and time; namely, maps, timetables, supply-status accounts, and morning reports. Adequate feedback was an implied capability in such a centralized command, knitted together by electrical communications.[81]

The electrical pathways established between echelons may be used in peacetime for practice. Moreover, they did not require the presence of troops for their usefulness. The processes of professionalization became available on a vastly broader scale for generals, who were denied field experience by reason of the expense of mobilizing an entire national army and the unpredictability of political consequences. War making acquired the potential of becoming more of an intellectual exercise when continuous practice was made possible. Only information inputs to the communications system need be changed. Knowing of their uninterrupted ability to send orders and receive reports to any location, generals could map the locations of their subordinate units and chart their performance as one might investigate a physical problem in a laboratory. The action would be as remote from the generals as a chemical reaction in a beaker was from the chemists. Explaining this new military function of electrical communications, Squier recommended that no delay should be permitted in instituting the telegraphic war game.

> The army which at the outbreak of war has developed this method [of command and control] to the utmost, will possess an insuperable advantage over an opposing army of equally brave and trained men, but who are subject to the blind and semi-independent control of individual leaders.[82]

Squier suggested that senior officers actually move into the field with their subordinate commanders and sufficient signalmen to maintain communications. No longer in a classroom, but on the actual terrain, they would respond to situations transmitted to them by telegraph and react as if they had actual troops under them. To Squier's knowledge, such an extension of the war game to actual terrain with generals and senior officers subordinated by telegraph to remote central authority had not yet been tried by any army.[83] He urged an innovative role for the U.S. Army in the development of military practice, believing that Americans should not content themselves with emulating foreign models.[84]

The game he suggested involved a large invading Red Army of 100,000 men, established on the Atlantic coast, near Fort Monroe. The reinforced defending Blue Army would move against the aggressors. The scale of the exercise was beyond anything previously contemplated. He recommended employing armies comprising nearly a quarter-million men from all combat arms. While the generals would be in the field, they would not possess freedom to move at will. His suggestion complemented Root's school program by utilizing War College students as planners and executors of the problem. Officers about to graduate from the War College would act as chiefs of staff, umpires, or hold other important positions as part of the problem. The entire problem would be directed from the War College and the headquarters of the Army in the field by the President of the War College or the Chief of Staff of the Army. Hourly situations would be telegraphed to field commanders calling for immediate creative responses. Because of reinforcements on one side or the other, offensive and defensive roles of Blue and Red Armies could be reversed without prior knowledge of the field commander. Of course, each commander had the responsibility of informing the War College of his moves so they might be reviewed and criticized.

Squier's scheme also envisioned war games against an imaginary invading force. Their actions and dispositions would be telegraphed to a defending force, which would take appropriate measures. These, too, would be scrutinized by the exercise team.

This graduating exercise for officers of the War College went beyond current conceptions of the employment of electrical communications in controlling large, dispersed armies. It introduced a new functional relationship between the signal services and the main structured elements of military organization. By 1917, the usefulness—the necessity—of electrical communications was well accepted throughout most of the Army. But its development in the Air Service had achieved a state no more sophisticated than that which had existed in the Army in 1898. Without radiotelephone links between airplane and airplane, and ground, the airplane would fail to develop as an effective military weapon.

Squier's order to hasten the development of airborne radiotelephones proceeded from a profound understanding of the relationship between effective communications and successful military operations. While no one man alone should receive credit for the airplane radiotelephones, Squier represented a Signal Corps tradition, which favored such an extension of telephony to aircraft. The radiotelephone was not just another gimcrack or pet device to be installed on aircraft.

The development of voice-commanded squadrons issued from the timely conjunction of technical advances and a doctrine of centralized command and control. Generous allotments, liberal working arrangements, and encouragement spur astonishing achievements in undeveloped technical areas, particularly in vacuum tubes. The establishment of Fort Monmouth provided a center for scientific research in electrical communications and signaling for the military, world famous even in the present time. The doctrine was articulated by the very man who had access to the necessary resources in industry and government. The fortunate conjunction of technology and doctrine was presided over by a man who represented both trends in his own thinking.

This important innovation in all aircraft operations, for control, navigational assistance, and communications, came from technical officers—primarily from technical officers and engineers who had a vision of aircraft operating in concert. The potential of military aircraft, thus conceived, differed markedly from the popular, romantic image of fighter pilot with scarf wrapped around his neck and flying in the breeze. The image shared much in common with the popular view of cavalry officers in the 19th century. He made a *coup d'oeil* at the aerial battlefield, mounted his winged steed, and engaged the enemy. Men like Culver and Squier persisted in their vision that an "army which ... has developed this method to the utmost, will possess an insuperable advantage over an opposing army of equally brave and trained men, but who are subject to the blind and semi-independent control of individual leaders."[85]

Claims are common that the most significant weapons to come out of World War I were the tank and the airplane. It is important to amend the claim with the observation that the tank and the airplane attained their full significance as weapons only after being equipped with radiotelephones. Squier could, thus, rightly regard their development as one of the most creative innovations of the war.

15
Retirement Years and Legacy

In 1916, Squier took charge of the Aviation Section at a time of internal dissension and external criticism. Unwelcomed by dissatisfied, even bitter, young flying officers, Squier set out immediately to establish industrial cooperation, research institutions, and, above all, an effective air service. Without necessary cooperation and sympathy from the General Staff, he undertook a task that was probably impossible for any one man to perform alone. Perhaps if he had been more conventional as an Army officer, the General Staff would have been more willing to assist him. But his eclecticism, his scientific reputation, his unorthodox behavior, and his habituation to cutting red tape and threading his way through bureaucracies probably separated him from his colleagues. Still, however unacceptable it might have been to contemporaries and superiors, Squier's creativity, willingness to assume responsibility, and ability to envision the largest dimensions of research demands in communications and aviation suited him to the challenge of institutionalizing science within the Army. In the short run those qualities were viewed as evidence of poor administrative ability. In the long run they may be interpreted as essential attributes for laying the foundations of scientific research and development in the Army.

Squier spent five years in the Army after the signing of the Armistice. He devoted most of his efforts to electrical communications, although he served on various NACA committees until 1920. Within a year after the war, AT&T made rapid progress in the development of multiplex telephony and telegraphy. Because of a change of heart or the poor advice of the Judge Advocate General in 1910, Squier tried to obtain a share in the immense commercial value of his multiplex patents.

Radio Affairs

Squier retired from the Army in December 1923. His last years in the service were devoted to matters concerning military, national, and interna-

205

Left to right: Television pioneer Francis C. Jenkins, David Sarnoff, George O. Squier, Secretary of Commerce Herbert Hoover, October 7, 1924 (courtesy CECOM Historical Office).

tional radio policies. He served as War Department representative on a committee formed by the Secretary of State to study ways of improving electrical communications in the Far East.[1] Later, at the invitation of the National Academy of Sciences and the National Research Council, he served on the executive committee of the American Section of the International Union of Scientific Radio Telegraphy.[2]

In 1921, he headed the United States delegation to the International Wireless Conference in Paris, which was convened to decide on the international allocation of frequencies for radiotelephony and radiotelegraphy. The United States position that certain wave bands should be reserved for radiotelephony carried the conference. Most delegates believed that radio would never supersede submarine cables.[3] Four months after returning from Europe, Squier resigned from the Army.

Awards, Decorations, and Recognition

Despite the public acrimony that dogged his leadership of the Air Service, Squier received American and foreign recognition for his contributions to the war and to science. The Franklin Institute awarded him its most prized Franklin Medal. The inscription recognized

> his valuable contributions to physical science, his important and varied inventions in multiplex telephony and telegraphy and in ocean cabling, and his eminent success in organizing and directing the Air and Signal Services of the U.S. Army in the World War.⁴

The National Academy of Sciences elected Squier a member of its prestigious body.⁵ He soon became a charter member of the Engineering Section of the Academy.⁶ In London, Field Marshall Sir Douglas Haig recommended that Squier be knighted for bravery, referring to Squier's tour as attaché in London and in the trenches in France.⁷ Thus he became Sir George Owen Squier, Knight Commander of the Order of St Michael and St George (KCMG). The King of Italy appointed him a Commander of the Order of the Crown, and France selected him as a Commander of the Legion of Honor. The United States awarded him the Distinguished Service Medal.⁸

In addition to the medals and awards of governments were the expressions of good will from respected colleagues. Squier's aeronautical counterpart in the Navy during the war, Admiral David Taylor, wrote a generous introduction

George Squier (left) at the White House with his sister, Mary, and her husband, Dr. E.H. Parker (courtesy CECOM Historical Office).

to his book, *Wings of War*, on the Army's role in the great enterprise of wartime aviation. In appreciative recognition, Secretary of War Baker wrote to Taylor, wondering

> sometimes whether all the smoke which impatience and misjudgment created about this exploit will ever clear away, so as to permit men like Squier. ... to emerge to the public view crowned with all the honors that their loyalty intelligence, zeal and success deserve.[9]

Charles Marvin, Chief of the Weather Bureau, sent a few lines to Squier, telling him:

> There has never been any doubt in my mind that time at least would demonstrate the great merit and importance of the work in aeronautics performed under your direction.[10]

The premier pioneer of American flight, Orville Wright, told Squier:

> I have always felt that the work done by you and your department would be appreciated in time; but nevertheless it is an outrage to one's feelings to see the criticism that you have had to endure from those who were too ignorant to appreciate what you were doing.[11]

One of the finest tributes to Squier as a scientist came in 1923 when the Massachusetts Institute of Technology asked him to be one of the speakers and the representative of the federal government at the inauguration of its new president, Samuel W. Stratton, former Chief of the Bureau of Standards. When the Secretary of War heard that "from all the scientists connected with the Federal government, an Army officer has been chosen..." he took the unusual step of placing Squier on military orders as the representative of the federal government to MIT.[12]

Major General George O. Squier USA (Ret.), formal portrait, 1932 (courtesy CECOM Historical Office).

Squier Sues AT&T

Squier entered a new phase of his business enterprises in the closing years of his Army career. The war, it will be recalled, interrupted Squier's personal involvement in the development of his multiplex patents in Europe. With the press of wartime duties subsiding, Squier returned to his old interest in their commercial exploitation. Not long after the Armistice was signed, the American Telephone and Telegraph Company succeeded in perfecting multiplex telephony with the aid of Campbell's wave filter and DeForest's vacuum tube. Squier looked excitedly at their progress. He wrote to his friend Robert Hadfield in England that the Bell System had made "wonderful strides in developing and applying [multiplex] telephony and telegraphy and 'wired-wireless' lines are now coming into practical use here." As many as five telephone conversations were simultaneously carried on a single pair of wires.[13]

From all appearances, however, Squier might have shared nothing in the American development of a system which he intimately associated with his own inventiveness, because the four mother patents taken out in 1910 were dedicated to the public. Still, the enormous number of patents issued to civilian scientists in uniform or employed by the government during the war prompted a reinterpretation of the 3 March 1883 Act, which authorized the grant of patents to any "officer of the government" (except Patent Office employees). So, an Army and Navy Patent Board was convened to study the matter. They soon realized that an interpretation of the phrase "any other person in the United States" was essential to explaining how broadly the dedication to the public clause of the 1883 Act should be construed. Thus, the Patent Board asked for a ruling from the Judge Advocate General of the Army. Acting Judge Advocate General S.T. Ansell declared on 30 November 1918 that the key phrase meant any other "like person in the United States," that is, any other person performing work for the government.[14]

Two weeks later, the president of AT&T, Theodore N. Vail, began laying the groundwork for a defense against any possible claims from Squier that the public dedication of 1910 patents did not apply to corporations or private persons. In a letter to Postmaster General Albert S. Burleson, Vail said that, while no substantial practical results had been achieved in multiplex telephony before recent AT&T demonstrations, the work of earlier investigators was naturally suggestive in the successful solution of the problem. He declared he had in mind the work of Squier.[15] It was not clear from the evidence that Squier had definitely decided at this point to pursue the question. In instructing his attorney, R. Randolph Hicks of Satterlee, Canfield, and Stone in New York City, on the sale of his Canadian multiplex patents, he said he preferred to have Canadian purchasers. He explained that, in the United States, he had already given the patents to the public.[16] He may well have wanted to keep the issue of domestic rights to private working of his American patent. This interpretation makes sense in light of correspondence three months later in which Hicks remarked to Squier about the illogicality of AT&T's claim that their multiplex system was different from Squier's while Western Electric, to all intents and purposes a part of AT&T, was buying Squier's rights in England.[17] Whatever the state of negotiation, they were suspended until September 1919, when federal control over telephones reverted to AT&T following the expiration of governmental wartime powers.[18]

By October 1919, AT&T made it quite clear that they intended to resist Squier by relying on the reasoning that antecedent patents on the multiplex principle existed—and that AT&T had purchased them. But to avoid a troublesome suit, AT&T offered to settle, on a nuisance basis, for $15,000.[19] Squier's lawyer then recommended filing an infringement suit against AT&T.

In his reply to Hicks's recommendation, Squier reported personal knowledge of similar cases in which AT&T initially offered to settle for a few thousand dollars and later paid over $100,000 when the inventor stood firm. He informed Hicks that General Electric might soon be approaching him about purchase of his multiplex rights.[20] Meanwhile, the case came to John J. Carty's attention. Upon leaving the Signal Corps, having risen to the rank of Brigadier General, Carty became a vice-president of AT&T. In a personal conversation with Squier, Carty agreed to take over the negotiations himself.[21] Carty's taking charge failed to move them along toward a speedier conclusion, however. For long periods of time no communications passed between the parties. Then, in March 1920, the Attorney General of the United States concurred in the opinion of the Judge Advocate General of the Army on the interpretation of that crucial clause regarding "any other person" in the 1883 Act.[22] This opinion further bolstered Squier's position, but not enough to give potential purchasers in the United States, Canada, Great Britain, and Germany the confidence that they would not be paying for rights that AT&T would obtain free of charge. For the next year and a half, negotiations with AT&T bumped along.

Then, in November 1921, Squier and AT&T came to an agreement. Squier would bring an infringement suit against AT&T to test the validity of his patents. If Squier could sustain his position in court then the following companies would contribute to the purchase of rights: General Electric, Western Electric, United Fruit Company, Westinghouse Electric Manufacturing Company, and the Radio Corporation of America. The agreement required Squier to "institute and prosecute vigorously" an action against AT&T. AT&T, for its part, paid $100,000 to Squier to begin proceedings and promised to pay another $750,000 if Squier won in court.[23] It was assumed that legal fees would take about one-half of the $100,000. The difference belonged to Squier, win or lose.

15. Retirement Years and Legacy   211

Five months later Squier instituted proceedings in U.S. District Court in New York. The *New York Times* announced the suit with the observation that the case will "likely ... become one of first importance to the wireless industry."[24]

While the case was being considered in court, Squier applied the "wired-wireless" concept to power lines, transmitting radio programs along ordinary electric lines. He designed a receiver which plugged directly into a wall socket. In early 1923, the North American Company purchased the rights to Squier's multiplex patents used in this connection. North American installed a radio transmitting service over the electric lighting lines in Milwaukee, Wisconsin. A similar service was later operated in Cleveland, Ohio, and New York City. Like present-day cable-TV services, the central transmitting station received and rebroadcast a variety of programs. A stockbroker circular describing the offering of stock in North American prophesied the dawning of an entirely new public utility.[25]

In the long interval during which the case was at trial, Squier's bargaining power eroded overseas. Nevertheless, as far as available records indicate, he received $100,000 from the Canadian government and $75,000 from the British government for his patent rights.

Then, on 4 September 1924, the court ruled against Squier. Without judging the merits of the infringement arguments, the presiding judge said that Squier had made it abundantly clear that his patents were freely available to the public. Typical of such statements noted by the court was part of an address before the Telephone Society of Washington in which Squier declared: "We wish it understood that this work is absolutely free to the public and that the patents were free to any person or corporation who wishes to use them."[26]

While Squier personally stood to lose a great deal of money, he argued that the issue was larger than himself. He decided to pursue the suit because it was important for the government to establish the rights, or lack of rights, of scientists in public service: "Strangely enough this question has never been threshed out in the courts before and it has never been decided what rights these men have in their inventions." As for himself, Squier said, "Let me make it plain that personally I feel no grievance in this matter. I am not a poor, downtrodden army officer with no place to lay his head." The issue was over the future of inventive research in the government service.[27]

Thus Squier immediately appealed the decision, pledging to take his case to the Supreme Court, if necessary, to secure the rights of inventors in government employment. Nine months later, he lost his appeal. The appellate court observed how:

pride in the army and his own corps and a very human liking for the "limelight" all combined to produce in Maj. Gen. Squier a desire, quite sincere at the time, to do exactly what the subtitle of his patent indicated and have it "dedicated to the public."[28]

Thus, whatever he might now feel about his donation, the act was irrevocable. As the court acidly remarked:

> Of course the very existence of this suit shows that the fit of public spirit has passed.... We are satisfied that plaintiff wished and was proud to be generous.[29]

Squier did indicate once in an interview with a *New York Times* reporter that he was given no choice in dedicating the multiplex patents to the public. Squier was "impelled to renounce all rights in his patent in favor of the public." The reporter quoted Squier in what sounded like a time-worn Army homily, "When an army man begins to think about money, he begins to forget the army."[30] A decade later, Squier indicated that the Judge Advocate General had misled him on his rights in 1910.[31]

Squier's suit drew attention to the relationship between corporations and individual inventors. The independent inventor enjoyed a revered and romantic image among a public strongly sympathetic to their plight of an unequal footing with corporations. Most large corporations, so the myth taught, rested on the creative genius of a single man. In the case of AT&T, Bell was the cornerstone. Although the view was certainly simplistic by 1925, it had a firm grip on the public's imagination. Thus, the corporate heirs of individual inventors should refrain from mistreating independent inventors. A prominent geologist and petroleum engineer, Edwin Hopkins, wrote the *New York Times* that "the beneficiaries of Dr. Bell today do not shine in a favorable light when they dip into their war chests, to fight independent inventors." There was also a question of fairness. Hopkins stated that, while AT&T might well be criticized for paying royalties to Squier since his invention was made with public money while in government service, the "company must still suffer loss of prestige and public approbation through appropriating such a valuable invention without remuneration."[32] AT&T did, of course, satisfy some of Hopkins's objections, without public knowledge.

Squier dismissed any further appeals. But, in one important sense, Squier succeeded because Congress legislatively addressed the question. In 1928, a new law redefined the rights of inventors in public service.[33] Squier expressed his delight and satisfaction with the legislation to Secretary of War Dwight Davis, who had spearheaded the drive for reform. He praised it as a model for all industries and thought the government could now attract young inventive talent, despite low salaries.[34] A *New York Times* editorialist concurred in Squier's optimism and remarked that it "may well be that for once Uncle Sam has proved himself to be more far seeing than the private business man or manufacturing corporation."[35]

15. Retirement Years and Legacy 213

As noted in Chapter 8, Professor Mischa Schwartz's recent investigation proved that AT&T's use of carrier multiplexing was based on Squier's invention, despite the technically incorrect and self-serving claims of AT&T executives to the contrary.[36] It is clear that, in the interest of simple fairness, AT&T should have compensated Squier appropriately for the invention that made possible the huge and highly profitable expansion of the telephone system. Had they done so, Squier would not have embarked on the complex legal odyssey that brought him little but an unfair stain on his reputation.

Muzak

Squier's 1910 patent on carrier multiplexing, which led to the modern telephone system, also provided an effective means of audio transmission over electrical power lines, which could carry higher frequencies modulated by audio signals. In contrast to wireless radio, transmitting music through the system Squier named "wired wireless" ensured higher signal quality, regardless

Major General George O. Squier with his bust by sculptor Moses Wainer Dykaar (courtesy CECOM Historical Office).

of atmospheric or solar conditions. The technology led to the first countrywide communications network, allowing the simultaneous delivery of programs through utility lines to remote radio-transmitting stations. Squier, however, was not satisfied with the structure of radio, in which programs were funded by intrusive commercials. He envisioned a new nationwide network, supported by subscription fees, that would also make unnecessary the commercials and program interruptions from sponsors (a concept like today's satellite radio). Squier approached the North American Company, then the nation's largest utility company, to transmit music over their lines. North American agreed to license the necessary patents from Squier and formed Wired Radio, Incorporated, in 1922. The company was not a success. It proved difficult to convince consumers to subscribe to a service they seemingly got for free from commercial radio. Looking to increase business, Squier came up with a catchy brand name for the venture in 1934. He combined the word *music* with the camera/film company's Kod*ak*, and came up with "Muzak." In the same year, North American formed the Muzak Corporation to transmit music directly to homes in Cleveland.

Muzak enjoyed its first success, however, in an entirely different market, sending music to the workplace. As Chief Signal Officer, Squier used music to increase the productivity of his secretaries. Afterwards, he investigated ways that music could soothe the nerves of employees while increasing their output. Muzak soon proved effective in locations beyond the office or factory floor. As skyscrapers climbed higher in American cities, building owners employed Muzak to calm anxious elevator riders, quickly earning its programs the name "elevator music."

Squier did not live long enough to see his vision become a great commercial success, but he received stock and was otherwise compensated for the use of his patents and the advice he gave the new company. Benefiting from the findings of industrial psychology, the company went on to become ubiquitous in the factory environment and in professional offices. Muzak was also widely used as background music for supermarket shoppers.

A Country Place for Country People

Squier had a philanthropic side of his personality which the appellate court's ruling hardly suggested. Sometime around the turn of the century he began buying property surrounding his grandfather's farm. In time, he acquired a large holding of land which he reserved for community use. He called it the "Poor Man's Country Club." Squier built facilities at his own expense for boating, fishing, baseball, golf, hunting in season, and outdoor camping. A local

15. *Retirement Years and Legacy* 215

Alden Hills Golf Club, Squier Park, Dryden, Michigan (courtesy Dryden Historical Society).

family maintained the club in exchange for free lodging. All people were invited to use the club without charge.[37] By 1934, over 60,000 guests visited annually. After the war he constructed a convalescent home for veterans, which he called Forest Hall, and donated it to the government. He occasionally lived in a house on the property that he called Burnside House, after his Philippine command. It held his library, papers, and memorabilia. During Christmas holidays 1933, the house burned to the ground, with the resulting loss of nearly all his library. Squier directed in his will that the entire country club be donated to Lapeer County, Michigan, after his death. In 1937, the county dedicated the site as the General George O. Squier Country Club. It is still in existence.

Squier's Legacy

Squier materially helped in institutionalizing scientific research and development in the Army. The last of his contributions occurred with the passage of a new patent law in 1928. The law assured personal rights in valuable inventions produced by persons employed by the government. Thus, the institutionalization process that characterized Squier's career was rounded out with the provision of legal rights for scientists, engineers, and inventors working in government facilities.

Squier's career was one that presented the challenges and opportunities of a new scientific and technological age to the Army. His personal problems of professional military and scientific growth prefigured the institutional problems of reconciling two increasingly professionalized and divergent traditions. As he worked out his own difficulties and strove to realize a vision of science in the service of the Army, he also laid the groundwork for some of the still-great scientific research and development institutions in the American government.

Assessment of Squier's impact upon the Army can be made more accurately today than 90 years ago. In the period following World War I, Squier became the scapegoat of the aviation scandal. Pershing's evaluation of Squier five years after the war evidenced the unattractive reputation Squier still carried. After his retirement from the Army, Squier's judgment faltered. An unscrupulous promoter persuaded him to lend his name to the solicitation of funds to commission a bust of General Pershing by the sculptor Moses Dykaar.[38] The bust was delivered and is currently in the Smithsonian Institution, but the funds raised were much greater than those needed. The promoter, Alfred Layton, absconded with the funds and was indicted for fraud in 1934. Squier was mortified and arranged to reimburse his Army colleagues and many prominent contributors, which was eventually accomplished by his estate after he died. This episode and minor episodes of eccentric behavior in the years just before his death, in March 1934, at the age of 69, detracted further from his reputation and deprived him of recognition and respect from his professional colleagues.

Viewed from the perspective of ninety years, however, encompassing several major wars in which victory depended as much on scientific and technological skills as on martial prowess, Squier's objectives were sound and beneficial to the U.S. Army. The validity of his basic premises that science and technology can multiply the strength of a military force well beyond its mere numbers and that access to science and technology are essential to modern military forces has been demonstrated. In addition to publicizing the need for integrating professional science and technology into the Army, he established some of the important research institutions for exploiting what are still two of man's most exciting technical challenges—electrical communications and powered flight.

Appendix: Technical Information

Squier's career was suffused with technological innovation, both by him and by other prolific inventors. Squier authored (or co-authored) many technical papers and patented over 30 inventions, something he continued to do almost until the day he died. Rather than burden the general reader with technical details, some are included here for those who are interested.

The Polarizing Photochronograph

The polarizing photochronograph, jointly patented by Squier and Crehore, consisted of two sections, a transmitter and a receiver. The transmitter contained an arc lamp and a Nicol prism (which only transmits light polarized in a single plane). The light transmitted through this prism was confined to the vertical plane. The receiver contained another Nicol prism, which was rotated 90 degrees so it would only transmit light confined to the horizontal plane. In this "crossed" position, no light would be transmitted through both of them.

The transmitter also contained a clear glass cylinder filled with carbon bisulfide liquid surrounded by a coil connected externally to wires. Carbon bisulphide will change the polarization of light passing through it in the presence of a magnetic field, such as would be produced by current flowing through the coil. When there was current in the coil, the plane of the light was rotated away from the vertical by the carbon bisulphide and some light would be passed through the prism in the receiver.

The coil, when connected, led to a series of screens placed in the path of a projectile's trajectory. As the projectile penetrated each screen the circuit connected to that screen was momentarily broken, the coil de-energized, and the prisms returned to the crossed position. Light was thereby prevented from emerging from the analyzer. The interruption of the light when the circuit was broken was detected by a photographic plate seated in a flywheel, which was attached to a rotating motor shaft. When the coil in the transmitter was energized, light emerged from the analyzer and struck the rotating photographic plate. When the circuit was interrupted by a projectile, current ceased to flow, the coil lost its field, the plane of polarization reverted to a "crossed" position.

A record of the projectile's passage would thus be traced upon the film in the receiver. A timing record, derived from tuning forks, was simultaneously projected upon the film. The polarizing photochronograph, therefore, constituted a camera with a massless shutter system, the first of its kind. (While physical mass no longer influenced operation of the shutter, electrical and magnetic inertia could not be ignored. Squier and Crehore's investigations indicated, however, that electromagnetic effects were insignificant.[1])

Figure A-1 is a photograph of the polarizing photochronograph manufactured by the Warner-Swayze Company.

Figure A-1. Polarizing photochronograph, manufactured by Warner-Swayze (courtesy Dryden Historical Society).

The Synchronograph and Its Application to Telegraphy

With the arrival of Pupin's high-speed alternator, Squier and Crehore had a transmitter capable of increasing signaling speeds from 1,200 words per minute to 3,000 words per minute, 12 times faster than rapid telegraphic systems of the period. The transmitter differed from the standard Morse system by using alternating current instead of direct current. Thus, the messages received by the Synchronograph took the form of a sine wave in which some half cycles were purposely suppressed to indicate the dots and dashes of the Morse code. Figure A-2 shows how the sample message TEN was handled by the Synchronograph and recorded on film or chemically treated paper.

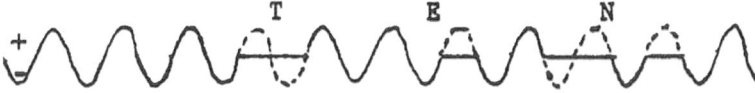

Figure A-2. Alternating current telegraph code.

Two suppressed half cycles represented a dash, and one suppressed half cycle represented a dot. Letters were separated by a full cycle. Dotted lines indicate suppressed half cycles.

The method by which the half cycles were generated was very ingenious, since it is difficult to interrupt the sine wave at precisely the point at which the wave passes through the zero point. It came to Crehore quite unexpectedly after months of thinking about the problem.[2] It is not possible to manipulate a key at high speed to open and close a circuit hundreds of times per second at the exact instant when the voltage is naturally zero. It came to him that the proper place to manipulate such a voltage controller, where the circuit must be made and broken at distinct points of phase, is at the generator itself.

The following simple example shows how a single half-wave is generated. The Morse code for any word or sentence may be formed by the appropriate combination of full- and half-wave cycles.

In Figure A-3, S represents the shaft of a two-pole alternating current generator, which drives wheel W through the gears M and N. The circumference of this wheel is one con-

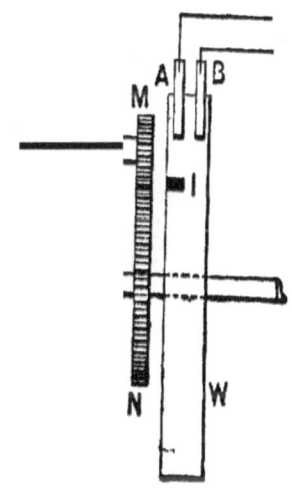

Figure A-3. Half-cycle generator.

tinuous conductor, presenting a smooth surface for brushes to bear upon. If the periphery of the wheel is divided into two equal parts, and is geared to run at the speed of the generator shaft, each division corresponds to one-half cycle of the current produced by the generator. Upon the wheel bear two brushes A and B carried by a brush-holder, which is capable of adjustment. These two brushes are connected in series with the line, so that the current which passes in at one brush, is conducted through the wheel to the other brush, and thence to the line.

Since the line current used is from the generator, the current is synchronously operated upon by the wheel. If both brushes remain continually in contact with the wheel, the current transmitted would have the regular sine form represented in Figure A-4.

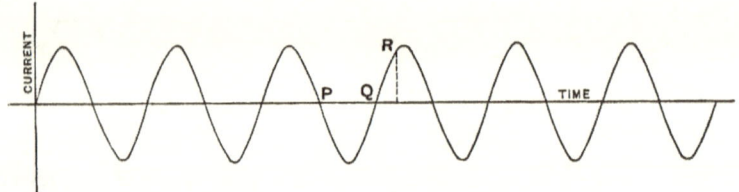

Figure A-4. Sine wave produced by generator.

If half of the circumference of the wheel is covered by paper and the brush A adjusted to ride on and off this insulation just as the voltage is changing from one-half cycle to the next (points P and Q in the figure), while the brush B is in continuous contact with the wheel, the half cycle represented by the section covered will be suppressed, without any sparking, because the voltage is zero, even if the potential used is high. There will be sparking at any other point, e.g. point R in the figure. In practice, the brush A is easily adjusted to this point by moving it slightly, backward or forward around the circumference of the wheel until the sparking ceases. This adjustment once made, the brush is fixed in position and so remains. In each succeeding revolution of the wheel, this cycle of operations is exactly repeated, and the current sent over the line would have every half cycle omitted.

If the generator has ten poles, it will produce five full-cycle sine waves for each revolution of the generator shaft. If the wheel is geared to run at one-fourth the speed of the shaft, one rotation of the wheel corresponds to 20 full-cycle sine waves. If the wheel is divided by paper into 40 equal parts, the current will have every half cycle omitted. For transmission of information, the connection between the brush and the wheel is interrupted not by paper attached to the wheel, but by a tape perforated according to a code and moving at a speed in synchronization with the wheel. The coded intelligence is transmitted over the line by the position of the complete and suppressed half cycles, as shown in Figure A-2.

Wireless Technology

From 1890 to approximately 1914, wireless meant Wireless Telegraphy (W/T), using Morse code. The technology did not support wireless telephony (voice). W/T used spark transmitters, which created a spark by interrupting the flow of current in a coil, just as an ignition coil works in an automobile. The spark transmitter is very inefficient; a low percentage of input power is actually radiated from the antenna. The first receivers used a "coherer," which was a glass tube full of iron filings that became a conductor of current in the presence of the electromagnetic field radiated by the transmitting antenna. The coherer was later supplanted by a magnetic, and then a crystal detector.

Inefficiency wasn't the principal problem of spark technology, however. Mutual interference between transmitters is the fundamental and inescapable problem. It is the reason spark gap transmitters have been illegal throughout the world since the late 1920s. In order to reduce mutual interference from spark gap transmitters, Marconi and others used the principle of "syntony," meaning an oscillation in the receiver's circuit, which is induced by an oscillation of the same frequency in the transmitter's circuit. The term "syntony" is taken from the sympathetic vibration induced in a tuning fork resonant at one tone, e.g., the note "A" on the piano (440 cycles per second), when a second tuning fork, resonant at the same tone, is struck in close proximity. Syntony between transmitter and receiver was the technical concept at the heart of Marconi's "four sevens" patent (British patent number 7,777, issued on 26 April 1900), which was the principal technical basis for the success of the Marconi Company.

Like the tuning fork, the spark gap transmitter excites a "resonant" circuit tuned for a certain frequency, and the receiving circuit is tuned to the same frequency. The electromagnetic wave radiated by the transmitter from multiple discharges of the spark looks like the acoustic wave radiated by the tuning fork, and is shown in Figure A-5. It is called a "damped sine wave" since the amplitude of the oscillation decreases with time while the frequency of oscillation (the number of cycles per second) remains the same. It can be shown mathematically and experimentally that such a wave contains higher frequencies than the single-resonant frequency of the transmitting and receiving circuits.[3] This "spillover" of higher frequencies is the reason spark-gap transmitters interfere with each other. It is fundamental to their operation to emit a series

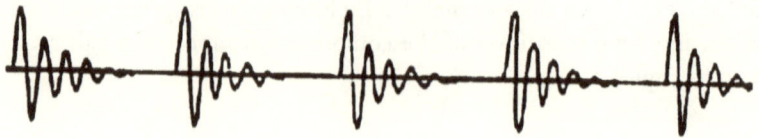

Figure A-5. Damped sine waves radiated by a spark gap transmitter.

of damped sine waves, one for each discharge of the spark. Clever design can minimize the effect, but cannot eliminate it.

Mutual interference severely limits the number of communication channels (transmitter/receiver pairs) that can be used simultaneously. During World War I, with spark technology, multiple transmitter/receiver pairs interfered with each other if they were less than 3,000 yards apart, despite the use of syntony to reduce interference. This was the major restriction on World War I wireless in an environment that requires significant communication capacity. Clearly, wireless based on spark technology was close to useless on a World War I battlefield.

Before the war, another technology which virtually eliminated the problem of mutual interference had been invented. Not only did the technology improve wireless telegraphy, it also offered the promise of wireless telephony. This was continuous wave (CW) transmission, which used a single frequency (called the carrier), as shown in Figure A-6, for each communication channel, like today's AM or FM broadcast radio. Use of a single frequency for transmission, rather than a damped sine wave, eliminates the "spillover" of extraneous frequencies from one channel to another and prevents mutual interference. Rather than the 3,000-yard separation required of spark transmitters, many CW transmitters using different frequencies could be operated in close proximity. This fact was well known to radio engineers before 1914, but could not be exploited until a compact source of CW was available.

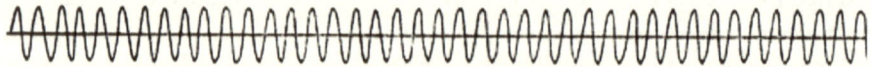

Figure A-6. Continuous sine wave radiated by a CW transmitter.

CW wireless depends on having a reliable single-frequency oscillator. It also requires devices to modulate (add information to) and amplify the signal. All three of these functions can be performed by electrical circuits containing three-element (cathode, grid, and anode) vacuum tubes, called triodes. The cathode, when heated, emits electrons into the vacuum; they travel to the anode, which is at a much higher voltage than the cathode. The grid is placed between the cathode and anode; the voltage on the grid accelerates or inhibits the electron flow. A small change in grid voltage leads to a much larger change in anode voltage, thus amplifying the grid voltage. Vacuum tubes were available before the war. Dr. Ambrose Fleming, of Marconi, patented a two-element vacuum tube, or diode, in 1904, calling it a "valve." (Britain continued to use this nomenclature for all vacuum tubes in the 20th century.) Building on Fleming's diode, Lee DeForest patented a three-element triode in 1906, calling it the "audion." (America did not continue to use this name.) Using the triode, Howard Armstrong patented the oscillator and regenerative amplifier in 1912.[4]

The availability of vacuum tubes was crucial to the timely application of this technology to World War I wireless. The triode is essential for: 1. Transmission, because in the appropriate circuit it is the source of sustained continuous oscillation, the "carrier frequency" of modern radio. It also is the only means of increasing the power of the signal delivered to the antenna; 2. Reception, because any high-frequency signal detected by a crystal or other means is weak, and must be amplified in order to be heard; and 3. Weight reduction, because the CW transmitter is much more efficient than a spark gap transmitter and uses less power.

The status of wireless technology in 1914–1918 was that the spark-gap wireless was inadequate for military requirements, and CW wireless promised to overcome the spark gap's inadequacies. The regenerative circuit, which allowed transmission and reception of continuous waves, was invented in 1912. When vacuum tubes were produced in sufficient quantities, as they were early in the war, the technology for building and deploying CW radiotelegraph and radiotelephone sets was available to all belligerents early in World War I.

Chapter Notes

Preface

1. General A. W. Greely won his second star after his transfer from the Signal Corps. A. W. Greely, *Reminiscences of Adventure and Service* (New York: C. Scribner's Sons, 1927), 219.
2. Paul Wilson Clark, "Major General George Owen Squier: Military Scientist," unpublished Ph.D. dissertation, Case-Western Reserve University, 1974.

Introduction

1. U.S. War Department, "Report of the Chief Signal Officer," *U.S. War Department Annual Reports, 1919* (Washington, D.C., 1919) XI, 262.
2. Paul Wilson Clark, "Early Impacts of Communications on Military Doctrine," *Proceedings of the IEEE*, September 1976, 1410.
3. Ibid., 1407.
4. George Owen Squier, "Multiplex Telephony and Telegraphy by Means of Electric Waves Guided by Wires," Professional Paper of the Signal Corps, U.S. Army, Washington, Government Printing Office, 1911.
5. Mike Bullock and Laurence Lyons, *Missed Signals on the Western Front—How the Slow Adoption of Wireless Restricted British Strategy and Operations in World War I* (Jefferson, NC: McFarland, 2010), 193–194.

Chapter 1

1. For genealogical information and many stories regarding George Squier's ancestors, the authors are indebted to the observations provided directly by Cathy D. Barry-Orth, and to her extensive *Squier-Atwell Family Tree—The Descendants of Samuel Squire*, available at URL http://familytreemaker.genealogy.com/users/b/a/r/Cathy-Diane-Barryorth/BOOK-0001/0000-0001.html. (Hereinafter referred to as *Family Tree*.)
2. Ibid.
3. Among other documents provided to us by Ms. Barry-Orth is the remarkable autobiography: *A Few Facts of My Own History*, by 2d year Cadet George Owen Squier, West Point, also available at the URL http://familytreemaker.genealogy.com/users/b/a/r/Cathy-Diane-Barryorth/BOOK-0001/0099-0001.html. Most of the details of his early life presented in this chapter are from this document. (Hereinafter referred to as *A Few Facts*.)
4. *Family Tree*.
5. *A Few Facts*.
6. Ibid.
7. Ibid.
8. Ibid.
9. Ibid.
10. Ibid.
11. Ibid
12. Ibid.
13. *Family Tree*.
14. *A Few Facts*.
15. Ibid.
16. Ibid.
17. Squier, Michigan, West Point Diary.
18. Ibid.
19. Squier, Monmouth, Accession 607, Speech by Mary Squier Parker to students of Dryden School, October 1937.
20. Association of Graduates of the United States Military Academy: *65th Annual Report* (Newburgh, NY: Moore Printing Company, June 11, 1934), 158.

21. A. Hunter Dupree, *Science in the Federal Government* (Cambridge: Harvard University Press, 1957), 184.
22. Stephen E. Ambrose, *Duty, Honor, Country: A History of West Point* (Baltimore: Johns Hopkins Press, 1966), 192.
23. Ibid., 200.
24. Ibid., 205.
25. Samuel E. Tillman, "The Academic History of the Military Academy," *The Centennial of the United States Military Academy at West Point, New York*, Vol. I of 2 (Washington: Government Printing Office, 1904), 269–270.
26. Ernest Dupuy, *Men of West Point* (New York: William Sloan Associates, 1951), 113.
27. *Centennial*, II, 318.
28. Extract, *Class and Conduct Reports of the Military Academy for the Annual Examination, 1887*, June 22, 1887, George Owen Squier Collection, Michigan Historical Society, University of Michigan at Ann Arbor. (Hereinafter referred to as Michigan Papers.)
29. A. E. Kennelly, "George Owen Squier," *National Academy of Science Biographical Memoirs*, Vol. 20, 151.
30. Squier, Michigan, *West Point Diary*.
31. Squier, Michigan, Request submitted 30 May 1887 and approved by Special Order No. 93, Headquarters, USMA, *Bound Notebook, West Point, 1883–1887*.
32. Association of Graduates, 158.
33. Squier, Michigan, *European Letters*, 1887, Lt. Squier to Julius Lauder, 25 July 1887.
34. George Owen Squier to Mary Squier, 25 July 1887, Michigan Papers.
35. George Owen Squier to Mary Squire, 1 August 1887, Michigan Papers.

Chapter 2

1. "Eugene Griffen," *Dictionary of American Biography*, 1928, VII, 48.
2. Frank C. Page, "Lieutenant Colonel George Owen Squier, USA, Inventor," *World's Work* (1916), 455.
3. When public property has been damaged, a board of survey is convened to determine extent of the damage, the degree of responsibility, and reimbursement due the government by the accountable party.
4. Notebook, Johns Hopkins University, Michigan Papers.
5. Squier, Michigan Papers, *Personal Letters and Orders, 1891–92*, Orders dated 15 July 1891, and a newspaper clipping dated 25 July 1891.
6. Squier, Michigan Papers, *Personal Letters and Orders, 1890–91*, Adjutant General, Ohio National Guard, through the Governor of Ohio to the Adjutant General of the U.S. Army, 9 July 1891.
7. Squier, Michigan Papers, *Personal Letters and Orders, 1891–92*, Lt. Col. Brush to Secretary of War, 14 August 1891.
8. George Owen Squier's Personal Folder (ACP), Old Military Records (NNO), National Archives, Washington, D.C. (Squier to Adjutant General of the Army, September 1889.) (Hereafter cited as Squier's ACP.)
9. 1st Lt. I. R. Williams, Commanding Officer, Battery G, 3d Art. to the Adjutant General, 17 September 1889, Squier's ACP.
10. Adjutant General to 1st Lt. Williams, 23 September 1889, Squier's ACP.
11. Squier to Adjutant General, 16 September 1889, Squier's ACP.
12. Ira Remsen, acting president of The Johns Hopkins University to the adjutant general, 20 March 1890, Squier's ACP.
13. Adjutant General to Squier, 31 March 1890, Squier's ACP.
14. Adjutant General to Commanding Officer, Artillery School, Fortress Monroe, 17 May 1890, Squier's ACP.
15. Letter, Colonel Royal T. Frank to Adjutant General, 19 May 1890, Squier's ACP.
16. Squier to Adjutant General, 22 May 1890, Squier's ACP.
17. Ibid., Endorsement to letter.
18. Emmet Dougherty, "Army's Greatest Inventor," *Popular Mechanics*, September 1927, 410.
19. President Gilman to Squier, no date, Michigan Papers.
20. Ralph W. Pope to Squier, endorsed by Louis Duncan, W. F. C. Hasson, and William J. A. Bliss, 19 May 1891, Michigan Papers.
21. Telegram, Adjutant General to Squier, 27 February 1891, Michigan Papers.
22. Special Orders No. 88, Headquarters of the Army, 18 April 1891, Michigan Papers.
23. Adjutant General to Squier, 21 August 1891, Michigan Papers.
24. Note from E. W. Bass, West Point, 26 January 1892; note from Bliss to Squier, 3 March 1892, Michigan Papers.
25. Captain G. C. Thurston to Squier, 18 April 1892, Michigan Papers.
26. Notebook. Artillery Notes, 1892–93, Michigan Papers.
27. Ibid.
28. Squier's thesis appeared in print during the month of his graduation: "Electro-Chemical Effects Due to Magnetization," *London, Edinburgh, and Dublin Philosophical Magazine and Journal of Science*, June 1893, p. 473ff.

29. Adjutant General to Squier, June 24, 1892, Michigan Papers.
30. 1st Endorsement, Squier to the Adjutant General, 1 June 1892, Squier's ACP.
31. 2d Endorsement, Squier to the adjutant general, 1 June 1892, Squier's ACP.
32. 3d Endorsement, June 24, 1892, letter, Squier to the adjutant general, 1 June 1892, Squier's ACP.
33. Special Order No. 95, Headquarters 3d Artillery, 31 August 1892, Michigan Papers.
34. Squier to post adjutant through the battery commanding officer, 6 October 1892, Michigan Papers.
35. Special Order No. 148, Headquarters 3d Artillery, 28 December 1892, Michigan Papers.
36. Association of Graduates of the United States Military Academy, "John Wilson Ruckmann," *Fifty-third Annual Report* (Newburgh, NY: Moore Printing Company, 1922), 133.
37. Association of Graduates of the United States Military Academy, "Henry Clarence Davis," *Sixty-third Annual Report* (Newburgh, NY: Moore Printing Company, 1932), 123.
38. *Centennial*, II, 328.
39. Association of Graduates of the United States Military Academy, "Charles Dyer Parkhurst," *Sixty-third Annual Report*, 94.
40. Squier, "Electricity and the Art of War," *Journal of the United States Artillery* II (January 1893), 99. (Hereinafter referred to as *Artillery Journal*.)
41. Squier served on the Committee of Direction and Publication as representative of the 3d Artillery. The committee assisted the editor, Lt. John W. Ruckmann, 1st artillery, in managing the publication *Artillery Journal*.
42. Association of Graduates, "Ruckmann," 133.
43. Maurice Matloff, ed., *American Military History* (Washington, D.C.: Government Printing Office, 1969), 290.
44. Russell F. Weigley, *History of the United States Army* (New York: Macmillan, 1967), 274.
45. Squier, "Electricity and the Art of War," 101.
46. Ibid.
47. Memorandum of the Adjutant General, 17 May 1893, Squier's ACP.
48. Mendenhall to Secretary of War, 16 May 1893, Squier's ACP.
49. Mendenhall to Secretary of War, May 1893, Squier's ACP.
50. Squier, "The International Electrical Congress of 1893, and its Artillery Lessons," *Artillery Journal* III (January 1894): 1.
51. Pure theory included electric waves, theories of electrolysis, electric conduction, magnetism, etc. Theory and practice included studies of dynamos, motors, storage batteries, measuring instruments, materials for standards, etc. Pure practice dealt with telegraphy, telephony, electric signaling, electric traction, transmission of power, systems of illumination, etc.
52. Squier, "On the Electrical Congress of 1893," *Artillery Journal* IV (January 1895): 154.
53. Squier, "The Electrical Congress and its Artillery Lessons," 6.
54. Of special interest were the electrically controlled 30-inch searchlight suspended from the Russian ironclad *Navarin* and the electrically operated turret on the French ironclad *Le Tonnant*.
55. Squier, "The Electrical Congress and its Artillery Lessons," 9.
56. Squier to adjutant general, 5 September 1893, Squier's ACP.
57. Ibid.
58. 3d Endorsement, Squier to Adjutant General, 5 September 1893, Squier's ACP.
59. 4th Endorsement, Squier to Adjutant General, 5 September 1893, Squier's ACP.
60. 5th Endorsement, Squier to Adjutant General, 5 September 1893, Squier's ACP.
61. 6th Endorsement, Squier to Adjutant General, 5 September 1893, Squier's ACP.
62. 10th Endorsement, Squier to Adjutant General, 5 September 1893; War Department Special Order No., 228, 4 October 1893, Squier's ACP.
63. Squier, "Some Tests of the Magnetic Qualities of Gun Steel," *Journal of the Military Service Institution*, XXIII (July 1898): 35; and *Artillery Journal* III (October 1894): 559.
64. Reported in *Philadelphia Magazine* (1873).
65. Supplied in his test case by the Bethlehem Iron Company of Bethlehem, PA.
66. Squier, "Some Tests of the Magnetic Qualities of Gun Steel," *Journal of the Military Service Institution*, 51.
67. Notebook, *The Physical Laboratory, 1893*, Michigan Papers.
68. Ibid.
69. Adjutant General to Commanding General, Department of the East, 18 May 1894, Squier's ACP.
70. War Department Special Order No. 118, 19 May 1894, Squier's ACP.
71. Notebook, *Physical Laboratory, 1893*, Michigan Papers. Squier's mentor Rowland also studied in Helmholtz's laboratory as a young man.
72. Letter and endorsements, Squier to Major General Commanding, June–July 1894, Squier's ACP.

73. Squier to post adjutant, Fort McHenry, 1 July 1894, Squier's ACP.

Chapter 3

1. Squier, "On Coast Artillery Fire Instruction," *Artillery Journal* III (April 1894): 247.
2. "Jump" is defined in Note 23.
3. Ibid.
4. Ibid.
5. Guns, per se, were certainly not new. The power, range, projectile dimensions, size of carriage, mounting and chase were unprecedented, however.
6. Albert C. Crehore, *Autobiography* (Gates Mills, OH: self-published, 1944), 37.
7. Albert C. Crehore, "A Reliable Method of Recording Variable Current Curves," *Transactions, American Institute of Electrical Engineers* XI (October 1894): 507–522.
8. Crehore, *Autobiography*, 48. Crehore and Squier, "Experiments with a New Polarizing Photo-Chronograph, Applied to the Measurement of the Velocity of Projectiles," *Artillery Journal* IV (July 1895): 409.
9. Roger R. Bruce, ed., *Seeing the Unseen: Dr. Harold E. Edgerton and the Wonders of Strobe Alley* (Rochester, NY: The Publishing Trust of George Eastman House, distributed by MIT Press, 1994).
10. Benjamin Robins, *Principles of Gunnery*, edited by Charles Hutton (London, 1805) 84, quoted in the *Encyclopædia Britannica*, 11th ed., VI, 302. Later modifications of the technique added multiple supporting cords, a method called geometric suspension. The method relies on the assumption that the time of collision, i.e., the time required for the bullet to cease its forward motion, is very short compared to the time of the swing of the pendulum, thereby assuring the presence of no external horizontal component of force influencing the system and the consequent conservation of momentum. The speed of the bullet is far too fast to measure directly, but the slow speed of a heavy block plus bullet is easily determined. If the mass of the bullet is represented by m, its velocity u, the mass of the pendulum block M, and the velocity of block plus bullet v, then velocity of the bullet can be derived from the relation, $mu = (m + M)v$ and $v = s/t$, where, s is the distance moved and t the time required for such movement. Thus, $u = (m + M)s/mt$.
11. The earliest forms of gun chronographs were probably those of Colonel Grobert, 1804, and Colonel Dabooz, 1818. Other investigators included Sir Charles Wheatstone, Professor Joseph Henry, A. J. A. Navez, Bashforth, Sir Andrew Noble, Le Boulenge, H. S. S. Watkin, and F. Jervis-Smith, who, with the sole exception of Henry, were all Europeans. The *Encyclopædia Britannica*, 11th ed., VI, 305 contains a remarkable bibliography on 19-century chronographic techniques in ballistics, physiology, and astronomy.
12. Many soldiers during World War I and the inter-war period knew Colonel Royal T. Frank's name well as belonging to one of America's concrete ships that plied the ocean between the West Coast and Hawaii. According to the last chief signal officer of the Army, Major General David P. Gibbs, who had sailed on it, she was tagged the *Rolling T*.
13. Crehore, *Autobiography*, 48.
14. Velocity at the very moment of departure from the gun's bore.
15. Crehore and Squier, "Experiments with a New Polarizing Photochronograph," 443.
16. Crehore and Squier, "The New Polarizing Photo-Chronograph at the U.S. Artillery School, Fort Monroe, Va., and Some Experiments with It," *Artillery Journal* VI (December 1896): 302.
17. Joseph Ames to Squier, 4 March 1895, Michigan Papers.
18. Squier to Crehore, 6 March 1895, Michigan Papers.
19. Crehore to Squier, 12 April 1895, Michigan Papers.
20. Squier's Account Book, 45, Michigan Papers.
21. Squier to Board of Ordnance and Fortification, 10 May 1895, and approval on 13 May 1895, Michigan Papers.
22. Extract, *Proceedings of the Board of Ordnance and Fortification*, 16 July 1895, and Squier's Account Book, 57, Michigan Papers.
23. Headquarters, U.S. Artillery School, General Orders No. 42, 2 July 1895, Michigan Papers.

In addition to their preparation for the August trials, Squier and Crehore explored the recoil motion of a gun, using photography. High-powered guns with long bores accentuated recoil motion and made complete knowledge of "jump" a necessity. Jump is defined as the difference in the angle of elevation at which the gun is laid and its angle when the projectile leaves the bore. Because angle of jump varies with each gun type, mounting, and elevation, tables are prepared to advise gunnery officers what correction is required in firing their particular gun. For the guns of today, gases produced during firing are exhausted through vents to the rear. Hence, recoil motion is not the factor it was in Squier's time.

Crehore and Squier, "Note on a Photographic Method of Determining the Complete Motion of a Gun during Recoil," *Artillery Journal* IV (July 1895): 470.

24. The idea of measuring muzzle velocities attracted Squier's attention as early as 1891. In a notebook entry for 30 October, he recorded the suggestion that a new form of chronograph stylus (reported in the *Philosophical Magazine*, London, July 1891) might be substituted for the Russell interrupter of the Schultz Chronoscope. Squier, Notebook, Physics, JHU, 1890–92, Michigan Papers.

25. Crehore and Squier, "Experimental Determination of the Motion of Projectiles inside the Bore of a Gun with the Polarizing Photochronograph," *Artillery Journal* V (June 1896): 326–327.

26. Ibid., 333.

27. Ibid., 351.

28. Crehore and Squier, "The New Polarizing Photochronograph at the U.S. Artillery School," 271.

29. Crehore and Squier, "The Synchronograph," *Transactions, American Institute of Electrical Engineers* XIV (April 1897): 116.

30. Crehore and Squier, "New Polarizing Photo-Chronograph," *Artillery Journal* VIII (August 1897): 301.

31. Extract from Proceedings of the Board of Ordnance and Fortification, 17 November 1896, Record Group 165, Chief of Staff File, National Archives, Washington, D.C. (Hereinafter referred to as BOF Files.)

32. Chief of Ordnance, Brig. General A. W. Flagler to Secretary of War, 21 April 1897, BOF Files.

33. Jennet Conant, *Tuxedo Park* (New York: Simon and Schuster, 2002), 32–33.

34. Devised by Charles Wheatstone, a Wheatstone bridge consists of four branches or legs, each one containing, in its simplest form, a resistive element. Its most useful application is as a balancing circuit.

A simple relationship exists between the four resistive elements. In the figure, $R_1R_4 = R_2R_3$. If three resistive values are known, the fourth can be readily calculated with remarkable accuracy. A galvanometer connected across points A and B indicates when a balance is achieved by indicating a null current through it. It was designed so that the magnitude of inductive reactance was much, much greater than resistance in the coil plus connecting wires.

35. Joseph John Thomson, *Notes on Recent Researches in Electricity and Magnetism* (Oxford: Clarendon Press, 1893).

36. Squier to Recorder, BOF, 12 December 1896, BOF Files. Crehore and Squier, "Discussion of the Currents in the Branches of a Wheatstone Bridge, where each Branch Contains Resistance and Inductance, and there is an Harmonic Impressed Electromotive Force," *Philosophical Magazine*, London 43 (March 1897): 161.

37. Squier to Recorder, BOF, 11 January 1897, BOF riles.

38. Crehore and Squier, "An Alternating Current Range and Position Finder," *Artillery Journal* VII (February 1897): 42.

39. Report of Committee selected by Board of Ordnance and Fortification, 28 February 1900, BOF Files.

40. Ibid. The use of light and shadow had formerly been a feature of the Smith-Crampton Range Finder, but the Board of Ordnance and Fortification had disapproved construction of that finder. Squier readily acknowledged indebtedness to Smith and Crampton and freely admitted conferring with Smith on improvements of the concept. The Committee selected by the Board said that if "the Board had disapproved utilization of the method in the earlier range finder it is believed it made a mistake."

41. BOF to Squier, 1 February 1897; and War Department Special Orders No. 31, 6 February 1897, Squier's ACP.

42. Squier to Adjutant General, 8 May 1897, Squier's ACP.

43. Report upon Search Light Installations at Fort Monroe, VA, 14 November 1896.

44. Ibid.

45. Squier to Recorder, BOF, 11 April 1898, BOF Files.

46. Application for Commissioned Officer in Engineers, 2 June 1898, Squier's ACP.

47. Charles Singer, ed., *A History of Technology*, Vol. 5 of 5 (Oxford: Clarendon Press, 1967), 228.

48. T. Commerfield Martin, editor of *The Electrical Engineer* to Squier, 22 November 1898, quoting a letter to him from Squier dated

10 December 1897, George Owen Squier Collection, United States Air Force Academy Library, Colorado. (Hereinafter referred to as USAFA Papers.)
49. Squier to Recorder, BOF, 12 February 1898, BOF Files.

Chapter 4

1. Squier to Recorder, Board of Ordnance and Fortification, 14 August 1896, BOF Files.
2. Crehore and Bedell already had a fruitful collaboration in studying alternating current theory. Their joint work, *Alternating Current* (New York: W. J. Johnston Co., 1895) was one of the first theoretical treatises on the subject published in America.
3. Squier to Board of Ordnance and Fortification, 4 April 1897, BOF Files.
4. Ibid.
5. Ibid.
6. Crehore and Squier, "The Synchronograph," *Artillery Journal* VIII (August 1897): 19.
7. Squier to Recorder, Board of Ordnance and Fortification, April 4, 1897, BOF Files. In modern parlance a Chief Electrician would be called the Chief (Electrical) Engineer.
8. Squier's Account Book, 5 May 1897, Michigan Papers.
9. Squier to Adjutant General, 25 May 1897 and approving endorsement, Office of the Adjutant General, Record Group 94, National Archives, Washington, D.C. (Hereinafter referred to as AGO Files.)
10. Squier to Adjutant General, 27 August 1897, AGO Files.
11. Part of the grand link utilized short submarine cables between some of the British Isles.
12. Greely to AGO, 25 April 1898, AGO Files.
13. A. W. Greely, *Reminiscences of Adventure and Service* (New York: Charles Scribner's Sons, 1927), iii.
14. President Franklin D. Roosevelt presented the Congressional Medal of Honor to Greely on his 91st birthday for the many services he performed for his country, in peace and war.
15. Max L. Marshall, *The Story of the U.S. Army Signal Corps* (New York: Franklin Watts, 1965), 121.
16. Greely to General Arthur MacArthur, 29 April 1905, Adolphus W. Greely Papers, Library of Congress. (Hereinafter referred to as Greely Papers.)
17. Robert V. Bruce, *Lincoln and the Tools of War* (Indianapolis: Bobbs-Merrill Co., 1956), 264. The Army had tolerated, even esteemed, officers who invented and patented devices or machines for the military. Major Thomas J. Rodman was perhaps the most notable example of a universally respected Army officer who held patents on cannons of his own design. Twice, however, the possession of these patents disqualified him from further consideration for promotion to Chief of the Ordnance Corps.
18. U.S. War Department, *Annual Reports, 1893*, Report of the Chief Signal Officer (Washington, D.C.: Government Printing Office, 1898), 901. (Hereinafter referred to as *CSO Report*, 1898.)
19. The Siphon recorder, invented in 1867, had replaced the mirror galvanometer, perfected by Lord Kelvin, which had originally made undersea cable transmission over trans-ocean distances a reality. The Siphon recorder provided a permanent record of the signals as received, and was in general use throughout the world.
20. *CSO Report*, 1898, Appendix 12, 988–991. Squier's Notebook in the University of Michigan Collection records more experiments on the Canso cable than are indicated in the Annual War Department Report.
21. U.S. Congress, Senate, Committee on Naval Affairs, *Construction of Telegraphic Cables between the United States, Hawaii, Guam, Philippine Islands, and Other Countries, and to Promote Commerce, Hearings*, S. Doc. 141, Serial 4231, 57th Cong., 1st Sess., 1902, 22. (Hereinafter referred to as Senate Document 141, Serial 4231.)
22. Greely to AGO, 22 October 1898, AGO Files.
23. Greely to Commissioner of Patents, 12 January 1899, Office of the Chief Signal Officer Files, Record Group 111, National Archives, Washington, D.C. (Hereinafter referred to as OCSO Files.) In all, four patent applications had been filed since October 1898. Greely's January letter requested that all be marked "special" under the provisions of Rule 63, section 1 of Rules of Practice in Office of Patents.
24. Squier "Influence of Submarine Cables," *Scientific American*, 20 April 1899, 21156.
25. U.S. War Department, *Annual Reports, 1899*, Report of the Chief Signal Officer (Washington, D.C.: Government Printing Office, 1899), 742. (Hereinafter referred to as *CSO Report*, 1899.)
26. General Thomas E. Eckert, President of Western Union Telegraph Company, to Greely, 15 September 1853, referring to a written request from Greely, dated 3 September 1898, USAFA Papers.
27. Diary, 1898, entry for 11 December, Monmouth Papers.

28. Squier to AGO, 23 February 1899, Squier's ACP.
29. Commanding Officer, Battery B, 3d Artillery, Fort Monroe, VA, to Alcatraz Island, 13 February 1899, Squier's ACP.
30. Squier to CSO, 14 February 1899, and endorsements, Squier's ACP.
31. Fitness Report, 1899, Squier's ACP.
32. Diary, 1899, entry for 15 January, Monmouth Papers.
33. Ibid.
34. Ibid., entry for 24 January.
35. Ibid., entry for 9 January.
36. Ibid., an unidentified newspaper clipping enclosed in the diary.
37. *CSO Report*, 1899, 754.
38. Ibid.
39. Greely to Recorder, Board of Ordnance and Fortification, 19 December 1900, BOF Files.
40. U.S. War Department, *Annual Reports*, 1900, Report of the Chief Signal Officer (Washington, D.C.: Government Printing Office, 1900), 993. (Hereinafter referred to as CSO Report, 1900.)
41. *CSO Report* 1899, 754.
42. Squier to CSO, 4 May 1899, Squier's ACP.
43. CSO to Recorder, Board of Ordnance and Fortification, 19 December 1900, BOF Files.
44. Greely to AGO, 26 April 1900, Squier's ACP.
45. Herbert Satterlee in later years served as Squier's patent attorney in New York City.
46. Thomas C. Piatt to Elihu Root, 28 May 1900, Squier's ACP.
47. Garfield to AGO, no date, Squier's ACP.
48. Greely to AGO, no date, Squier's ACP.
49. Stockton to Squier, 17 May 1900, AGO Files.
50. War Department SO #98, 26 April 1900, Squier's ACP.
51. Squier to AGO through the CSO, 21 May 1900, AGO Files.
52. Ibid., 1st endorsement, 22 May 1900, AGO Files.
53. Ibid.
54. Squier and Crehore referred to their system as "the sine wave system."
55. Crehore and Squier, "A Practical Transmitter," *Transactions, American Institute of Electrical Engineers* XVII (May 1900): 425.
56. Ibid., 438–39.
57. Ibid., 432.
58. The ionosphere was formerly called the Kennelly-Heaviside Layer, in recognition of one of his scientific investigations.
59. Crehore and Squier, "A Practical Transmitter," 440.
60. Ibid., 443. Comments rendered by Bedell and Delaney.
61. Ibid., 417.
62. Ibid., 416.
63. J. W. Freebody, *Telegraphy* (London: Sir Isaac Pitman & Sons, 1958), 118.
64. Crehore and Squier, "A Practical Transmitter," 421.
65. Ibid., 422.
66. Ibid., 396, 411–12.
67. Ibid., 441.
68. H. Nyquist "Certain Factors Affecting Telegraph Speed," *Bell System Technical Journal* 3 (April 1924): 324–46.
69. Crehore Autobiography, 69.
70. Martin Gardiner, *Fads and Fallacies in the Name of Science* (New York: Dover, 1952), 46.
71. Interview with Dr. I. I. Rabi by Thomas S. Kuhn at his New York home, 8 December 1963, at http://www.aip.org/history/ohilist/48 36.html.
72. Squier to AGO through CSO, 21 May 1900, AGO Files.
73. Ibid.
74. Greely to AGO, 26 May 1900, and War Department SO #125, 28 May 1900, Squier's ACP.
75. Stockton to Assistant Secretary of the Navy, 4 June 1900, 5th Endorsement, 25 June 1900, AGO Files.
76. Squier to AGO, 21 May 1900, 1st Endorsement, 22 May 1900, with attached schedule of events for week ending 14 July 1900, AGO Files.

Chapter 5

1. Greely to AGO, 29 November 1898, Squier's ACP.
2. *CSO Report* 1899, 743.
3. U.S. Navy Department, Annual Reports, 1899, Report of the Chief of the Bureau of Equipment (Washington, D.C.: Government Printing Office, 1899).
4. *CSO Report*, 1899, 741.
5. Haigh, *Cableships and Submarine Cables* (London: Adlard Coles, Ltd., 1968), 328, and *CSO Report*, 1399, 16.
6. Ibid.
7. CSO to AGO and AGO to Commanding Officer, Fort Myer, VA, 31 July 1900, AGO Files.
8. WD SO #190 series 1900, Squier's ACP.

9. Fitness Report, 1901, Squier's ACP and *CSO Report*, 1900, 989.
10. Historical sketch, no author, no date, OCSO Files.
11. Greely to General Arthur MacArthur, 28 January 1905, Greely Papers.
12. Ibid.
13. Greely to Secretary of War, 6 December 1900, AGO Files. Edgar Russell was promoted at the same time.
14. U.S. War Department, *Annual Reports*, 1901, Report of the Chief Signal Officer (Washington, D.C.: Government Printing Office, 1901), 931. (Hereinafter referred to as *CSO Report*, 1901.)
15. Fitness Report, 1901–1902, Squier's ACP.
16. Newspaper Clipping in Notebook, 1902, Michigan Papers.
17. U.S. War Department Annual Reports, 1902, Report of the Chief Signal (Washington, D.C.: Government Printing Office, 1902), 707. (Hereinafter referred to as *CSO Report*, 1902).
18. Greely to MacArthur, January 28, 1905, Greely Papers.
19. *CSO Report*, 1902, 707.
20. U.S. Congress, House, Committee on Interstate and Foreign Commerce, *Cables Between the United States and Hawaii, Guam, and Philippine Islands, Hearings*, H. Doc. 568, Serial 4401, 57th Cong., 1st Sess., 1902, 1–2. (Hereinafter referred to as House Document 568, Serial 4401.)
21. *Messages and Papers of the Presidents*, Vol. XV, 6647, cited in Leslie Tribolet, *International Aspects of Electrical Communications in the Pacific Area* (Cambridge: Oxford University Press, 1929), 180. (Hereinafter referred to as *International Aspects*.)
22. Ibid.
23. Ibid.
24. Tribolet, *International Aspects*, 182, and House Document 568, Serial 4401, 4.
25. It provided that

any telegraph company now organized or which may hereafter be organized under the laws of any State in this Union, shall have the right to construct, maintain and operate lines of telegraphs through and over any portion of the public domain of the United States, over and along any of the military of post roads of the United States, which have been or may hereafter be declared to be such by Act of Congress, and over, under or across the navigable streams of water of the United States.

Provided that before any telegraph company shall exercise any of the powers or privileges conferred by this Act, such company shall file their written acceptance with the Postmaster-General of the restrictions and obligations required by this Act.

That telegraphic communications between the several departments of the Government of the United States and their officers and agents shall, in their transmission over the lines of any said companies, have priority over all other business, and shall be sent at rates to be annually fixed by the Postmaster-General.

That the United States may at any time after the expiration of five years from the date of the passage of this Act, for postal, military or other purposes, purchase all the telegraph lines, property and effects of any and all of said companies, at an appraised value, to be ascertained by five competent, disinterested persons, two of whom shall be selected by the Postmaster-General of the United States, two by the company interested, and one by the four so previously selected.

[That the lines of such company shall not be constructed or maintained so as to obstruct the navigation of any waters of the United States, or interfere with the ordinary travel on military or post roads.

(Provisions of this act were cited in Tribolet, *International Aspects*, pp. 183–184.)

26. House Document 568, Serial 4401, 5.
27. Senate Document 141, Serial 4231, 46.
28. Tribolet, *International Aspects*, 70–80.
29. Senate Document 141, Serial 4231, 43.
30. House Document 568, Serial 4401, 8–11.
31. Tribolet ascribed to Congress a passive attitude vis-à-vis the Executive regarding control of cable landings in the period 1873–1921. I believe he fails to appreciate the extent of feeling against large corporations in the Progressive Era and the distrust they engendered in men like Corliss who used his committee to denounce Mackay's arrogant attitude about Congress's role in granting franchises. Corliss played an active role in attempting to prove the case for a government cable. He received less than full disclosure and honesty from the Commercial Pacific Cable Company in dealing with those who wanted a private firm to handle the Pacific cable project.
32. Senate Document 141, Serial 4231, 22–23.
33. Garfield to Root, 4 February 1902, Squier's ACP.
34. *CSO Report*, 1900, 988.
35. Squier to AGO, 1 July 1902, Squier's ACP. Corliss asked Squier to be present "in connection with the Congressional consideration of the proposed Trans-Pacific Cable."

36. Cablegram, AGO to Squier, 30 March 1902, Squier's ACP.
37. Fitness Reports, 1901–1902, Squier's ACP.
38. Squier to AGO, 1 July 1902, Squier's ACP. Greely was on a tour of inspection in Alaska during this period, although he did give testimony during the February House hearings.
39. Squier to acting CSO, 10 July 1902, Michigan Papers.
40. Telegram, Corliss to Root, 11 July 1902, Squier's ACP.
41. Squier to AGO, 14 July 1902, Squier's ACP.
42. Senate Document 141, Serial 4231, 25.
43. Garfield to Root, 14 July 1902, Squier's ACP. Acting CSO Edgar Russell approved the request.
44. U.S. President of the United States, *In matter of application of Commercial Pacific Cable Company for permission to land on shores of United States, Hawaiian Islands, Midway Islands, Guam, and Philippine Islands, telegraph cable to be laid between United States and Philippine Islands, and to China*, S. Doc. 24, Serial 4417, 57th Cong., 2nd Sess., 1902.
45. U.S. Congress, Senate, Committee on Interstate Commerce, *Cable Landing Licenses, Hearings*, before a subcommittee of the Committee on Interstate Commerce, Senate, on S. Bill 4301, a bill to prevent the unauthorized landing of submarine cables in the United States, 66th Cong., 3d Sess., 1921, 270.
46. Tribolet, *International Aspects*, 242.

Chapter 6

1. Philip C. Jessup, *Elihu Root* (New York: Dodd, Mead, 1938), 260.
2. Greely to MacArthur, 28 January 1905, Greely Papers.
3. Colonel F. K. Ward, 7th Cavalry, to Brigadier General James Allen, 27 June 1907, OCSO Files.
4. Fitness Report, 1907, Squier's ACP.
5. Crehore to Squier, 29 January 1905, Michigan Papers.
6. Lt. Gen. David Gibbs, last Chief Signal Officer of the Army, private interview, Colorado Springs, CO, February 1971.
7. Edward O. Purtee, *History of the Army Air Service, 1907–1926* (Unpublished study, Wright-Patterson AFB, 1948), 20.
8. Since then, the Army has captured a great deal of the Air Force's aviation mission. By almost any measure the Army presently operates an air service as large as the Air Force.

9. War Department General Order No. 145, 16 August 1906.
10. Squier to School Secretary, 16 August 1906, AGO Files.
11. Ibid., 2d Endorsement.
12. Ibid., 3d Endorsement, 24 December 1906.
13. Extract of Proceedings of the Academic Boards of the Infantry and Cavalry School, Signal School, and Army Staff College, convened pursuant to the verbal orders of the Commandant, 13 February 1907, Fort Leavenworth, KS, 2, AGO Files.
14. Ibid.
15. Squier to School Secretary, 20 December 1906, Endorsements 11 and 12, AGO Files.
16. Alfred Hurley, *Billy Mitchell* (New York: Franklin Watts, 1964), 11.
17. Assistant Commandant, *Annual Report*, U.S. Signal School, 1906, 2, OCSO Files.
18. Hurley, *Billy Mitchell*, 11.
19. Curtis Le May, "U.S. Air Force: Power for Peace," *National Geographic*, September 1965, 291.
20. Alfred Goldberg, *A History of the United States Air Force, 1907–1957* (Princeton: Van Nostrand, 1957), 3. (Hereinafter referred to as *A History*.)
21. Ibid., 2.
22. Juliette Hennessey, *The United States Army Air Arm April 1868 to April 1917* (Mobile, AL: Gunther AFB, 1958) 25. (Hereinafter referred to as *Army Air Arm*.)
23. Charles Chandler and Frank Lahm, *How Our Army Grew Wings* (New York: Ronald Press, 1943); Grover Loening, *Our Wings Grew Faster* (Garden City, NY: Doubleday Doran, 1935); Grover Loening, *Takeoff into Greatness* (New York: G. P. Putnam's Sons, 1968).
24. Marvin McFarland, ed., *The Papers of Wilbur and Orville Wright, 1906–1948*, 4 volumes (New York: McGraw-Hill 1953), II, 684. (Hereinafter referred to as *Papers of Wright*.)
25. Squier to Greely, 22 January 1906, OCSO Files.
26. Ibid.
27. In his first fitness report of Squier, Commandant J. Franklin Bell recommended him for Chief Signal Officer of an army in the field or of the Army. He judged him as "one of the ablest men of his day and age." Squier's ACP.
28. Virgal to Greely, 16 May 1902, OCSO Files.
29. Greely to Fisher, 21 May 1902, OCSO Files. Greely's letter is astonishing. Not to have known about the Western Society of Engineers indicates incredible incompetence in engineering matters, a judgment not in keeping with his

many achievements. It was one of the oldest and largest engineering societies in the United States. Established at Chicago in 1869, The Western Society of Engineers was the chief professional engineering group in the Midwest for years. Many outstanding engineers presented their papers to the WSE, including Octave Chanute's own pioneering studies in aeronautics.

30. Greely to Russell, 24 June 1902, OCSO Files.
31. Carroll Glines, *The Compact History of the United States Air Force* (New York: Hawthorn Books, 1965), 47.
32. Squier to Allen, 19 June 1906, OCSO Files.
33. George Squier, *Field Equipment of Signal Troops* (Fort Leavenworth, KS: U.S. Army Signal School, 1907), 32–33. This was first presented as a speech to the Staff College in 1906.
34. Hennessey, *Army Air Arm*, 25.
35. Alvin Josephy, Jr., ed., *The American Heritage History of Flight* (New York: American Heritage, 1962), 120.
36. Chandler and Lahm, How Our Army Grew Wings, 135–140; Fred Kelly, *The Wright Brothers* (New York: Harcourt, Brace, 1943), 121–27.
37. A patent covering the Wrights' airplane was issued by the Patent Officer on 22 May 1906. The Aero Club of America published in the spring of 1906 an account of the 1904–1905 Wright flights. It is conceivable that Squier drew some of his information from the Aero Club publication. Such a publication does not necessarily explain his confident assurance of the fact of powered flight.
38. Goldberg, *A History*, 3.
39. Hennessey, *Army Air Arm*, 25.
40. Squier, *Field Equipment of Signal Troops*, 33.

Chapter 7

1. John F. Victory, first civilian employee of the National Advisory Committee on Aeronautics, private interview, Tucson, AZ, November 1970.
2. Charles Chandler and Frank Lahm, *How Our Army Grew Wings* (New York: Ronald Press, 1943), 80.
3. Several authors have placed Squier in Europe in the summer and fall of 1907, although the record of his assignments and correspondence does not support this. The error probably stems from a House of Representatives document (dated 1927) entitled "Pioneer Aviators." Some of the authors who have perpetuated this story are: Elsbeth Freudenthal, *Flight into History* (Norman: University of Oklahoma Press, 1949), 161; Royal Frey, *Evolution of Maintenance Engineering, 1907–1920* (unpublished study, Maxwell Air Force Base, Alabama, undated), 5; and Irving Holley, *Ideas and Weapons* (New Haven, CT: Yale University Press, 1953), 27. Frey states that issuance of the specification for the first Army airplane resulted from two events: Squier's return from Europe and Wilbur Wright's meeting with General Crozier of the Ordnance Department in November 1907, upon his own return from Europe. Frey is probably correct about the immediate events, but we believe that Squier returned from New York instead of Europe.
4. Chandler and Lahm, *How Our Army Grew Wings*, 80–81; CSO to Lahm, 16 August 1907, OCSO Files.
5. Chief, Military Intelligence Division, to CSO, 28 August 1907, OCSO Files.
6. CSO to Secretary of War, 16 October 1907, and Patent Commissioner Edward Moore to CSO, 26 October 1907, OCSO Files.
7. Chandler and Lahm, *How Our Army Grew Wings*, 78.
8. *New York Herald*, newspaper clipping, 15 October 1907, Chief of Staff Files, Record Group 165, National Archives, Washington, D.C. (Hereinafter referred to as C/S Files.)
9. Squier to CSO, 15 October 1907, OCSO Files.
10. Carroll Glines, *The Compact History of the Air Force*, 46.
11. CSO to Recorder, Board of Ordnance and Fortification, 10 October 1907, BOF Files.
12. Ibid.
13. A&M Motor Company to Squier, 11 November 1907, OCSO Files.
14. Curtiss to Squier, 11 November 1907, OCSO Files. The Wright airplane accepted by the Signal Corps the following year developed 30.6 horsepower, according to Chandler, *How Our Army Grew Wings*, 301.
15. Squier to Allen, 6 November 1907, OCSO Files.
16. Ibid.
17. Allen to CSO, Department of the East, and Squier, 16 November 1907, OCSO Files.
18. Hutchison to Dewey, 18 November 1907, OCSO Files.
19. Allen to Squier, 20 November 1907, OCSO Files.
20. Allen to Squier, 20 November 1907, OCSO Files. This was a second letter, written the same day.
21. Squier to Allen, 22 November 1907, OCSO Files.

Notes. Chapter 7   235

22. Ibid.
23. Ibid.
24. Allen to Squier, 20 November 1907, OCSO Files.
25. Allen to Adjutant General, 30 November 1907, OCSO Files.
26. Wilbur Wright to Octave Chanute, 3 December 1907, quoted in McFarland, *Papers of Wright*, II, 835.
27. Chanute to Wilbur Wright, 4 December 1907, Ibid.
28. Edward Purtee, *History of the Army Air Service, 1907–1926*, Appendix A, paragraph 2: "The operations of this division are strictly confidential, and no information will be given out by any party except through the Chief Signal Officer of the Army or his authorized representative." Army support of the Langley airplane, which failed to fly after two trials in 1903, was the subject of considerable public criticism and ridicule. The perceived need for secrecy is hardly surprising.
29. Orville Wright to Major Harold S. Martin, Air Service's Technical Data Section at McCook Field, 2 December 1920, McFarland, *Papers of Wright*, II, 1128.
30. Allen to Crozier, with tentative specifications attached, 5 December 1907, OCSO Files. Allen's preoccupation with motor reliability is evident in the fifth provision, which required that the airplane "should be provided with some device to permit a safe descent in case of accidental stopping of propelling and lifting machinery."
31. Squier to Bell, 9 December 1907, and reply from Charles Cox (Bell's secretary), 16 December 1907, OCSO Files.
32. Herring to Russell, 16 December 1907, OCSO Files.
33. Kimball to Russell, 18 December 1907, OCSO Files.
34. Squier to Hewitt, 9 December 1907, OCSO Files.
35. Squier to Hutton, 9 December 1907, OCSO Files.
36. Allen to Wright, 26 December 1907, OCSO Files.
37. Cited in Hennessey, *Army Air Arm*, 27.
38. Ibid.
39. "An Invaluable Contribution to Popular Knowledge," *Flight* 1 (1909): 112.
40. U.S. War Department, *Annual Reports*, Report of the Chief Signal Officer (Washington, D.C.: Government Printing Office, 1908), 211.
41. Orville Wright to Wilbur Wright, 23 August 1908, McFarland, *Papers of Wright*, II, 914–915.
42. Ibid.

43. Dr. Oliver Wendell Holmes, private interview, Washington, D.C., 10 February 1971.
44. Signal Corps Log of Wrights' Aeroplane, 3 September 1908, McFarland, *Papers of Wright*, II, 197. Orville Wright's diary of the 1908 Fort Myer trials was lost many years ago. The collected Wright papers substituted the official Signal Corps "Log of Wright's Aero plane" for the missing diary.
45. Ibid.
46. The flight of an airplane had been publicly observed the previous March when a plane designed and built by the Aerial Experimental Association, under the leadership of Alexander Graham Bell, competed for the Scientific American Trophy at Mineola Field, New York.
47. *Washington Post*, 10 September 1908, 2.
48. Ibid. The claim that Lahm was the first Army officer in the world to fly in an airplane has greater institutional significance than importance as a fact. The claim requires interpretation. Stephen Tillman stated in *Man Unafraid* that Thomas Selfridge, whom Allen sent in the fall of 1907 to work with Bell's Aerial Experiment Association, was on record as having flown the *White Wing* on 19 May 1908. Squier himself claimed throughout his life to have been the first Army officer to fly with Orville Wright. It is difficult to imagine a contractor who earnestly desired selling his government an invention inviting a junior officer to be the first military man to fly in a Wright airplane, notwithstanding Orville's friendship with Lahm's father. Doing so would be an obvious breach of military protocol. Squier was keenly aware of the event's historical importance; he was that kind of person. Thus, it seems unlikely that he would have gone second unless he had flown earlier, just as he stated to a reporter in 1927. Even so, the institutional implications provide more insight to the process of self-identification. Only Foulois and Lahm took up careers in the Air Corps. Selfridge died early, and Squier was too clearly identified with the Army establishment to be allowed to take a place with other heroes of aeronautics as he never became a pilot. The repetition of only one of several possible claims reflects a serious ignorance of the period and the operation of a powerful institutional bias.
49. Signal Corps Log, 12 September 1908, quoted in McFarland, *Papers of Wright*, II, 922.
50. *Washington Post*, 13 September 1908, 1.
51. *Washington Post*, 12 September 1908, 2.
52. *Washington Post*, 10 September 1908, 1.
53. Bishop Milton Wright's Diary, 17 September 1908, quoted in McFarland, *Papers of Wright*, II, 925.
54. *Washington Post*, 18 September 1908, 2.

55. Proceedings of the Aeronautical Board of the Signal Corps, which convened at Fort Myer at 10:15 a.m., September 18, 1908, for The Purpose of Investigating and Reporting upon the Cause of the Accident to the Wright Aeroplane which Resulted in the Death of First Lieutenant Thomas E. Selfridge, First Field Artillery, Major C. McK. Saltzman presiding, AGO Files.
56. Telegram, Squier to Allen, 18 September 1908, OCSO Files.
57. *Washington Post*, 19 September 1908, 2.
58. Orville to Wilbur, 14 November 1908, McFarland, *Papers of Wright*, II, 938.
59. Entry 18067, OCSO Files.
60. An editor of *Colliers* magazine asked Squier to write a piece on the sensation one feels when flying in an airplane. Squier declined on the ground that impartiality in his position as President of the Board of Aeronautics must be strictly observed. As much as he would enjoy writing such an article it would be inappropriate for a signed piece which expressed enthusiasm for Wright's airplane to appear publicly before judging other entries.
61. Jones to Squier, 19 September 1908, OCSO Files.
62. Squier to Jones, 21 September 1908, OCSO Files.
63. Squier to French, 3 October 1908, OCSO Files.
64. Squier to AGO, 8 October 1908, and an enclosure titled "Service in the Signal Corps of the Army," AGO Files.
65. Memo for Aeronautical Board by Squier, 7 October 1908, OCSO Files.
66. Proceedings of the Aeronautical Board, 8 October 1908, OCSO Files.
67. CSO to AGO, 28 November 1908, AGO Files.
68. George Squier, "The Present Status of Aeronautics," *Engineering News* LX (December 1908): 639.
69. Ibid., 646.
70. Squier, "Present Status of Military Aeronautics," *Flight* (London) (February 1909): 121.
71. Report on Investigation in Europe made in compliance with CSO letter of 16 August 1907, on the subject of dirigible balloons, etc., 13 February 1908, OCSO Files. Lahm found all foreign dirigibles, excepting the *Lebaudy*, unsuitable for military purposes. He suggested the Aeronautical Division supervise construction of its own dirigible. Lahm to CSO, 11 January 1908, OCSO Files.
72. George Squier, "The Present Status of Aeronautics," *Annual Report*, Smithsonian Institution, 1908 (Washington, D.C.: Government Printing Office, 1909), 117.

73. It should be noted that reliable field radios had not yet been developed.
74. Squier, "The Present Status of Aeronautics," 639–40.
75. Ibid., 639.
76. Ibid.
77. Ibid., 640.
78. Airships meant dirigibles to some and either dirigibles or airplanes to others, depending in part upon the context. The Wrights, as mentioned previously, relied on this public confusion over terms to maintain the secrecy of their work.
79. Squier, "The Present Status of Aeronautics," 646.
80. Editorial," *Engineering News* XXXVI (July 1908): 126.
81. Squier, "Present Status of Military Aeronautics," 112.
82. Ibid., 646.
83. J. B. Scott, *The Hague Conventions and Declarations of 1899 and 1907* (New York: Oxford University Press, 1915), 100. The Hague Convention (IV) Respecting the Laws and Customs of War on Land.
84. He raised no question about the prohibition against the "attack or bombardment, by whatever means, of towns, villages, dwellings or buildings which are undefended." Ibid., Article 25, Section II, Chapter I. This is entirely in accord with his quoted sentiments.
85. August Post said, "There is more valuable information contained in Major Squier's paper than in any other work I have seen on the subject." Post to Allen, 8 January 1909, OCSO Files.
86. Squier to Secretary, Aero Club of America, 2 November 1908, OCSO Files.
87. Squier to Aeronautical Division, 2 November 1908, OCSO Files.
88. Squier, Zahm, and Brockett served together for many years on a Smithsonian committee for bibliographic publications.
89. Walcott to Squier, 9 February 1909, Smithsonian Institution Archives, Washington, D.C. (Hereinafter referred to as Smithsonian Papers.) Entry 20238 in OCSO Files is filled with letters thanking him for the courtesy of a copy of his paper. Included are letters of appreciation from General Baden-Powell of Great Britain and Major von Parseval of Germany, both important pioneers of aeronautics.
90. Bell wrote Walcott from Nova Scotia in early December 1908: "The Wright Brothers are being deservedly honored in Europe. Cannot America do anything for them? Why should not the Smithsonian Institution give a Langley Medal to encourage aviation?" Bell to Walcott, 5 December 1908, Smithsonian Papers.

91. Walcott to Senator Henry Cabot Lodge, 3 February 1909, Smithsonian Papers. Lodge and Bell were regents of the Smithsonian Institution.
92. Bell to Walcott, 7 January 1909, Smithsonian Papers.
93. Brashear was the maker of fine instruments, who had made some of Squier's instruments a decade earlier. Pickering came from the Harvard Astronomical Observatory, and Newcomb had earned an enviable record as astronomer and government scientist, although he once ironically disclaimed the possibility of flight in heavier-than-air-craft.
94. Walcott to Bell, 10 January 1909, Smithsonian Papers.
95. Walcott to Squier, 13 January 1909, Smithsonian Papers.
96. Walcott to Chanute, 13 January 1909, Correspondence of Octave Chanute, Col. 12, Library of Congress Manuscript Collection. (Hereinafter referred to as Chanute Papers.)
97. Walcott to Lodge, 3 February 1909, Smithsonian Papers.
98. Means to Chanute, 25 January 1909, Smithsonian Papers.
99. Chanute to Means, 27 January 1909, Smithsonian Papers.
100. Squier to Chanute, 27 February 1909, Chanute Papers.
101. Walcott to Lodge, 3 February 1909, Smithsonian Papers.
102. Means to Walcott, 4 February 1909, Smithsonian Papers.
103. Chanute to Means, 27 January 1909, Smithsonian Papers.
104. Walcott to Squier, 10 February 1909, Smithsonian Papers. *Aerodromics* is the term Walcott tried unsuccessfully to substitute for aeronautics in honor of Langley, who had first used it in his own work.
105. Chanute to Squier, 4 March 1909, Chanute Papers.
106. Walcott to Wrights, 17 February 1909, Smithsonian Papers.
107. Many cables passed between Walcott and W. I. Adams, U.S. Ambassador to France, and Schleicher Freres, the firm that also struck the Smithsonian's Hodgkin's Medal.
108. Wilbur Wright to Walcott, 29 November 1909, Smithsonian Papers.
109. McFarland, *Papers of Wright*, 1093 (footnote).
110. Ibid., 1092.
111. Greely was now retired from the Army. Squier asserted membership in the Cosmos Club, but their records fail to substantiate this.
112. Allen to Colonel Frank Baker, 19 December 1908, OCSO Files. When Baker asked Allen for a speaker on aeronautics for a Yale night, Allen replied that no one was available and that Squier had already "presented the official views of this office on the subject of aeronautics. It is about the last word that can be said for the present."
113. Memorandum, Allen to General Edwards, 8 May 1909, OCSO Files. He prophetically said, "Packard Automobile Company could do as much or more toward the building of that engine than any other concern." Packard engineers were instrumental in the design and development of the Liberty engine, developed during World War I.
114. CSO to General Jones, 27 December 1909, OCSO Files.
115. Walcott to Squier, 6 April 1909, Smithsonian Papers.
116. Walcott to Squier, 25 May 1909, Smithsonian Papers.
117. Commencing 1 July 1910.
118. Memorandum, Allen to Assistant Secretary of War, 3 June 1909, Squier's Personal Folder, OCSO Files.
119. Office of Chief Signal Officer memo No. 18, 21 June 1909, OCSO Files.
120. CSO to Charles Day, 20 August 1909, OCSO Files.
121. Richard C. Drachmann, President of MIT, to CSO, 16 November 1909; T. T. Tuttle to CSO, 1 March 1910; and L. O. Howard, Permanent Secretary of the AAAS, to Secretary of War, 23 November 1909, OCSO Files.
122. Curtiss to Secretary of War, 19 November 1909, AGO Files. Beck's first run-in with the Army occurred when he wrote an article for a popular publication in which he disparaged a regular army colonel. At Fort Leavenworth he quibbled about the scoring of rifle targets in an attempt to qualify for a special marksmanship award. Squier himself intervened in that particular affair with a report unfavorable to Beck. Later, his father, a retired general officer, intruded with a special request that Beck be assigned to a staff school. Although his point average disqualified him, his father's persistence resulted in high-level attention and disgruntlement. During his career he received transparent assistance from congressmen in obtaining desired assignments and avoiding unpleasant duties. Nevertheless, in some respects, he is the prototype of the new officer in aviation after about 1911, when he was transferred to College Park for aviation training. Combat arms men were displacing Signal Corps technical men in the company grades. They often possessed considerable political clout. The father of the most fa-

mous of the early aviators, Billy Mitchell, had been a U.S. senator from Wisconsin. Their willingness to frequent the halls of Congress stands in marked contrast to Allen's diffidence in aeronautical matters after mid-1909. The capture process was underway.

123. CSO to Recorder, BOF, 13 October 1909, and 19 October 1909, BOF Files.

124. CSO to Recorder, BOF, 29 October 1909, BOF Files.

125. McFarland, *Papers of Wright*, 1092-1093.

126. One of the first items of business Squier tackled during the opening months of World War I was establishment of cross-licensing agreements between airplane manufacturers modeled after those in the automotive industry. With regard to Wright's behavior, Squier wrote: "I have another case of a fundamental patent, which I happen to know about personally, i.e., The Wright Bros. Aeroplane Patent. They had invented the aeroplane and taken out the patent for it, and they formed the idea that the world was willing to wait until they got around to [devising] an engine and different kinds of aeroplanes suited to different needs. Mr. Curtiss, and many others in America, went to them very generously, with a view of obtaining royalty arrangements, so that they could go ahead and develop aeroplanes, under the Wright patent. The Wright Bros. refused all these advances, and remaining at a small town in Ohio worked away on their engine. In the meantime, Mr. Curtiss, failing to get permission from Mr. Wright, went ahead anyway, to make aeroplanes, with the result that Mr. Curtis is to-day easily worth $5,000,000 and has the largest aeroplane factory in the world; whereas, Mr. Wright is still in Dayton, Ohio, working away to improve his engine, and he is, more or less, a broken-hearted man, because he did not realize that he had a fundamental thing which the world demanded." Squier to Wilkins, 28 September 1915, Michigan Papers.

127. Any disinclination on Allen's part is understandable in the face of General Staff indifference to airplanes. Even a year later, in a carefully argued position paper delivered before the Army War College, Brigadier General W.W. Wotherspoon, President of the College, stated that the net results of his inquiry of aeronautics in war showed there is no good reason to believe airplanes will be used for war purposes. He believed, moreover, that there were many reasons for judging that their use will be less extensive than enthusiasts believe. He concluded that "we should study the whole question from a conservative standpoint and deduce from this study, as aerial matters develop, the war purposes to which we can put this new device." Quoted from "Aeronautics," a lecture by Brigadier General W.W. Wotherspoon, 13 October 1910, Chief of Staff File, War College Division, Record Group 165, National Archives, Washington, D.C. (Hereinafter referred to as WCD Files.)

128. The outbreak of hostilities on the Mexican border in 1911 brought a hasty aeronautical appropriation of $25,000 from Congress as part of a $125,000 authorization for fiscal year 1912. New aircraft were ordered and estimates for refurbishing the 1909 Wright airplane requested from the Wright factory. The company quoted a price of $2,000, but recommended against repairs because of the many improvements developed over the previous two years, which would be impossible to incorporate in the older model. (Chandler and Lahm, *How Our Army Grew Wings*, 187.) On 8 May 1911, the Chief Signal Officer offered the *Wright Flyer*, the world's first military airplane, to the Smithsonian for its historical collection, illustrating the growth of aeronautics. The arrangement with the Institution had already been concluded verbally and their willingness to accept the tender understood by Allen in advance. (CSO to Walcott, 8 May 1911, OCSO Files.) The transaction bears all the marks of Squier, particularly in view of his close relationship with Walcott. The actual transference and associated correspondence all carry Squier's originating initials. Squier's sense of history and his inventor's concern with priority are signified in this gift to the Smithsonian. Moreover, he was the official who accepted the Army's first airplane.

129. Bell delivered the address on 10 February 1910; Chanute, Chairman of the Selection Committee, was out of town.

130. Remarks appended to a letter from Walcott to Zahm, 25 February 1910, Smithsonian Papers.

131. Walcott to William Dinwiddie, recounting his remarks of the occasion, 12 February 1910, Smithsonian Papers.

132. Bell to Walcott, 2 May 1910, Smithsonian Papers.

133. The Wireless Board to which Squier refers was established by the President on 24 June 1904. Members were Greely, Rear Admiral Henry N. Manney, Lieutenant Commander Joseph L. Janye, and Professor Willis L. Moore of the Weather Bureau, according to William Loeb, Jr., Secretary to the President, OCSO Files. Cooperation between the Signal Corps and the Weather Bureau also extended to balloons. Willis Moore agreed to let Allen's men use the gas generator on Mount Weather for

Notes. Chapter 8   239

supplying hydrogen during dirigible trials in spring and summer of 1908. [Allen to Moore, 27 November 1907, and Moore to Allen, 30 November 1907, OCSO Files.]

134. Squier to Bell, 1 May 1910, Smithsonian Papers.

135. Ibid.

136. Bell to Walcott, 2 May 1910, Smithsonian Papers.

137. CSO to Chief of Second Section, General Staff of the Army, 12 March 1909, with Lord Rayleigh's Interim Report of the Advisory Committee for Aeronautics, 24 July 1909, OCSO Files.

138. Charles Cox (Bell's secretary) to Squier, 3 May 1910, OCSO Files. Rayleigh, Chairman of the British Committee, was one of Squier's old friends from his visits to London in the late 1890s.

139. Bell to Walcott, 2 May 1910, Smithsonian Papers.

140. Walcott to Bell, 4 May 1910, Smithsonian Papers.

141. McFarland, *Papers of Wright*, 1096 (footnote).

142. Squier continued his aeronautical activities and Smithsonian associations throughout 1910. He attended the celebrated Harvard-Boston Aero Meet on official orders at Boston. A $5,000 prize was offered "to the competitor who is most successful in dropping improvised bombs most accurately down the funnel of the miniature warship." Ward Webster, Secretary to J. V. Martin, Manager, Harvard-Boston Aero Club and War Department Special Orders No. 204, 31 August 1910. The meet was held from 3 September through the 13th at Soldier's Field, Boston. Walcott asked Squier to review the *Langley Memoir* and render his opinion as to its value as a publication in the *Smithsonian Contributions to Knowledge*. Charles Manley, Langley's former assistant, directed preparation of the memoir. (Walcott to Squier, 23 November 23, 1910.) Squier, Zahm, and Bell were later asked to comment on the proposed inscription on the Langley Tablet to be installed at the Smithsonian and intended to memorialize Langley's early investigations in flight. (Walcott to Squier, 20 December 1910, Smithsonian Papers.)

Chapter 8

1. George H. Clark, *The Life of John Stone* (San Diego: Frye and Smith, 1946), 117–118. (Hereinafter referred to as Stone.)

2. Rexmond C. Cochrane, *Measures for Progress* (Washington, D.C.: U.S. Department of Commerce, 1966), 141.

3. Clark, Stone, 118, and "Ernest R. Cram Obituary," *Proceedings, Institute of Radio Engineers* XXXIX (November 1951): 1469.

4. CSO to Secretary of War, 18 October 1909, Bureau of Standards, Record Group 136, National Archives, Washington, D.C. (Hereinafter referred to as NBS Files.) Allen asked if laboratory space could be provided "in making scientific research in methods of transmitting intelligence, particularly in the subject of wireless telephony." The Bureau enthusiastically replied that "the cooperation requested is not only advisable on the part of the Bureau, but very desirable since it will undoubtedly be of great benefit to both its work and that of the Signal Corps." (Secretary of Commerce and Labor to Secretary of War, 26 October 1909, NBS Files.)

5. Cochrane, *Measures for Progress*, 431 (footnote).

6. CSO Report, 1908, 29.

7. Radiotelephony means the radio transmission of voice; radiotelegraphy means the radio transmission of telegraphic signals.

8. Memorandum, Allen to President, Board of Ordnance and Fortification, 30 October 1908, OCSO Files. Squier's initials on the front page signify he was the author.

9. Vacuum cleaners, hand drills, and other small household appliances produce broad band transmissions from their sparking motors. Their signals are heard on nearby radios. Sometimes the effects of their "noisy" transmissions are observable on the household television set.

10. Memorandum, Allen to President, Board of Ordnance and Fortification, 30 October 1908, OCSO Files.

11. Ibid.

12. Allen to Chief of Coast Artillery, 31 October 1908, Squier's Personal Folder, Office of the Chief Signal Officer, Record Group 111, National Archives, Washington, D.C. (Hereinafter referred to as Squier's Personal Folder.)

13. The Signal Corps received for radio research alone an amount of money equal to ten times the entire Signal Corps budget of just a decade earlier. This increase reflects the growth potential inherent in radio for commercial and military purposes and the impact of new responsibilities acquired in Latin America and the Pacific. Such a sum for research and development stood as a tribute to Greely's foresight in defining a mission for the Signal Corps and the men he selected to accomplish it. Allen and Squier were working together again on making useful radios for the Army.

14. U.S. War Department, *Annual Reports,*

1909, Report of the Chief Signal Officer (Washington, D.C.: Government Printing Office, 1909), 249. (Hereinafter referred to as *CSO Report*, 1909.)

15. U.S. Congress, Senate, *Multiplex Telegraphy and Telephony*, S. Doc. 172, Serial 8712, 69th Cong., 2d. Sess., 9 December 1926, 5.

16. Contract signed by Charles DeF. Chandler, 28 September 1909, Squier's Personal Folder. This machine was one of only two in the entire country and cost the Signal Corps $10,995. A variable transformer (cost: $1,390) adapted the alternator for slower-working frequencies, which were necessary for Squier's planned experiments. The machine had 600 poles, developed 20,000 RPM, and operated from a line voltage of 110 volts. It was capable of producing a high frequency of 100 KHz and a low frequency near 20 KHz. Other equipment purchased from Fessenden included a 2 KW radio transmitter (probably Fessenden's rotary spark gap transmitter) and a radio receiver.

17. Cochrane, *Measures for Progress*, 40.
18. Ibid., 68.
19. Ibid., 40.
20. U.S. War Department, *Annual Reports*, 1910, Report of the Chief Signal Officer (Washington, D.C.: Government Printing Office, 1910), 647. (Hereinafter referred to as *CSO Report*, 1910.)

21. Culver, "Guided Wave," 307. As late as 1912, DeForest's claims for his triode tube were under challenge as a fraud in the courts. (Lee DeForest, *Father of Radio* [Chicago: Wilcox & Follett, 1950], 285–286.) Triodes later became essential in long-distance wire telephony. Squier's application of the triode to wire communications was subsequently adopted as a major innovation in the improvement of long-distance wire telephony.

22. "General Squier's Wired Wireless System," unknown author, perhaps E. R. Cram, expert in radio with the Signal Corps, undated, found in Clark's *Radioana*, located in the Division of Electricity and Nuclear Energy, Smithsonian Institution, Washington, D.C. (Herein after referred to as Clark's *Radioana*.) Squier wrote DeForest several years later claiming to have first used the vacuum tube in a physical circuit. Its use in 1910 for the purpose claimed is also recorded in Squier's notebook of the multiplex experiments.

23. The National Electric Signaling Company sent H. W. Sparks to train Squier's men in operating the 100 KHz alternator. John H. Kelman, Superintendent of Construction, National Electric Signaling Co., Brant Rock, Mass. to Squier, 5 April 1910. As the end of the fiscal year approached, Alexanderson expressed concern to Kelman about achieving positive results. He said, "General Allen is extremely anxious to be able to show results of this work before June so as to justify the congress [sic] of giving another appropriation and I therefore think that everythink [sic] should be done to make as good deliveries as possible." Report of E. F. Alexanderson, representative of General Electric, to J. H. Kelman, 21 May 1910, Squier's Personal Folder.

24. Sen. Doc. 172, Serial 8712, op. cit., 4. Resonance curves trace the relation between frequency on a line or in a circuit and the current flowing through it. For all circuits capable of resonance there will be ore frequency at which a maximum or a minimum of current will flow, according to the way the circuit elements are connected. While the shape of a resonance curve is generally parabolic, Squier found asymmetry, a result confirmed by Bela Gati, a telephone engineer of high standing. Bela Gati, "Discussion on 'Multiplex Telephony and Telegraphy by Means of Electric Waves Guided by Wires,'" Chicago, 28 June 1911," *Transactions, American Institute of Electrical Engineers* 30 (August 1911): 1679.

25. Squier, "The Unification of Communication Engineering," *Scientific American* (August 1922): 85–86; E. H. Colpitts and O. B. Blackwell, "Carrier Current Telephony and Telegraphy," *Electrical Engineering* 20 (April 1921): 304. Uniqueness of Squier's system and its fundamental importance to the carrier telephony techniques subsequently developed in the laboratories of AT&T, are claimed by Charles A. Culver in "Guided Wave Telephony," *Journal of the Franklin Institute* 191 (March 1921): 301. John Stone, whose own multiplex experiments in AT&T laboratories in 1894 were cited as evidence that Squier had failed to achieve anything new in his multiplex system (Colpitts and Blackwell, "Carrier Current," p. 303), hailed Squier's achievement: "A new art has been born to us. The infant art of high-frequency multiplex telephony is the latest addition to our brood of young electric arts." John Stone, "The Practical Aspects of the Propagation of High Frequency Electric Waves along Wires," *Journal of the Franklin Institute* 174 (October 1912): 353. (Hereinafter referred to as "Electric Waves.")

26. Notebook, 1910, Michigan Papers. Abbreviations and punctuation are quoted directly from the notebook.

27. Sen. Doc. 172, Serial 8712, op. cit., 6.
28. Ibid., 27
29. Allen to AGO, 3 November 1910, Squier's Personal Folder.
30. The use of radio frequencies for the

transmission of voice through space was already known. In 1906 engineers at Telefunken modulated a von Arco arc by voice. In the same year Reginald Fessenden experimented with voice and music radio broadcasts from Brant Rock, Massachusetts. The carrier was obtained from an 80,000 Hertz Alexanderson alternator. Several other cases of voice transmissions by radio before 1910 are recorded.

31. Squier, *Multiplex Telephony and Telegraphy*, A Professional Paper of the Signal Corps, U.S. Army (Washington, D.C.: Government Printing Office, 1911), 71.
32. Ibid., 74.
33. Ibid., 82.
34. Ibid., 84.
35. Ibid., 91.
36. Stone, "Electric Waves," 373.
37. Squier, *Multiplex Telephony and Telegraphy*, 97.
38. Sen. Doc. 172, Serial 8712, op. cit., 27.
39. Stone, "Electric Waves," 343.
40. "A System of Multiplex Telephony," *Telephony, Telegraphy, and Wireless* 58 (January 1911): 97.
41. Frank B. Jewett, "Discussion on 'Multiplex Telephony and Telegraphy by Means of Electric Waves Guided by Wires,'" *Transactions, American Institute of Electrical Engineers*, op. cit., 1666. (Hereinafter referred to as "Discussion on Multiplex.")
42. Sen. Doc. 172, Serial l8712, op. cit., 11.
43. Memorandum, C/S Wood to AGO, 8 March 1911, and War Department Special Order No. 55, 8 March 1911, Squier's ACP; and Sen. Doc. 172, op. cit., 10.
44. There were two other difficulties. The first (and more important) was the power source, namely the Alexanderson high-frequency alternator; there were but two in the country. Their cost and limited number made practical applications expensive. Later, when Alexanderson introduced alternators with iron cores, a concept which Fessenden adamantly refused to adopt in alternators of his own design some radio stations adopted them for producing the radio wave carrier. L. S. Howeth, *History of Communications-Electronics in the United States Navy* (Washington, D.C.: Government Printing Office, 1963), 137. (Hereinafter referred to as *Communications-Electronics*.) The concept of using an alternator to produce radio waves was not defective; it was simply superseded by the advent of triode (three electrode vacuum tubes) oscillator circuits. G.G. Blake, *History of Radio Telegraphy and Telephony* (London: Chapman & Hall, Ltd., 1928), 137. The second difficulty relates to the number of independent conversations that could be multiplexed on a single line. Squier showed from sensitivity curves taken on the cable that the interval between the frequencies of each of the receiving sets must be greater than 44 percent and that each receiving set would fail if stray currents of 50 percent of normal operating value existed. (Squier, *Multiplex Telephony and Telegraphy*, 57.) Such broad band width requirements represented a potential problem until the introduction of filter circuits by Edwin Colpitts, O. B. Blackwell, and George Campbell in America, and K. W. Wagner in Germany. E. Mallett, "Wired Wireless," *The Wireless World and Radio Review*, 6 May 1922, 169.

45. Jewett, "Discussion on 'Multiplex,'" 1667–1669.
46. Ibid., 1667.
47. Ibid., 1666.
48. M. Schwartz, "Origins of Carrier Multiplexing: Major George Owen Squier and AT&T," History of Communications Column, *IEEE Communications Magazine* 46, no. 5 (May 2008): 20–24.
49. W. J. Blenheim to Squier, 23 October 1913, Michigan Papers.
50. MacLaurin discusses the work of Alexanderson on continuous wave transmission without a mention of Squier's experiments, which used one of the only two Alexanderson alternators in the United States. Rupert MacLaurin, *Invention and Innovation in the Radio Industry* (New York: Macmillan, 1949), 93–95. Moreover, he stated that Fessenden was the "first important American inventor to experiment with wireless" (59–60). MacLaurin's omission of Army inventors, particularly Squier, illustrates one of the problems in the history of technology, which often focuses upon inventive activity within the established institutions of academia and industry. Military services do not simply buy equipment off the shelves of industry. A far more complex process occurs between military procurers, military research organizations, industrial representatives, and industrial researchers. Writing histories of military technical services is a neglected field within military history, so it seems reasonable that civilian historians of technology should also overlook the contributions of military inventors and scientists. Much of the Squier story to date reveals a fascination within the Army for research activity and a willingness to support innovative institutional arrangements for supporting research work. While a parallel, but much larger, effort was proceeding apace in civilian research laboratories, the Army, too, had men, whom James Brittain portrayed as proficient in mathematical physics, and interested in advancing the application of science in

research laboratory settings. James Brittain, "The Introduction of the Loading Coil: George A. Campbell and Michael I. Pupin," *Technology and Culture* 11 (January 1970): 36.
51. Culver, "Guided-wave Telephony," 305.
52. Telegram, AGO to Commanding General of the Maneuver Division, 13 July 1911, and War Department Special Order No. 162, 13 July 1911, Squier's ACP.
53. Jessup, *Elihu Root* I, 243–251.

Chapter 9

1. Memorandum, Allen to C/S, 18 January 1912, Squier's ACP. Orders were issued a week later reassigning Squier, effective in March.
2. Memorandum, Alien to C/S, 15 February 1912, WCD Files.
3. Wood to Mills, 15 February 1912, WCD Files.
4. Burton Hendrick, *The Life and Letters of Walter H. Page* 3 vols. (Garden City, NY: Doubleday, Page & Co., 1922), III, 202.
5. Post Wheeler and Hallie Erminie Rives, *Dome of Many Coloured Glass* (Garden City, NY: Doubleday, 1955), 300. Quoted in Alfred Vagts, The *Military Attaché* (Princeton: Princeton University Press, 1967), 112.
6. AGO to Squier, 18 March 1912, WCD Files.
7. Vagts, *The Military Attaché*, 270.
8. Acting Secretary of State Huntington Wilson to Secretary of War, 30 April 1912, AGO Files.
9. Public Law No. 238, passed 23 July 1912.
10. Benjamin S. Cable to Secretary of State, 8 May 1912, AGO Files.
11. D. W. Todd, "The International Radio Telegraphic Conference of London," *Journal of the American Society of Naval Engineers* XXVI (August 1912): 1330.
12. *London Times*, 2 July 1912, 11.
13. Acting Secretary of State Alvee A. Adee to Secretary of War, 31 October 1912, Squier's ACP.
14. Squier may have come to Norman's attention as a member of the U.S. delegation to the Radio Telegraphic Conference or as one of the speakers at the British Association for the Advancement of Science meeting in September. General consensus at the meeting was that a great deal more needed to be known on the scientific side before radio telephony could compete successfully with radio telegraphy. The differences between nighttime and daytime transmission characteristics occupied most attendants' attentions. The most popular explanations for the differences were advanced by Lord Rayleigh, who stressed the irregular conductivity in the surface of the earth, and William Eccles, who emphasized the role of ionized atmospheric layers. Other speakers were J. A. Fleming, Sylvanus P. Thompson, A. E. Kennelly, and A. G. Webster (The *London Times*, 7 September 1912, 3). Norman may also have learned of Squier from word of his selection as winner of the Elliot Cresson Gold Medal, awarded by the Franklin Institute for his outstanding achievement in multiplex telephony. Secretary of State to Secretary of War, 18 November 1912, Squier's ACP, and Franklin Institute Gold Medal Report No. 2507, 5 June 1912, USAFA Papers.
15. Adee to Secretary of War, 6 November 1912, Squier's ACP.
16. Squier granted an 11-day option on 4 June 1912 to the Marconi Wireless Telegraph Company, Ltd., of London, Michigan Papers.
17. Eccles to Scriven, CSO, 17 November 1913, OCSO Files.
18. Eccles to CSO, 1 January 1915, OCSO Files.
19. Eccles to CSO, 17 November 1915, and Scriven to Eccles, 8 July 1915, OCSO Files. Radio strays were detected in a telephone receiver and consisted of three types: 1) More or less pronounced rattling or grinding noise ["grinders"]; 2) Sharp isolated knocks ["clicks"]; and 3) Buzzing or frying noise ["hum" or "sizzle"].
20. Report of the Aerodynamical Laboratory Commission, created 19 December 1912, Smithsonian Papers.
21. CSO to R.E. Heinselman (Lawyers COOP Publishing Co.), 9 June 1913, OCSO Files.
22. Ibid.
23. Zahm to Squier, 6 June 1913, Smithsonian Papers. "I am enclosing herewith a prospectus which for the present is confidential, but which I trust will interest you as embodying an ideal presented to the Institution by yourself three years ago."
24. McFarland, *Papers of Wright* II, 1096 (footnote).
25. Zahm to Squier, 6 June 1913, Smithsonian Papers.
26. Report of Aerodynamical Committee, 1913, Smithsonian Papers. R. B. Owens, another of Squier's lifelong friends from Johns Hopkins, present Secretary of the Franklin Institute, served on the Sub-Committee on Publication and Dissemination of Aeronautical Information. Zahm also chaired Owen's committee. (Report of the Committee on Publication and Dissem-

ination of Aeronautical Information, Smithsonian Papers.)
27. Zahm to Squier, 17 July 1913, Smithsonian Papers.
28. Secretary of the Navy to Secretary of War, 12 November 1913, and 1st Endorsement by Reber, 2 November 1913, OCSO Files.
29. Telegram, Sec/WCD to Squier, 18 December 1913, WCD Files.
30. Squier to WCD, 5 February 1914, WCD Files.
31. Memorandum, Acting C/S to AGO, 21 November 1912, WCD Files. Because of the advanced state of the war and Turkey's indisposition to permit more military observers, the Secretary of State thought the plan inadvisable. Secretary of State to Secretary of War, 3 December 1912, WCD Files.
32. Secretary of State to Secretary of War (to CSO), 6 December 1912, OCSO Files.
33. Squier to CSO, 9 July 1912, OCSO Files. Rayleigh still headed the Advisory Committee. Squier described him as "the most distinguished living physicist in England."
34. Squier to WCD, 8 January 1914, OCSO Files. Foremost among those reported was a Ripograph, capable of simultaneously recording nine measurable indices of airplane performance.
35. Memorandum, CSO to President, Army War College, 20 May 1914, OCSO Files.
36. Contract between Muirhead and Squier, Michigan Papers.
37. Ibid. The set was made in accordance with British Patent No. 30003, issued in 1910.
38. Charge Slip, Military Information Division, Chief of Staff File, Record Group 165, National Archives, Washington, D.C. (Hereinafter referred to as MID Files.)
39. Director, A. S. S. to CSO, 29 January 1914, OCSO Files.
40. Russell to Wiedman, 2 February 1914, OCSO Files.
41. Copy of Agreement and letter from Squier to John Waterbury, 17 September 1914, Michigan Papers.
42. Squier to John Waterbury, 17 September 1914, Michigan Papers.
43. "A New Portable Wireless Receiving Set for Use on either Regular Commercial Telephone Lines or Wireless Antennae," October 1914, typescript (unattributed, but almost certainly Squier's), Michigan Papers.
44. Squier to Waterbury, 28 August 1914, Michigan Papers.
45. Squier to Waterbury, 17 September 1914. He asked $50,000 at date of transfer as a guarantee his patents would be developed, and 20 percent of net profits from working the patents. In case there is a dispute, a third party would decide (Squier recommended Bion J. Arnold, a distinguished electrical engineer known to be fair and just), Michigan Papers.
46. John J. Carty to Waterbury, 29 December 1914, Michigan Papers.
47. Lee DeForest to Squier, 3 January 1915, Michigan Papers.
48. Charles Bright, *Telegraphy Aeronautics and War* (London: Constable & Co., 1918), p. 52.
49. Ibid., p. 245. Substitution of machine or automatic devices speeded transmission between cable terminals. Some of the important ones were Brown's drum relay, Muirhead's cold-wire relay, and Huertley's hot-wire magnifier. Magnifiers received weak cable signals and reproduced them on a local siphon recorder in amplified form. Huertley's magnifier represented a notable improvement over the carbon-water relay invented by E. Raymond-Barker. Huertley used a signal-actuating coil to move platinum wires heated by an electric current in and out of a cold air blast in order to influence the balance of a Wheatstone bridge, which was connected to the Siphon Recorder. His hot-wire magnifier improved, receiving speed by 35 to 40 percent over direct connections from cable to Siphon Recorder. John Gott of the United States started inverse current working about 1883. In his system, dots and dashes were of equal duration, but of different amplitudes. Raymond-Barker and Orling, whose system also used a jet of water to vary resistance in a local circuit connected to a Siphon Recorder, adapted their receiving sets to the Gott transmitting device. The great British cable theorist H. W. Malcolm analyzed cable arrival currents and devised a method of introducing artificial cable loading, whereby the lead element of the arrival curve was improved. Artificial loading made the leading edge of the arrival curve more abrupt to resemble the onset of a square wave and made the tail negligibly short.
50. W. A. O'Meara, "The Practical Problem of Telegraph Transmission from a New Angle of View," *The Royal Engineers Journal* XXII (October 1915): 184.
51. George Squier, "On an Unbroken Alternating Current for Cable Telegraphy," *Proceedings, Physical Society of London* XXVII (August 1915). U.S. Patent No. 1,233,519. *London Times*, June 30, 1915, 145b.
52. Nyquist, op. cit.
53. Bright, *Telegraphy, Aeronautics and War*, 52.
54. O'Meara, "New Angle of View," 192.

55. Squier to A. B. Crossman, an official of Muirhead's, 1 May 1915, Michigan Papers.
56. Squier to Walcott, 15 November 1915, Smithsonian Papers.
57. Letter of Assignment, 25 June 1915. Patent specifications were filed in the United States on 14 June 1915, Michigan Papers.
58. A. Baxendale to Squier, 9 May 1915, Michigan Papers.
59. R. A. Hadfield to Henry B. Jackson, First Sea Lord of the Admiralty, 27 September 1915, Michigan Papers. According to Howeth, it was Jackson who first commenced radio experiments in the British Navy in 1895. Howeth, *Communications Electronics*, 45.
60. Squier to Wilkins, 28 September 1915, Michigan Papers.
61. Ibid.
62. Squier to Baxendale, 4 October 1915, Michigan Papers.
63. Squier to Baxendale, 5 November 1915, Michigan Papers.
64. Squier to Wilkins, 15 January 1916, Michigan Papers.
65. His sine-wave system of telegraphy is described in British Patent No. 22,265. Squier's remarks before the Royal Institution were delivered on 20 January 1915.
66. Memorandum relating to the trial of the Sine-Wave Telegraph System invented by Lt. Col. G. O. Squier of the U.S. Army and developed by Messrs. Muirhead & Co., Ltd., London, undated, unsigned, Michigan Papers.
67. Squier, "On an Unbroken Alternating Current for Cable Telegraphy," 312.
68. Squier to DeForest, 1 March 1916, Michigan Papers.
69. Squier to DeForest, 29 January 1916, Michigan Papers.
70. DeForest to Squier, 16 February 1916, Michigan Papers.
71. AGO to Squier, 25 March 1916, WCD Files.

Chapter 10

1. George M. Trevelyan, *Grey of Falloden* (London, 1937), 102, as quoted in Ernest R. May, *The World War and American Isolation, 1914–1917* (Cambridge: Harvard University Press, 1959), 8. (Hereinafter referred to as *American Isolation*.)
2. Ibid.
3. Lloyd George, *War Memoirs*, II, 751, quoted in Philip Magnus, *Kitchener: Portrait of an Imperialist* (London: Butler and Tanner, Ltd., 1958), 285.
4. May, *American Isolation*, 11.
5. Lindley Garrison to Walter Page, 14 January 1914, WCD Files.
6. Telegram, Page to Garrison, 12 January 1914, Squier's ACP.
7. Garrison to Page, 14 January 1914, WCD Files.
8. Memorandum for the chief of staff, Subject: Military Observers with Foreign Armies, 3 August 1914, AGO Files.
9. Memoranda for the CSO from the Aviation Section Chief, 23 November 1914, 2 January 1915, and 20 March 1915, OCSO Files.
10. Secretary of the War College Division to Squier, 29 August 1914, WCD Files. Squier's notification reads: "Your function still remains that of general observer to note and report mobilization plans, changes in general policy, movements of troops, political exigencies, methods of procurement of general supplies, the supplies themselves, tests and charges in arms and equipment, recruiting to fill vacancies, organization of new units, and other information that could not be obtained in the field."
11. Secretary of state to secretary of war, 25 August 1914, AGO Files.
12. Telegram, Squier to chief of staff, 14 September 1914, AGO Files.
13. Ibid.
14. Chief of staff to Squier, 15 September 1914, AGO Files.
15. Squier, "Personal and Unofficial War Diary of the Military Attaché, American Embassy, London, with British Army in the Field," Communiqué from Page to secretary of state, 6 January 1915, George O. Squier papers, 1914–1961, U.S. Army Heritage and Education Center, Carlisle, PA. (Hereinafter referred to as "War Diary.")
16. Squier to Chief/WCD, 20 November 1914, WCD Files.
17. Hendrick, *The Life and Letters of Walter H. Page*, III, 205.
18. Magnus, *Kitchener: Portrait of an Imperialist*, 375.
19. Samuel Bemis, *A Diplomatic History of the United States* (New York: Henry Holt, 1953), 590 (footnote).
20. Charles C. E. Callwell, *Field-Marshall Sir Henry Wilson: His Life and Diaries*, 2 vols. (New York: Charles Scribner's Sons, 1927), I, 196 (footnote). Callwell served as Kitchener's Director of Military Operations.
21. Vagts, *The Military Attaché*, 266. Vagts reported that attaches of neutrals were escorted over the Maine battlefields of 1914, where they listened to lectures on the highlights of action by members of the French General Staff. Thus

the neutrals received guided tours in the absence of hostile action, which made for better propaganda than military history (271).
22. F. W. Honeycutt, "Report of Trip to British Headquarters in the Field," 3 December 1914, 3, WCD Files. (Hereinafter referred to as "Report of Trip to British Headquarters.")
23. Of course, there may have been others for whom no record survives. A great purge of attaché reports occurred in the years between world wars, so any record of attaché activity is, of necessity, partial. From memoranda and reports to the War College Division it appears that Squier's visit was indeed the only trip of an attaché permitted by the British prior to American entry into the war. On the eastern front, an American attaché succeeded in reaching the combat zones while acting in a fictional capacity. His circumstances were similar to Squier's in that the Russians wanted to exclude all attachés of neutral nations from the front, but allowed the American attaché to proceed as a personal favor. The attaché, 1st Lieutenant Sherman Miles, son of former Major General Commanding Nelson A. Miles, tells in his own words how he succeeded in joining Russian troops in the field.

[Unable to obtain permission in St. Petersburg] I left Petrograd for Warsaw on January 3d. In Warsaw I got permission from the Generalissimo through the intercession of a friend at headquarters to remain. I am now under the following status—the Generalissimo has given instructions that I am to be shown what Red Cross organizations in the group of armies about Warsaw I want to see. Under this authority, as long as it lasts, I can see everything I want to see.

This status is given me by favor, and its fictitious form is due to the desire of the Russians to keep all other neutral attachés from the front. In consideration of this, I request that my presence at the Russian front be kept confidential, and that anything I may be able to send in during the continuation of the war be kept strictly within official hands.
Miles succeeded in remaining throughout the entire year of 1915. Unfortunately, his surviving reports reflect superficial analysis and insubstantial data. Report, Miles to Chief/WCD, 23 January 1915, WCD Files.
24. Squier, "War Diary," 6.
25. Honeycutt, "Report of Trip to British Headquarters," 3.
26. Hendrick, *The Life and Letters of Walter H. Page*, III, 206.
27. Squier, "War Diary," 3.
28. Ibid., p. 55.

29. David L. Woods, *A History of Communications Techniques* (Orlando: Martin-Marietta, 1965), 220.
30. Purtee, *History of the Army Air Service*, 9.
31. Ibid., 75.
32. Ibid., 93.
33. His official report survived the purge of attaché reports mentioned earlier. However, the numerous attachments, including manuals, pamphlets, orders, illustrations, maps, and interview notes, apparently fell victim. Observer forms were completed on every combat and service arm, in many cases by the British officers themselves, as a favor to Squier. General Wilson, subchief to Sir Archibald Murray, chief of staff to 4th Division, looked through the War College forms and laughingly told Squier that General Macomb [Chief of the War College division] "evidently expects you to be a poobah and I advise you to send him a full set of our service manuals to begin with." Squier, "War Diary," 59.
34. Ibid., 105.
35. Ibid., 77.
36. Ibid., 36.
37. Honeycutt, "Report of Trip to British Headquarters," 24. The French first used motor transport to increase mobility of their forces in the first Battle of the Maine in 1914, when they put Paris taxicabs into service at the front.
38. Squier, "War Diary," 88.
39. Ibid., 29.
40. Ibid., 28-29.
41. Ibid., 29.
42. Rowland also conducted a study on the alleged German use of dum-dum bullets. He deprecated the idea that either side had used them, a practice strictly forbidden by the Hague Convention. Squier remarked: "But why split hairs about alleged violations of the Hague Convention, when I saw the Indian Corps making up the most infernal machines for tearing great chunks out of the men, by putting bolt nuts, jagged pieces of rusty iron and small stones into gun cotton bombs, made of jam tins and shrapnel cases, to be hurled by hand into the enemy's trenches a few yards away." Ibid., 30.
43. Ibid., 110.
44. Telegram, Page to SecState, 6 January 1915, Monmouth Papers.
45. Viscount Grey, *Twenty Five Years*, II, 110, cited in Ray Baker, *Woodrow Wilson, Life and Letters*, 8 vols. (Garden City, NY: Doubleday, Doran & Co., 1935), V, 202. (Hereinafter referred to as *Wilson*.)
46. Hermann Hagedorn, *Leonard Wood: A Biography* (New York: Harper & Bros., 1931), 150; and Baker, *Wilson*, 263.

47. Charles Seymour, *The Intimate Papers of Colonel House*, 4 vols. (Boston: Houghton-Mifflin, 1926), I, 388.
48. Hendrick, *Life and Letters of Walter H. Page*, III, 170.
49. Ibid., 175.
50. Fitness Report, 1913, Squier's ACP.
51. Squier, "War Diary," Staff Duties Section, 6, WCD Files.
52. Ibid., 8.
53. Ibid. A commander could cover 150 miles of his sector in an automobile without seriously affecting his stamina. On horseback, a 30-mile trip would reduce his efficiency significantly for several hours after arrival at his destination.
54. *The Navy and Army Illustrated* stated contemporaneously with Squier's comments: "The principal use and value of the airplane is in the work of scouting, observing, and reporting. Bomb dropping is a sphere of aeroplane work which has little importance except where it is used for the destruction of airships." Quoted in Woods, *Communications Techniques*, The Chief Signal Officer, George P. Scriven, said about bombing in his 1914 annual report: "These experiments were interrupted and had to be indefinitely postponed when the detachment was sent to Galveston." *CSO Report*, 1914, 24. Bombing received no mention in the CSO Report for the following year.
55. Squier, Reports of Military Attache on British army in the Field, 7, WCD Files.
56. Ibid.
57. Squier to Walcott, 15 November 1915, Smithsonian Papers.
58. Walcott to Squier, 7 December 1915, Michigan Papers.
59. Lloyd Morris and Kendall Smith, *Ceiling Unlimited: The Story of American Aviation from Kitty Hawk to Supersonics* (New York: Macmillan, 1953), 149–154.
60. Memoranda, Reber to Scriven, 23 November 1914, 28 January 1915, and 20 March 1915, OCSO Files.
61. Memoranda, Chief of Staff Scott to AGO, 27 November 1914, and 2 April 1915, AGO Files.
62. Charge Slip, 12–15 March 1915, Miscellany, Chief of Staff, Record Group 165, National Archives, Washington, D.C. (Hereinafter referred to as Miscellany, C/S Files.)
63. C/WCD to Squier, 31 October 1914, WCD Files.
64. Squier to C/WCD, 11 January 1916, WCD Files.
65. Squier to C/WCD, 2 November 1915, WCD Files.
66. Transmittal File No. 7242, WCD Files; Squier to C/WCD, 3 June 1914, WCD Files; Squier to C/WCD, 19 January 1916, WCD Files.
67. Squier to C/WCD, 8 December 1914, OCSO Files.
68. Memorandum, Major Wm. S. Graves, Secretary to the General Staff, undated, Squier's ACP.
69. Memorandum, Brigadier General H. L. Scott, Chief of Staff, to AGO, 2 December 1914, Squier's ACP.
70. Telegram, Italian Minister of Foreign Affairs to U.S. Secretary of State, 31 May 1915, AGO Files.
71. Memorandum, C/WCD to C/S, 19 February 1916, WCD Files.
72. Secretary of War Newton D. Baker to Defense Reports Committee, 27 September 1916, answering a protest telegram, AGO Files.
73. H. J. Creedy to Squier, 5 October 1915, WCD Files.
74. Squier to C/WCD, 2 November 1915, WCD Files.
75. Squier to C/WCD, 3 March 1916, AGO Files.
76. Squier to W. M. Wright, Adjutant General, 20 March 1917, Squier's ACP.
77. Hendrick, *Life and Letters of Walter H. Page*, III, 207.
78. See No. 23.
79. Macomb to Squier, 10 April 1916, Michigan Papers.

Chapter 11

1. The Aviation Section superseded the Aeronautical Division in 1914.
2. Squier to William Eccles, 14 April 1916, Michigan Papers.
3. Statement of 1st Lt. H. A. Dargue before a Special Committee of the General Staff, 11 May 1916, AGO Files.
4. Ibid.
5. Manuscript Scriven to AGO, 3d Endorsement, February 1916, OCSO Files.
6. Colonel Goodier doubtlessly acted with the concern of a father who must have also known of the Calliaferro incident of the previous year. Loss of confidence in the quality of aircraft manufacture markedly increased when investigators found defective ribs and spars in the wings of the airplane that had crashed, killing Lieutenant Calliaferro.
7. Mitchell was on the General Staff at the time. Scriven asked for him "in consideration of the shortage of experienced officers and of the

great amount of work which is falling upon this office due to present conditions" until the arrival of Squier. Scriven to AGO, 31 March 1916, AGO Files.

8. Benjamin D. Foulois and Carrol V. Glines, *From the Wright Brothers to the Astronauts: The Memoirs of Major General Benjamin D. Foulois* (New York: McGraw-Hill, 1968), 126–128. (Hereinafter referred to as *Memoirs*.)

9. "Above the Ground and on It: Our American Ace Extraordinary," *Air Travel* (May 1916): 353.

10. Joseph O. Maugorgne, Jr., private interview, Atlanta, Georgia, February 1971.

11. Efficiency Report, 16 May 1916–31 December 1916, Squier's ACP.

12. Memorandum, Chief of WCD to Chief of Staff, 21 March 1916, WCD Files.

13. Scott to AGO, 28 October 1916, AGO Files.

14. OCSO, War Department, *Equipment for Aero Units of the Aviation Section Signal Corps* (Washington, D.C.: Government Printing Office, 1916), 7.

15. Memorandum, Brigadier General Charles Treat, Acting Chief of the War College Division, 17 November 1916, AGO Files.

16. Ibid.

17. C/Ordnance to AGO, 8 February 1917, AGO Files.

18. Public Law #271, 3 March 1915, 63d Congress.

19. Dupree, *Science in the Federal Government*, p. 291.

20. U.S. Bureau of Aircraft Production, *History of the Bureau of Aircraft Production*, 8 vols. (Dayton, OH: Wright Patterson Air Force Base, 1951), III, 557–60. (Hereinafter referred to as *BAP*.)

21. Dupree, *Science in the Federal Government*, p. 286.

22. Ibid.

23. Memorandum, Chief of Staff H. L. Scott to Secretary of War, 26 February 1915, WCD Files.

24. Scriven to Advisory Committee on Aeronautics, 16 April 1915, OCSO Files.

25. Hennessey, *Army Air Arm*, p. 128.

26. *BAP III*, 557–560.

27. The officer in charge of the Aviation Section, Samuel Reber, replied that experiments were conducted at the Signal Corps School at San Diego, but admitted that "the laboratory equipment is insufficient owing to lack of suitable appropriation." (Reber to Secretary, NACA, 22 September 1915, OCSO Files.) The experiments to which Reber referred derived in large part, if not altogether, from a request made by Scriven to Walcott in late 1913. Scriven asked for a member of the Langley Aerodynamical Laboratory to visit San Diego and give him a report on the general subject of the safety of the various types of machines being used by the Signal Corps. (Scriven to Walcott, 6 February 1913, OCSO Files.) Walcott assigned Alfred Zahm to the task. Zahm reported later that the machines appeared to be structurally sound, but stresses in vital parts should be determined while in actual operation in order to calculate a factor of safety. It also appeared that air frames lacked uniform strength. He investigated the effect gyroscopic torque of the propeller had on pilots of single-engine craft. He found it too small to endanger a trained aviator. He made recommendations on the correction of the combined tendency of the center of pressure to move rearward and the action of propeller thrust upon the center of gravity which created the tendency toward unexpected dives. He noted that this tendency created "pilot apprehension." (Zahm to Scriven, 23 January 1914, OCSO Files.) Zahm soon inaugurated a series of conferences on "aeromechanics" and airplane design at San Diego. Upon hearing of the lectures, Scriven wired school officials instructions to "furnish him [Zahm] every facility you can and see that every officer attends each daily lecture." (Telegram, Scriven to San Diego School, 4 December 1913, OCSO Files.) Zahm explained to his audience the results obtained by leading experimentalists in aeronautical laboratories and factories, obviously drawing upon his summer experiences in Europe (mentioned in Chapter Eight).

28. *BAP* III, pp. 561–562. The subcommittees formed were: (1) Relation of the atmosphere to aeronautics; (2) Standardization and investigation of materials; (3) Aeronautical nomenclature; (4) Radiator design; (5) Motive Power; (6) Specifications for aeronautic instruments; (7) Design, construction, and navigation of aircraft; (8) Site for experimental field; (9) Government relations; and (10) Bibliography of Aeronautics.

29. Walcott to Squier, 16 March 1916, Michigan Papers.

30. Dupree, *Science in the Federal Government*, p. 307.

31. Ibid., p. 306.

32. Squier to Walcott, 17 April 1916, Smithsonian Papers.

33. Newton D. Baker to Squier, 26 May 1916, Michigan Papers, Presidential Commission, dated 31 May 1916, Squier's ACP.

34. Memorandum, Subcommittee on Applied Aerodynamics and Aeronautical Standards, undated, 1916, Smithsonian Papers.

35. Proceedings, Public Session of Executive Committee of National Advisory Committee for Aeronautics, held at Smithsonian Institution, 8 June 1916, p. 1, National Advisory Committee for Aeronautics, Record Group 255, National Archives, Washington, D.C. (Hereinafter referred to as NACA Files.)
36. Ibid., 42.
37. Ibid., 9.
38. Squier's interest was realized in development of the famous Liberty engine. Squier "Liberty Aircraft Engine," *Transactions. American Institute of Electrical Engineers*, XXXVIII (February 1919): 104–111; Squier, "Liberty Engine One of Main Wartime Achievements," *U.S. Air Services*, 16 March 1931, 38; Squier, *Aeronautics in the United States at the Signing of the Armistice*, November 11, 1918, an address before the American Institute of Electrical Engineers (New York: AIEE, 1919), 54–61.
39. *BAP* III, 563.
40. Rexmond C. Cochrane, *Measures for Progress: A History of the National Bureau of Standards*, 159–89.
41. John F. Victory claimed that the terms *airplane* and *Liberty engine* were due to Squier's influence. Squier believed *aeroplane* to be poor English and persuaded the rest of the nomenclature subcommittee to recommend *airplane* to American lexicographers. Many people referred to the Liberty engines as motors. After some discussion, Squier turned to an automobile manufacturer familiar with shipping rates and asked him which class—motor or engine—traveled more cheaply. When he was told that the answer was engines, Squier recommended that, henceforth, the official designation should be Liberty engine.
42. The formation of the Subcommittee on Power Plants "to take up the development of motive power for aircraft," followed an Executive Committee session and several meetings with engine manufacturers in late May and early June 1916. The Committee consisted of S. W. Stratton, Chairman; Captain Mark L. Bristol, U.S.N.; and Squier. Walcott to J. Daniels, 10 June 1916, Smithsonian Papers.
43. *New York Times*, October 6, 1916, 5.
44. Squier, "Scientific Research for National Defense as Illustrated by the Problems of Aeronautics," *Journal of the Franklin Institute*, CLXXXIII (January 1917), 40. The same article appeared in *Nature* XCVIII (February 1917), 440ff and *Proceedings, National Academy of Sciences* II (December, 1916): 740ff. *Aviation and Aeronautical Engineering* I (December 1916), 287ff.
45. Ibid., 38.

46. Henry Woodhouse to Walcott, 13 January 1917, Smithsonian Papers.
47. The articles penned by Henry Woodhouse attacked Billy Mitchell, whom Squier publicly defended. An investigation was ordered by Baker. After Mitchell was cleared, Alan R. Hawley of the Aero Club of America denounced the entire Aviation Section in a series of articles in *Aerial Age*, a publication owned by Woodhouse. The Secretary of War answered with a curt reference to a press bulletin on the affair. (Philip J. Roosevelt, Military Editor of *Aviation*, to Secretary Baker, 16 October 1916, and Joseph P. Tracy, on behalf of Baker, to Roosevelt, October 1916, AGO Files.) When Benjamin Foulois came to Washington to discuss the First Aero Squadron, he and Squier held their meetings in the Army-Navy Club downtown, because Squier "did not trust [his assistant] Mitchell and did not want him to know what we were discussing." (Foulois, *Memoirs*, 138.) Foulois had doubts as to Mitchell's "knowledge, experience, and accomplishments." When he finally came to Washington as Squier's deputy—Mitchell failed to retain this post—Foulois said he found Mitchell uncommonly good at putting words on paper, but the quality of his actions were deficient. Mitchell's "failure to anticipate and plan for the purchase of land for airfields in the best flying areas in the United States was enough to convince Squier that Mitchell lacked the necessary training and experience to develop the basic precepts, principles, and policies required to meet all of our military air war plans." (Foulois, *Memoirs*, 140.) Squier described Mitchell in his Efficiency Report for the year as "specially fitted for field service with mounted troops." (Efficiency Report of William Mitchell, 3 April–31 December 1916, AGO Files.) In January 1917, Squier recommended Mitchell's assignment to France to conduct a study of aircraft development, employment and personnel training. He was given $40,000 to purchase French aircraft. (Squier to AGO, January 1917, and Secretary of State to Secretary of War, 28 February 1917, AGO Files.)
48. Walcott to Michael Pupin, 26 January 1917, Smithsonian Papers.
49. The establishment of an aviation experimental field is discussed in Chapter 11.
50. There is a fine account of the Wright-Curtiss feud in C. R. Roseberry, *Glenn Curtiss: Pioneer of Flight* (Garden City, NY: Doubleday, 1972), 332–63.
51. Howard Mingos, *The Birth of an Industry* (New York: W. B. Conkey, 1930), 12–13. Mingos was engaged in the early activities of the Manufacturers Aircraft Association and the Aeronautical Chamber of Commerce.

52. Secretary of the Navy Josephus Daniels wrote a letter to the NACA in which he described the situation: "Various combinations are threatening all other airplane and seaplane companies with suits for infringement of patents. It is difficult to get orders filled because some companies will not expend any more money on their plans for fear that suits brought against them will force them out of business. To protect themselves, in case they are forced to pay large license fees, the companies have greatly increased the sale prices of their products. As the Army and Navy are the principal purchasers of aircraft in this country they are bearing the brunt of this levy." Secretary of War Newton D. Baker seconded his fellow secretary and added his own wish that "a just and equitable solution to all concerned may be reached, which will apply not only to this department but to all other departments of the government purchasing airplanes." Quoted in Mingos, *The Birth of an Industry*, 14–15.

53. "Minutes of Subcommittee on Governmental Relations' Meeting, 24 February 1917, Smithsonian Papers. The organization mentioned in these discussions should not be confused with the Aircraft Manufacturers Association, formed on 13 February 1917. This organization made no attempt to solve the patent crisis until asked to intervene after America entered the war. Their efforts were directed toward improvements of the airplane and education of the public about the future of airplanes. (Manufacturers Aircraft Association, Inc., *Aircraft Year Book*, 1919 [New York: Manufacturers Aircraft Association, Inc., 1919], 32.) Confusion in names between the Aircraft Manufacturers Association and its successor, the Manufacturers Aircraft Association, has led several authors in the literature to ascribe confused founding dates to the same organization. Some indeed have failed to distinguish between the two organizations, treating them as one and the same.

54. Minutes of Subcommittee on Patents, NACA, 22 March 1917, Smithsonian Papers.

55. *Aircraft Year Book*, 1919, 37; Roseberry, *Glenn Curtiss*, 361.

56. National Advisory Committee for Aeronautics, *Annual Report*, 1917 (Washington, D.C.: Government Printing Office, 1918), 18. (Hereinafter referred to as *NACA Report*, 1917.)

57. *Aircraft Year Book*, 1919, 33.

58. Minutes of Subcommittee on Patents, NACA, 24 April 1917, Smithsonian Papers.

59. Foulois and Glines, *Memoirs*, 143.

60. U.S. War Department, *Annual Reports*, Report of the Chief Signal Officer (Washington, D.C.: Government Printing Office, 1918), 1074. (Hereinafter referred to as CSO Report, 1918.)

61. Foulois and Glines, *Memoirs*, 143. Mitchell, in Europe, appears to have helped draft the telegram.

62. Manufacturers Aircraft Association, *Aircraft Year Book*, 1919, 35.

63. *NACA Report*, 1917, 19.

64. Charter Members were: Aeromarine Plane and Motor Company, The Burgess Company, Curtiss Aeroplane and Motor Corporation, L. W. F. Engineering Company, Standard Aircraft Corporation, Sturtevant Aeroplane Company, Thomas-Morse Aircraft Corporation, and Wright-Martin Aircraft Corporation. A few months later the Dayton Wright Airplane Company, Fisher Body Corporation, Wright-Martin Aircraft Company of California (formerly the Glenn L. Martin Company) became members. *Aircraft Year Book, 1919*, 39. This achievement of bringing peace to the aircraft industry and of providing for orderly, cooperative use of patents belies the gloomy assessment of progress in the aviation industry during World War I, as rendered by John B. Rae in his *Climb to Greatness: The American Aircraft Industry, 1920–1960* (Cambridge: Massachusetts Institute of Technology Press, 1968), 1–2.

65. W. Durand to Walcott, 9 August 1917, Smithsonian Papers. A copy of the cross-license agreement may be found in the same location.

66. Mingos, *The Birth of an Industry*, 28.

67. *BAP* I, 46.

68. Mingos, *The Birth of an Industry*, 29.

69. Minutes of the Executive Committee of the NACA, 2 October 1917, quoted in *BAP* I, 52.

70. Grover Loening, *Take Off into Greatness* (New York: Putnam's Sons, 1968), 100–03.

71. Robert A. Millikan, *The Autobiography of Robert A. Millikan* (London: MacDonald, Ltd., 1951), 166. (Hereinafter referred to as *Autobiography*.)

72. It is not our intention to explore the "aircraft production scandal." It needs examination, but not in this work, which tries to develop other aspects of Squier's career in the Army.

73. Foulois and Glines, *Memoirs*, 145–147.

74. House Military Affairs Committee Report cited in Ibid., 147.

75. "The General Staff, and bureau officers, at Washington at this time are as a whole very poor—with some marked exceptions like Generals Squier and Crawder and Col. Van Deman." Theodore Roosevelt to Archibald Roosevelt, *The Letters of Theodore Roosevelt*, 8 volumes, edited by Elting E. Morison (Cambridge: Harvard University Press, 1954), VIII, 1241.

76. Isaac F. Marcosson, *Colonel Deeds: Industrial Builder* (New York: Dodd, Mead, 1947), 219–220. Deeds esteemed Squier highly, considering "no man in the Army possessed such a many-sided equipment. Scholar and scientist as well as professional soldier, he had a great gift of brilliant analysis. Short, dapper, keen-eyed, he talked and walked decisively. He was what people called 'a born leader.' Men followed him implicitly because of his passionate sincerity ... for years prior to our entry into the war he had lived and breathed aviation. He chafed at Congressional disregard of his persistent pleas for larger appropriations for the aircraft which he regarded as vitally essential, if we were to play our full part in the conflict." Clearly, this is a more sympathetic evaluation of Squier's role than most. It is important, though, because of the close working relationship that existed between Squier and Deeds. To many other Army officers, Squier appeared aloof, if not secretive. Deeds knew him as few others did. One senses that Squier was more at ease with civilians. In this crucial period the only support he could count on came from civilians, in government science and industry. This represented another difficulty in institutionalizing science in the Army. Support for programs developed in a scientific-industrial environment necessarily depends heavily on support without the Army's assistance.

77. Millikan, *Autobiography*, 166.

78. Theodore MacFarlane Knappen, *Wings of War* (New York: Knickerbocker Press, 1920), 25.

79. U.S. War Department, *Annual Reports*, Report of the Chief Signal Officer (Washington, D.C.: Government Printing Office, 1919) 892. (Hereinafter referred to as *CSO Report*, 1919.)

80. Ibid., 886.

81. Historical Section, The Army War College, *The Signal Corps and Air Service*, Monograph No. 16 (Washington, D.C.: Government Printing Office, 1922), 4. Squier turned over 16,084 officers, 147,932 enlisted men and 9,838 civilian employees to the newly created Division of Military and Bureau of Aircraft Production just 14 months later.

82. Army Times, eds., *A History of the U.S. Signal Corps* (New York: G. P. Putnam's Sons, 1961), 99. Except Mitchell in Europe, of course, who paid $1,500 for private flying lessons in order to qualify as a military aviator. The logic for his career of being the only field grade officer with wings was, according to John F. Victory, pointed out to Mitchell by Capt. Thomas DeWitt Milling. Victory Interview, *op. cit.*

83. *CSO Report*, 1918, 1073–74.

84. This account of Squier's rebuff came from John F. Victory, the first civilian employee of NACA, who was hired in 1915. He stated that Squier told this story to Walcott in his presence. Victory Interview, *op.cit.*

85. Ibid. What became of the arrangement is unknown.

86. Millikan, *Autobiography*, 167.

87. Benedict Crowell, Acting Secretary of War, to H. Snowden Marshall, 15 March 1918, Benedict Crowell Papers, Case Western Reserve University, Cleveland, OH. (Hereinafter referred to as Crowell Papers.)

88. The H. Snowden Marshall Report on Matters to be Investigated by Board Appointed to Investigate Air Service and Signal Corps in respect to arrangements for the Production of Airplanes for the United States Army, Crowell Papers, 5.

89. Ibid., 10.

90. Ibid., 17.

91. Ibid., 2.

92. Walcott to the President, 15 April 1918, Smithsonian Papers. We are indebted to Daniel Kevles, University of California at Berkeley, for sharing this letter with us.

93. *New York Times*, 25 April 1918, 1, and Executive Order #2863-A, Smithsonian Papers.

94. *New York Times*, 26 April 1918, 14.

95. J. S., Congress, House, Select Committee, *Expenditures in the War Department, Hearings*, before House Subcommittee No. 1 (Aviation), House of Representatives, on H.R. 637, Serial 7652, 66th Cong., 2d Sess., 1920, 2666.

96. Memorandum, Secretary of War to Army Chief of Staff, 17 May 1918, and Chief of Staff to Squier, 2 May 1918, Miscellany, C/S Files. A subsequent investigation, conducted by Chief Justice Hughes, found no "imputation of any kind upon Gen. Squier's loyalty or integrity." Therefore, Squier was informed, there was no necessity for a court of inquiry. (Chief of Staff to Squier, 16 November 1918, Miscellany, C/S Files.)

97. Squier to Baker, 14 November 1918, Papers of Newton D. Baker, 1918, Manuscript Division, Library of Congress, Washington, D.C.

98. *New York Times*, 6 April 1918, 14.

99. *New York Times*, July 1, 1917, VI, 3.

100. The priorities, in order, were training airplanes, aviators, and combat airplanes.

101. *CSO Report*, 1919, 893.

102. Goldberg, *A History of the United States Air Force, 1907–1957*, 15.

103. Leonard P. Ayres, *The War with Germany: A Statistical Summary* (Washington, D.C.: Government Printing Office, 1919), 85. (Hereinafter referred to as *Statistical Summary*.)

104. Ibid., 87–89.

105. Squier, "Liberty Aircraft Engine," Transactions, American Institute of Electrical Engineers XXXVIII (February 1919): 104–107. Squier, "Liberty Engine One of America's Main Wartime Achievements," U.S. Air Services (March 1931): 38.
106. Ayres, Statistical Summary, 25.
107. Ibid., 86–87.
108. Charles J. Gross, "George Owen Squier and the Origins of American Military Aviation," The Journal of Military History (July 1990): 281–305.

Chapter 12

1. Squier, Aeronautics in the United States at the Signing of the Armistice, 11 November 1918, an address before the American Institute of Electrical Engineers (New York: AIEE, 1919), 62.
2. Thomas P. Hughes, *Elmer Sperry: Inventor and Engineer* (Baltimore: The Johns Hopkins Press, 1971), 262–63.
3. Bion J. Arnold, "The Secret Report on Automatic Carriers, Flying Bombs (F. B.), Aerial Torpedoes (A. T.)," 31 January 1919, 2. (Hereinafter referred to as Liberty Eagle Report.) The Liberty Eagle Report was classified Secret until 1958, when it was downgraded to Restricted. In about 1970 it was placed among open documents in OCSO Files.
4. CSO to C/S, 5 October 1918, in Exhibit E of Liberty Eagle Report.
5. Squier to Chairman, Aircraft Board, 26 November 1917, Exhibit B, Liberty Eagle Report.
6. Ibid.
7. Quoted in Liberty Eagle Report, 5.
8. H. H. Arnold, *Global Mission* (New York: Harper & Brothers, 1949), 76.
9. Liberty Eagle Report, 5.
10. T. A. Boyd, *Professional Amateur: The Biography of Charles Franklin Kettering* (New York: E. P. Dutton, 1957), 109.
11. Frank Harris to Squier, 22 October 1918, Exhibit Q, Liberty Eagle Report.
12. Secretary of War Newton D. Baker disclosed the development of the Liberty Eagle on 24 March 1919 at Fort Worth. In a speech favoring the League of Nations, Baker told his audience that what was "regarded as one of the most destructive weapons invented during the war," had been placed in the secret archives of the War Department. Baker and Kettering both hoped it would always remain there. See "Automatic Air Bomber," *Army-Navy Journal*, Serial No. 2909 (1919): 1314–1516.

13. Charles D. Walcott, ed., *Above the French Lines, Letters of Stuart Walcott, American Aviator: July 4, 1917 to December 8, 1917* (Princeton: Princeton University Press, 1918).
14. Esther C. Goddard and G. Edward Pendray, eds., *The Papers of Robert H. Goddard* (New York: McGraw-Hill, 1970), p. 210. (Hereinafter referred to as *Papers of Goddard*.)
15. Goddard to Col. L. A. Codd, 13 July 1938, Goddard and Pendray, *Papers of Goddard*, III, 1172.
16. Walcott to Squier, 30 April 1918, Ibid., I, 224.
17. H. W. Dorsey to Goddard, 18 May 1918, Ibid., I, 226.
18. Goddard to Walcott, 18 May 1918, Ibid.
19. Shinkle to Goddard, 27 May 1918, Ibid., I, 228.
20. Goddard to Walcott, 18 May 1918, I, 227.
21. Shinkle to Rockwood, 27 May 1918, Ibid., I, 228.
22. Memorandum, Saltzman to Acting Chief of Ordnance, 27 May 1918, Ibid.
23. R. H. Goddard to Smithsonian Institution, 29 May 1918, Ibid., I, 227.
24. Ibid.
25. Walcott to Squier, 31 May 1918, Ibid., I, 232.
26. Walcott to R. S. Woodward, President, Carnegie Institution of Washington, 1 June 1918, and Woodward to Walcott, 3 June 1918, Ibid., I, 233.
27. R. H. Goddard's diary, En Route to California, 5–10 June 1918, Ibid., I, 234.
28. Squier to C/Ord, 19 July 1918, Ibid., I, 249.
29. H. W. Dorsey to C. G. Abbot, 22 August 1918, Ibid., I, 277, and Telegram, Squier to Goddard at Pasadena, 27 August 1918, OCSO Files.
30. Squier to C/Ord, 22 August 1918, Goddard and Pendray, *Papers of Goddard*, I, 277.
31. R. H. Goddard's diary, 13, 14 September 1918, Ibid., I, 285.
32. Hale to General Clark C. Williams, Ibid.
33. Abbot to Squier, 18 October 1918, Ibid., I, 291.
34. R.H. Goddard's diary, 7 November 1918, Ibid., I, 299.
35. C. G. Abbot to R. H. Goddard, 26 March 1918, Ibid., I, 315.

Chapter 13

1. George P. Scriven, *The Service of Information*, Circular No. 8, Office of the Chief Sig-

nal Officer (Washington, D.C.: Government Printing Office, 1915), 13.

2. National Research Council, *Council of National Defense* (Washington, D.C.: Government Printing Office, 1917), 1.

3. CSO Report, 1919, 888.

4. In October 1916, Squier addressed the following letter to John J. Carty:

It is the desire of the Signal Corps to attract to its Officers Reserve Corps the very best engineering talent in the country interested in our line of work. The American Telephone and Telegraph Company, Western Electric Company, Western Union Telegraph Company, and Postal Telegraph Company being engaged in work almost identical with the Signal Corps, it is very desirable that all these organizations be drawn upon in order that their wonderful resources can be utilized if necessary by the Government in any great national crisis.

The remarkable thing about this letter and the arrangements it produced was its timing. There was more than enough work for one man in that position. Yet he here appeared to be doing Chief Signal Officer Scriven's job, too. Squier, *Telling the World* (Baltimore: Williams and Wilkins, 1933), 145.

5. J. J. Carty accepted a commission as major in the Signal Corps Officers' Reserve Corps and took his oath at Governor's Island New York on 30 January 1917. (Carty to AGO 30 January 1917, AGO Files.) Squier called him to active duty one week after the declaration of war. (Squier to Carty, 14 April 1917, AGO Files.)

6. CSO Report, 1919, 888.

7. Ibid., 1228.

8. Thomas Marley Camfield, *Psychologists at War: The History of American Psychology and the First World War*, unpublished Ph.D. dissertation, The University of Texas at Austin, August 1969, 137. (Hereinafter referred to as *American Psychology*.)

9. Squier, "Scientific Research for National Defense as Illustrated by the Problems of Aeronautics," *Journal of the Franklin Institute*, CLXXXIII (January 1917): 37.

10. Camfield, *American Psychology*, 137.

11. *The Signal Corps and the Air Service*, Monograph No. 16 (Washington, D.C.: Government Printing Office, 1920), 50.

12. Camfield, American *Psychology*, 137.

13. Robert M. Yerkes, "Robert Means Yerkes," Volume II of *A History of Psychology in Autobiography*, edited by Carl Murchison, 4 volumes (Worcester: Clark University Press, 1930), 277.

14. Millikan, *Autobiography*, 167.

15. Ibid., 165.

16. *BAP* II, 403.

17. Squier to George E. Hale, 2 July 1917, in National Academy of Sciences, *NAS Report*, 1917, 47.

18. Dupree, Science *in the Federal Government*, 314. The acquisition of NRC research resources was even greater than Dupree suggests. The *Annual Report for 1917* of the National Academy of Sciences remarked laconically that "the work of the committees of physics, mathematics, astronomy, optical glass, and navigation and nautical instruments is now concentrated in the division of physics and related sciences, which also comprises the work of the Science and Research Division of the United States Signal Corps." (p. 65). More indicative of the influence Squier had to exert on the NRC are the budget figures for the period. From all funding sources in the period 1 September 1916 to 1 March 1918, the Signal Corps allotted $110,000 of the NRC appropriation of $270,000. While such research expenditures seem modest by contemporary standards, control of 41 percent of the budget was significant. And Squier appeared to appreciate his influence in the way he phrased his letter to Hale, mentioned in endnote 17. It is an old military custom to recommend to superiors and to suggest to subordinates. His choice of "suggest" in reference to Millikan applying for a commission may indicate how Squier perceived his role.

19. *CSO Report*, 1919, 1105.

20. Quoted in *BAP* II, 404.

21. Ibid.

22. Millikan, *Autobiography*, 174.

23. *BAP* II, 411–412.

24. *CSO Report*, 1919, 1105–06.

25. Cooperating organizations, institutions and laboratories were: a) *Government Organizations*: Wilbur Wright and McCook Fields, Dayton, OH; Aberdeen Proving Ground; Sandy Hook Proving Ground; Medical Research Laboratory, Hazelhurst Field, Mineola, NY; U.S. Navy Yard, Washington, D.C.; Naval Air Station, Hampton Roads; Bureau of Standards; U.S. Weather Bureau; Bureau of Chemistry; Bureau of Mines; b) *Educational and Research Institutions*: Johns Hopkins University; Carnegie Institute, Pittsburgh, PA; University of Chicago; Columbia University; University of Illinois; University of Minnesota; University of Wisconsin; Department of Terrestrial Magnetism, Carnegie Institution of Washington; and c) *Commercial and Industrial Laboratories*: General Electric Company, Lynn, MA; General Electric Company, Nela Research Laboratory, Cleveland, OH; Western Electric Company; Burke

and James Company; Bristol Company, Waterbury, CT.
26. Memorandum to Adjutant General, December 1917, NBS Files.
27. Stratton to Squier, 20 December 1917, NBS Files.
28. Stratton to Millikan, 27 December 1917, NBS Files.
29. Illustrative of this point is a telegram from Deeds to H. E. Talbott, who received all confidential telegrams for the leadership of Dayton-Wright Airplane Company at Dayton, OH: "Gen. Squier went direct to Detroit. Will probably spend Fourth at his old home in Michigan. May be in Dayton Thursday or Friday arriving there from Detroit or from Champaign, Ill. Harold, Kettering, and Wright can take care of him. He will be interested in the Dayton-Wright factory and laboratory, Orville Wright laboratory, and especially Mr. Kettering's views on scientific subjects. In general he is highly technical." (Deeds to Talbott, 3 July 1917, in Hearings before the Subcommittee No. 1 [Aviation] of Select Committee on Expenditures in the War Department, 66th Cong., Ser. 9, vol. 3 at 3883, Exhibit J, [1920].) In 1917 Squier also served as an editor of the *Journal of the Franklin Institute*.
30. Squier had a penchant for composing creeds and tributes for various groups. When he was selected as a member of the Court of Honor in the Boy Scouts of America in 1917, he wrote a creed for them. After the war he wrote a tribute to the telephone girls in appreciation of their valuable switchboard services in France and the United States.
31. Paul W. Clark, "Millikan and the Rice-Webster Gun Controversy," a paper presented before the annual meeting of the Society for the History of Technology, December, 1971.
32. Statement made by L. W. Stotesbury of the Inspector General's Office Relative to Charges made by one E. L. Rice, 6 September 1918, Archives of the National Academy of Sciences and the National Research Council, Washington, D.C. (Hereinafter cited as NAS/NRC Archives.)
33. Memorandum for the Adjutant General, prepared by Major General Frank McIntyre. Subject: Investigation concerning the National Research Council charges by E. L. Rice, 25 October 1918, WCD Files.
34. Ibid.
35. Ibid.
36. *BAP* II, 383-84.
37. Memo, Baker to G. S. Scott, no date, AGO Files.
38. Memo, Squier to CSO, 18 July 1916, AGO Files.

39. John F. Victory, "Memorandum on the National Advisory Committee for Aeronautics and Its Relations with the War Department at Langley Field, Virginia," NACA Files. (Hereinafter referred to as Victory Report.)
40. *BAP* II, 384-85.
41. Mitchell, Acting OIC A/S, to Scriven, 30 August 1916, AGO Files, and SO #216, 15 September 1916, *BAP* II, 385.
42. *New York Times*, 4 September 1916, 4. The noted engineer Henry Souther headed the Inspection Corps.
43. Victory Report, Appendix A, 2.
44. Ibid., 7.
45. *BAP* II, 385, and Appendix A, 2, Victory Report. Members of the subcommittee were C. D. Walcott, C. F. Marvin, head of the Weather Bureau, and S.W. Stratton, head of the Bureau of Standards.
46. Victory Report, Appendix A, 3, 9 November 1916.
47. Ibid. Military members were Marshall, Clark, Milling, and Souther. Naval representatives were Hunsaker and Tower.
48. Ibid.
49. *BAP* II, 386.
50. Victory Report, Appendix A, 4, 23 November 1916.
51. *BAP* II, 388.
52. Ibid., II, 389, 1 December 1916.
53. Memo Joseph Kuhn to C/S, 5 May 1917. Some of the remaining $10,000 in the original appropriation was used to purchase adjacent tracts of land (AGO Files).
54. Victory Report, Appendix A, 5.
55. Ibid., 7.
56. The Navy failed to establish itself as the proving ground during the war. There was no record to account why, according to W. G. Child, Office of Naval Operations. After the war, when Squier no longer had responsibility for Army aviation, the War Department refused permission for the Navy to establish an experimental station at the proving ground. W. G. Child to Captain Craven, 17 June 1920, NACA Files.
57. Victory Report, 7.
58. The peculiar status of NACA as the proving ground was formalized in writing for the first time in April 1919, when Major General Charles T. Menoher, Director of Air Services, assigned Plot 16 to the NACA in writing. Upon approval of the Secretary of War, the assignment became official.
59. Memo, Tasker Bliss to chief of Staff, 4 January 1917, WCD Files.
60. AGO to CSO, 4 January 1917, 1st Ind., 5 January 1917, WCD Files.

61. Memo, Acting C/WCD, Colonel C. W. Kennedy, to C/S, 15 January 1917, WCD Files.
62. Radio telegraphy systems in use on aircraft permitted one-way communications, from air to ground. More of this development will be treated in the next chapter.
63. *New York* Times, 14 January 1917, Sec. I, 18.
64. Order, USAFA Papers.
65. *BAP* II, 390, and Interview with John F. Victory, *op. cit.*
66. Squier submitted the general plan of construction, prepared by Kahn, to the Adjutant General on 15 May 1917 for approval; Secretary Baker approved the plan a few days later. On 25 May 1917 the Aircraft Production Board, a creation of the war, recommended placement of the construction contract with the J. G. White Engineering Corporation (*BAP* II, 391).
67. Ibid., 21 May 1917.
68. Ibid., II, 393.
69. Office Memo No. 54, OCSO, 18 July 1917, cited in *BAP* II, 393.
70. Paragraph 1, War Department General Orders, no. 104, Series 1917 (Cited in *BAP* II, 394).
71. *CSO Report*, 1919, 889.
72. *Fort Monmouth History and Place Names, 1917–1961* (Fort Monmouth, NJ: Fort Monmouth Tradition Committee, 1961), 2. (Hereinafter referred to as *Fort Monmouth History*.)
73. General Order No. 1, Signal Corps Camp, Little Silver, NJ, 17 June 1917, quoted in *Fort Monmouth History*, 2.
74. The pigeon service was officially disbanded c. 1963. Their use continued, however, during the Vietnam conflict. They still represent the best of silent communications systems.
75. *CSO Report*, 1919, 1125.
76. Special Order No. 45, Camp Alfred Vail, 23 February 1918, cited in *Fort Monmouth History*, 8.
77. *Fort Monmouth History*, 77.
78. *A Concise History of Fort Monmouth, NJ and the U.S. Army Communications-Electronics Command (CECOM) Life Cycle Management Command* (Fall 2009): 80.
79. Fitness Report, 1923, Squier's Personal Folder, St. Louis Records Repository, St. Louis, MO. (Hereinafter referred to as St. Louis Papers.)

Chapter 14

1. The Appendix of this book provides the technical details necessary to understand the difference between spark and continuous wave technology, and why the latter was crucial to voice communications.
2. Mike Bullock and Laurence Lyons, *Missed Signals on the Western Front—How the Slow Adoption of Wireless Restricted British Strategy and Operations in World War I* (Jefferson, NC: McFarland, 2010).
3. Squier, *Aeronautics*, 62.
4. Historical Section, the Army War College, *The Signal Corps and Air Service: A Study of Their Expansion in the United States, 1917–1918* (Washington, D.C.: Government Printing Office, 1922), 99. (Handwritten dedication identifies author as Colonel Richard H. Fletcher, U.S. Army.) According to G. H. Clark, Harry Horton made the first communication from air to ground on 27 August 1910. George H. Clark's *Radioana*, Division of Electricity and Nuclear Energy, Smithsonian Institution, Washington, D.C. (Hereafter referred to as Clark's *Radioana*.)
5. Benedict Crowell, *America's Munitions, 1917–1918* (Washington, D.C.: Government Printing Office, 1919), 323.
6. Joseph O. Mauborgne, Jr., "My Early Work in Airplane Radio Development," Clark's *Radioana*, SRM 151 035. Clark reports that an Ensign Charles H. Maddox, U.S. Navy, first achieved two-way communication in the summer of 1912 (SRM 100 483). Mauborgne identified the pilot of the radio-equipped Army airplane as H. H. Arnold, who much later became Chief of Staff of the Air Force. The other pilot was Thomas DeW. Milling. Maddox's work was little appreciated by the Navy at the time. Not until 22 years later did he receive a letter of commendation from the Secretary of the Navy in recognition of his work. Howeth reports that little was achieved in the improvement of aircraft radio in the Navy until 1915. Pilots generally considered the additional weight detrimental to safety. They also considered its utilization "an undesirable personal burden." Howeth, *Communications Electronics*, 191.
7. Ibid.
8. Crowell, *America's Munitions*, 323, and Joseph C. Mauborgne, Jr., to George H. Clark, 29 January 1941, Clark's *Radioana*, SRM 151 033. In this letter, Mauborgne purports to set the record straight on his artillery spotting and Philippine exploits which, in his opinion, the Press Bureau of the Air Corps had confused.
9. Mauborgne,"My Early Work in Airplane Radio Development."
10. Lloyd Espenscheid, "The Origin and Development of Radiotelephony," *Proceedings of the Institute of Radio Engineers* XXV (September 1937): 1106–07.

11. The original Wright design, in which motors and propellers were mounted behind the pilot, proved fatal to many pilots when their airplanes crashed. The position of the pilot between ground and engine was irresistible. The tractor types of aircraft, pioneered by Glenn Martin, pulled themselves through the air by propellers and engines mounted on the front. The great rush of wind then created in the cockpit bathed the pilot in noise.

12. The major engineering achievement in military aeronautics at North Island, San Diego, in 1916 was the successful development and installation of airborne radio by Captain Charles C. Culver. Two types were under test. The first was a short-range battery-type radio, weighing about 40 pounds and intended for a 20-mile transmitting range. The second type, a long-range radio, worked from a generator mounted with a two-bladed fan and set in the air stream. Transmissions from Santa Monica to San Diego, some 119 miles away, were successfully received. Another set under development used a self-exciting generator from a 1912 Signal Corps pack mule set; it delivered 110 volts at a frequency of 500 Hz and 250 watts. At 45 pounds, it weighed 15 pounds less than the other long-range set and could transmit over greater distances. Royal D. Frey, *Evaluation of Maintenance Engineering, 1907–1920*, typescript, Serial No. 5-1877-1A (Maxwell AFB, Alabama, no date), 66.

13. The development of tank radios shared much in common with airplane radios. The late installation of radiotelephone equipment in tanks also pertained. In the spring of 1918, the Signal Corps sent several officers to British units to test American aircraft radiotelephones. Captain S. C McCutchen wrote that "so far the British have used no tank wireless." For a time they used reels and wire. When shell fire failed to sever the wire, noise in the tank made hearing impossible. He recommended communicating in tanks the same way the Signal Corps did in aircraft. (Report on Radio Communications in British Tank Corps by S. G. McCutchen, Capt., S. C., 7 March 1918, Records of the National Academy of Sciences, Record Group 189, National Archives, Washington, D.C. Hereinafter referred to as NAS Files.) The recommendation merely gave approval to actual practice since another officer, Captain H. W. Webb, S. C, was installing Signal Corps Aircraft Interphones in British tanks. The whole experience of the Signal Corps in integrating individual tanks into a solid mass of coordinated armor was seen as applicable to tank doctrine. (Tests of Interphones for Tanks by H. W. Webb, Capt., S. C, 28 February 1918, NAS Files).

14. The Signal Corps in France constructed over 1,340 miles of telephone and telegraph pole lines carrying more than 30,000 miles of wire. The network extended all over France, radiating from five seaports. Some 273 telephone central stations were installed, maintaining contact with 8,000 outlying stations. At the peak workload 43,845 telegraph messages were sent daily (on the average there were 60 words per message). In addition, 72,000 local- and 5,000 long-distance telephone calls were placed. In a speech to the U.S. Veteran Signal Association in 1919, Lt. Col. John C. Moore said that "our telephone system was so well [built] and operated that General Foch would travel several miles out of his way to use it in preference to using a French or British System." (Scrapbooks of U.S. Veteran Signal Assoc, OCSO Files.)

15. Nugent H. Slaughter, "Radio Development during the World War," *Electrical World*, 15 February 1919, 311.

16. More attention was paid to French sets than British sets since American military units operated in French sectors. This necessitated American radio equipment resembling French apparatus. *CSO Report*, 1919, 1132.

17. Slaughter, "Radio Development during the World War," 312.

18. George Clark took notes of E. R. Cram's account of the improvement over existing mica condensers made by the Signal Corps in 1915. Cram recounted Major Russell's determination to keep the Signal Corps "a closed corporation as to design and manufacture.... Russell [in charge of Signal Corps research at this time] was very much against letting any commercial concern build condensers. Cram made a tour of New York to find condenser manufacturers, in 1915 or so, and did find some, but the idea of getting these to design condensers for the Signal Corps was summarily vetoed by Major Russell." [George H. Clark to George Lewis, Vice-President, International Telephone Development Co., Inc., 29 September, 1941, *Clark's Radioana*, CWC 101 089.

19. Crowell, *America's Munitions*, 324. This part of Crowell's work was written by one of the participants, C. C. Culver.

20. Squier, *Aeronautics*, 51.

21. Nugent A. Slaughter, G. Francis Gray, and John W. Stokes, "Radio-Telephone Development in WWI," *Electrical World* LXXIV (16 August 1919): 341.

22. *CSO Report*, 1919, 1146.

23. Ralph Brown, "War-Time Development of Vacuum Tubes," *Electrical World* LXXIII (22 February 1919): 359.

24. This decision represented a departure

from the French and British practice of using one type for both receiving and transmitting functions (*CSO Report*, 1919, 1147).

25. Prior to the war, vacuum tubes were in limited use as "repeaters" or relay amplifiers. Until just prior to the outbreak of war all vacuum tubes for the military had been obtained through one manufacturer. The anticipated demand for tubes prompted Squier to begin developing other sources of procurement. A conference was held between engineers of the Signal Corps, the General Electric Company, and the Western Electric Company. By the end of the war the DeForest Radio Telephone and Telegraph Company also manufactured tubes for the Signal Corps (*CSO Report*, 1919, 1098–1147).

26. Mr. Heising worked on transmitting circuits, Mr. W. E. Booth on the mechanical design, Mr. H. M. Stohler on the wind-driven generator, Mr. Nicols and Mr. Van Der Bijl on vacuum tubes, Mr. Newton on the interphone transmitter and helmet, and Mr. Oswald on the antenna investigation (*CSO Report*, 1919, 1142).

27. David L. Woods, *A History of Tactical Communication Techniques* (Orlando, FL: Martin-Marietta Corporation, 1965), 229.

28. Crowell, *America's Munitions*, 325, and CSO Report, 1919, 1142.

29. Ibid., 327. In 1918 the demonstrations were repeated on the front lawn of the White House. President and Mrs. Wilson tried for themselves the mysterious telephone that linked their thoughts with those of fliers high overhead. (George Raynor Thompson, "How Development of Radiotelephone [Voice Radio] in World War I Promoted the Radio Broadcast Industry in the 1920's," a private manuscript. Thompson served for many years as the Chief Historian of the Signal Corps.)

30. Ibid.

31. *CSO Report*, 1919, 1150.

32. One of the most significant engineers to work at the Signal Corps laboratory in Paris was Edwin H. Armstrong, recruited from the Physics Department at Columbia. In June 1918 he proposed a new type of radio circuit which would amplify incoming signals independent of their frequency. Contemporary amplifiers could handle only with great difficulty radio frequencies of 1000 KHz. Above 3000 KHz it was practically impossible to find an amplifier capable of delivering enough of the incoming signal to the first stage due to very small capacitive effects in tubes and wiring. Since capacitive reactance becomes vanishingly small to high-frequency signals, much of the incoming signal was uselessly dissipated. Small changes in the physical orientation of the set also caused changes in tuning at these high frequencies. Armstrong's invention reduced the high frequency at the input to a lower one, which the amplifier could handle with ease. Because the lower frequency was always the same one by design, the operation of the amplifier was simplified. With the range of frequencies above 1 MHz now available to radiotelephony, it was possible to use small antennae in the aircraft. Since high-frequency signals are more directional in their transmission paths than low-frequency signals, the security of aircraft communications was improved. Armstrong's heterodyne circuit assumed fundamental importance in postwar radio communications. (Lawrence Lessing, *Man of High Fidelity: Edwin Howard Armstrong* [Philadelphia: J.B. Lippincott, 1956], 87–101.)

Another significant electrical circuit perfected at the Signal Corps laboratory in Paris was the Master Oscillator Power Amplifier (MOPA) circuit. This circuit freed the radio set from dependence upon interaction of antenna electrical parameters and the radio detection and amplifier circuits. A form of the MOPA circuit had been tried earlier, in the 1915 radio trials at Montauk Point, NY, and Arlington, VA. Its perfection at Paris had special significance in operating tank radios. Because tank radios operated in a frequency range requiring long wire antennas which trailed behind the tank, they frequently changed length due to combat action. The MOPA circuit made it possible for tanks to stay on frequency when antennae were cut by shell fire or other accidents of combat.

Helpfulness and cooperation characterized relations between the Paris Signal Corps laboratory, the French military telegraphic service, and the scientific departments of the Sorbonne. When the war ended, berths aboard troop ships were at a premium. So arrangements were made to send the signalmen assigned to the laboratory in Paris to the Sorbonne for advanced instruction in radio. General Ferrie of the Etablissement Central de Materiel de la Radiotelegraphic Militaire directed the arrangements himself. The Allied Radio Engineering Class attended courses in circuits, vacuum tubes, radio direction finding, etc., from March to July 1919. The 50 or so Americans posed for a picture near the Eiffel Tower upon graduation. Squier was in Paris at this time, too, and had his photograph taken with General Ferrie on the observation platform. The entry of these men trained in the science of radio in Europe and America sparked engineering advances for the next decade (Marshall, *U.S. Army Signal Corps*, 166). Fred Lack and William MacDonald of Hazeltine Corporation both agreed that for "some years after that war, the entire entertainment industry, radio, slid along on the equipment and the know-how of the military communications effort of World War I." (MacDonald and Lack Interview, *op. cit.*)

33. Crowell, *America's Munitions*, 327.
34. Ibid., 328.
35. Brown, "Vacuum Tubes," 360. The VT-1 departed from the usual tungsten filament and developed a coated filament which yielded longer working hours, in excess of 500 hours.
36. Later, the General Electric Company introduced the "type P" tube, which was capable of producing 100 watts of high-frequency power. Fifteen hundred volts were required with approximately 0.2 amperes of current on the filament. Additional experimental forms of the "type P" tubes successfully delivered a quarter kilowatt of high-frequency power.
37. Crowell, *America's Munitions*, 329.
38. Brown, "Vacuum Tubes," 361.
39. Slaughter, *et al*, "Radio-Telephone Development," 343.
40. *BAP* IV, 2241. The development of voice-commanded flying continued at Lake Charles until 6 August when a storm destroyed the station. In September, voice-commanded training resumed at several other flying fields. The use of radiotelephones in flying courses reduced training periods by two-thirds of the previous time. By exercising control of pilots in the air, accidents were virtually eliminated. In November and December a fleet of 204 airplanes was commanded by voice control in the skies over San Diego. Radiotelephone-trained pilots began arriving in France in October 1918. They had no opportunity to function in combat as voice-commanded squadrons, but they never operated in any other capacity in subsequent conflicts.
41. Espenscheid, "Radiotelephone," 1109. The French did use a telephone set in their captive balloons, which were used as aerial-observation posts. The telephone cable was attached to the tethering cable of the balloon. *Projet De Notice sur le Materiel et les liaisons Telephoniques des Compagnies D'Aerostation*, Report #48/1, 5 March 1917, 6 (French military manual in the Gimble Aeronautical Collection, USAF).
42. Interviews with William A. MacDonald and Fred R. Lack at the Hazeltine, Long Island (Little Neck) Corporation factory, 4 April and 20 May 1960. Conducted by George Raynor Thompson, former Chief Historian of the U.S. Army Signal Corps.
43. Arthur E. Kennelly, "Advances in Signaling Contributed during the War," in *The New World of Science*, edited by Robert M. Yerkes (New York: Books for Libraries Press, 1969), 237.

44. General Drum told the House Committee on Military Affairs that during the Meuse Argonne battle in 1918 Billy Mitchell had 150 airplanes high in the air, striking enemy airfields. Meanwhile, the Germans were fighting over American front lines bombing and strafing infantrymen. Statement of Brigadier General Hugh A. Drum, Assistant Chief of Staff, before the House Committee on Military Affairs with regard to HR 10147 ("To Create a Department of Aeronautics," February 1925, OCSO Files).
45. *CSO Report*, 1898, 876.
46. *Traditions of the Signal Corps*, unpublished draft manuscript (Fort Monmouth, NJ: Signal Corps Museum, April, 1959), 110.
47. Greely, *Reminiscences*, 185.
48. Ibid., 178.
49. Ibid., 182–85.
50. This event was, surprisingly, omitted by David Kahn in his comprehensive book *The Codebreakers*. The most sensational cryptologic achievement in the 1890s, according to Kahn, was decipherment of the Panizzardi telegram which indicated the innocence of the unfortunate Captain A. Dreyfus. David Kahn, *The Codebreakers* (New York: Macmillan Co., 1968), 254–62.
51. Greely, *Reminiscences*, 185.
52. Howeth, *Communications-Electronics*, 65.
53. Elting E. Morison, *Men, Machines, and Modern Times* (Cambridge: MIT Press, 1966), 98–122.
54. *The Instructions of Frederick the Great for his Generals, 1747*, translated by Thomas R. Phillips (Harrisburg, PA: Military Service Publishing Co., 1944), 342.
55. Ibid., 410, 429.
56. Karl von Clausewitz, *Principles of War* (Harrisburg, PA: Military Service Publishing Co., 1944), 342.
57. Helmuth K. B. von Moltke, *War Doctrines*, quoted in Samuel C. Meyer, *The Impact of Signal Communications Upon the Art of War, 1860-1942* (unpublished B.A. dissertation, Princeton University, 1943), 19. (Hereinafter referred to as *Signal Communications*.)
58. T. Bentley Mott, trans., *The Personal Memoirs of Joffre, Field Marshal of the French Army*, 2 vols. (New York: Harper & Brothers, 1932), I, 259.
59. Samuel C. Meyer, *Signal Communications*, 17. We are indebted to Mr. Meyer for the broad outline of the previous argument concerning 19th-century command and control conceptual development.
60. Ibid., 17.
61. Matloff, *American Military History*, 19.

62. Greely, *Reminiscences*, 185.
63. Ibid., 186. Present authors' own emphasis. Greely had only $800 when war was declared. The great electrical companies of the country—General Electric, Western Union Telegraph, and the Mexican Telegraph—furnished the means and men on credit until Congress belatedly appropriated Signal Corps war funds.
64. Ibid., 186. Greely said, "I decided to force electrical facilities upon the Army."
65. In a series of letters to the Secretary of War, General Russell A. Alger, Greely pleaded for an allotment of $40,000 just prior to the war to correct an alarming lack of fire-control systems in major U.S. harbors. Such major ports as Boston, New York, and San Francisco, and principal cities including Philadelphia and Washington were without electrical communication with their own outlying military forts. Emphasizing unity of control, Greely alone urged that electrical communications, allowing harbor defense to conduct operation in unison, immediately under the direction of the commanding officer. Greely to Secretary of War, 11 March, 22 March, and 24 March 1898, OCSO Files.
66. Albert C. Crehore and George Owen Squier, "A Transmitter Using the Sine Wave for Cable Telegraphy; and Measurements with Alternating Currents Upon an Atlantic Cable," *Transactions: American Institute of Electrical Engineers* XVII, No. 25 (May 1900): 399.
67. Squier, "The United States Pacific Cable," *Independent*, LII (February 1900): 362.
68. Squier, "Pacific Cable," *Journal U.S. Artillery* (March-April 1900): 172.
69. Squier, "The Influence of Submarine Cables upon Military and Naval Supremacy," *Proceedings of the U.S. Naval Institute* XXVI (1900): 235.
70. Squier, "Pacific Cable," *Journal U.S. Artillery* (March-April 1900): 172.
71. Squier, "Influence of Submarine Cables," *Scientific American* (April 20, 1901): 21156.
72. Ibid.
73. CSO Report, 1899, 742.
74. Report to Secretary of War, 1908, cited in Meyer, *Impact*, 17.
75. E. D. Peek, *The Necessity and Use of Electrical Communication on the Battle Field*, p. 129, cited in Meyer, 18.
76. War games probably began about 3,000 BC in China. The Chinese game Wei-Hai is still played today under the Japanese name *Go*. (Andrew Wilson, *The Bomb and the Computer: Wargaming from Ancient Chinese Mapboard to Atomic* Computer [New York: Delacorte Press, 1968], 1.) The modern war game originated in Prussia with Baron von Reisswitz, the King's War Counselor, i.e., Kriegs und Domanenrat, and his son, Lt. von Reisswitz, in 1811 [*Militar-Wochenblatt*, No. 402, 1824; Nos. 56 and 73, 1874. *Allgemeine Deutsche Biographie*. Leipzig, 1899. Vol. 28, 153–154. Cited in Alfred Hausrath, *Simulation and Security in War and Peace: The Role of War* Gaming (McLean, VA: Research Analysis Corporation, 1968), I, 1.3. An excellent guide to war-gaming literature from several countries may be found in the appendices.) Diffusion of the war game to other countries quickened after the 1870 Franco-Prussian War. Austrians, English, Italians, French, Japanese, Turks, and Russians soon adopted the war game as part of an officer's professional training. The first significant American work on war games appeared in 1883. W. R. Livermore, a major in the Corps of Engineers, called his game the American Kriegsspiel (John P. Young, *A Survey of Historical Developments in War Games* [Bethesda, MD: Operations Research Office, Johns Hopkins University, 1960], pp. 27–31). The American Kriegsspiel ran through several editions, the last appearing in 1898. When the Army Staff College began in 1904 at Fort Leavenworth, war gaming was in the curriculum. It became part of the regular course of instruction at the Army School of the Line in 1907. By 1911 war gaming was taught in post-graduate schools at all Army posts. The highly influential teacher of war games at Fort Leavenworth, Major Farrand Sayre, taught the subject during the years of Squier's assignment to the Signal School (Farrand Sayre, *Map Maneuvers and Tactical Rides* [Springfield, MA: Springfield Printing and Binding Company, 1911], vii]. Sayre likened tactical rides to tactical map maneuvers. They covered the same field of instruction and game rules but moved over the actual ground instead of the maneuver map (Sayre, *Map Maneuvers*, 160). Maneuvers employing only one force (one side maneuvers) could be conducted with ease, but two side maneuvers (two opposing forces) were difficult for one director to manage. Thus, Sayre advises, "When it is desired to conduct a two side maneuver as a tactical ride, the opposing forces should not be large; should be started in contact with each other and given missions which will ensure their remaining in contact." (Sayre, *Map Maneuvers*, 161.) Exercises were clearly conducted without regard for electrical communications. In one side exercises, Sayre makes frequent reference to verbal orders and meetings of exercise directors and commanders at such and such a place at a predetermined time (163). In two side exercises, he cautions that opposing sides must stay out of earshot, but no more than

50 or 60 yards apart (167). Sayre considered briefly large-scale exercises, termed two side staff rides, with exercise directors with each force. He rejected such exercises because of the expense of quartering and transporting officers. He disapproved of the long telegrams which must conclude the operations of each day. He also objected to the prolonged absence of senior officers from their offices (170).

77. It was widely held that Japan's success in 1904 derived in large measure from the training received in war games, according to E. A. Raymond and Harry W. Beer in "History of War Games," *Reserve Officer*, October and November 1938, respectively (Cited in Young, *War Games*, 30).

78. Memorandum, Squier to C/S, 12 October 1908, OCSO Files.

79. For years the British utilized their vast worldwide network of cables and telegraph lines to glean information about international and domestic business affairs, diplomatic instructions, and military status of virtually every nation in the world. The All-Red cable system provided the British with incomparable "listening posts" around the world. It's no wonder that an internationally minded United States should insist on American ownership and control of cables between the mainland and its colonies and possessions.

80. A fourth role of military electrical communications is an aggressive one. Intentional injection of false or misleading information into the enemy's communications system and use of radio for propaganda or morale impairment are but two examples of their combative use. This use of electrical communications has been attributed to Professor A. M. Low, *Modern Armaments* (London: John Gifford, Ltd., 1939), 226–29.

81. The evolution of the skill of returning information up the line for review and action actually took longer than the exercise of control down the chain of command. *The Battle of Jutland* (Philadelphia: DuFour Editions, 1964), by Geoffrey Bennett, illustrates concisely the problem of junior officers faced with a new kind of responsibility and failing to initiate adequate information for their superior officers.

82. Memorandum, Squier to C/S, 12 October 1908, OCSO Files.

83. Ibid.

84. Such pride seems justifiable in light of a British army decision two years later to remove electrical communications from the Royal Engineers and place the responsibility for them in a newly formed Signal Corps patterned after the U.S. Army Signal Corps.

85. Memorandum, Squier to C/S, 12 October 1908, OCSO Files.

Chapter 15

1. Secretary of War to Secretary of State, 31 July 1918, Miscellany, C/S Files.
2. Chronology, Squier's Personal Folder, St. Louis Papers.
3. *New York Times*, 23 August 1921, p. 21.
4. R. B. Owens, Secretary, Franklin Institute, to Squier, 8 April 1919, Michigan Papers.
5. Arthur E. Kennelly, "George Owen Squier," *National Academy of Science Biographical Memoirs* 20 (1934): 159.
6. H. L. Abbot to Squier, 20 October 1919, Michigan Papers.
7. Biographical Sketch, Accession 607, Fort Monmouth Papers, and correspondence in St. Louis Papers.
8. Ibid.
9. Newton D. Baker to D. W. Taylor, no date, OCSO Files. Baker sent Squier a copy of the correspondence.
10. Marvin to Squier, 12 February 1919, Michigan Papers.
11. Orville Wright to Squier, 8 February 1919, Michigan Papers.
12. Major Stephen Malmsley to Lt. Col. L. D. Gasser, Secretary to the General Staff, 11 May 1923, and Dwight F. Davis, Acting Secretary of War, to Squier, 15 May 1923, OCSO Files.
13. Squier to Hadfield, 8 March 1919, Michigan Papers.
14. S. T. Ansell to Army and Navy Patent Board, 30 November 1918, OCSO Files.
15. Vail to Burleson, 12 December 1918, Michigan Papers. The letter also appeared in the public press the following day; *New York Times*, 5.
16. Squier to Hicks, 23 January 1919, Michigan Papers.
17. Hicks to Squier, 26 March 1919, Michigan Papers.
18. Hicks to Squier, 5 October 1919, Michigan Papers.
19. Hicks to Squier, 5 October 1919, Michigan Papers.
20. Squier to Hicks, 17 October 1919, Michigan Papers.
21. Squier to Hicks, 29 October 1919, Michigan Papers.
22. *Official Opinions of the Attorneys General of the United States*, volume 32, p. 145, available at https://archive.org/details/officialopinion01statgoog.

23. Memorandum of Agreement, November 1921, Michigan Papers.
24. *New York Times*, 16 March 1922, 15.
25. Circular published by F. B. Keech and Co., 16 February 1923, OCSO Files.
26. *New York Times*, 4 September 1924, 9.
27. *New York Times*, 1 October 1924, 36.
28. Squier v. American Telephone and Telegraph Co., 7F (2d) 831.
29. Ibid.
30. *New York Times*, 25 December 1919, 1.
31. *New York Times*, 5 July 1928, 18.
32. *New York Times*, 31 May 1925, Sect. 9, 12.
33. HR, 6103.
34. *New York Times*, 25 June 1928, 12.
35. *New York Times*, 5 July 1928, 18.
36. M. Schwartz, op. cit.
37. "A Country Club for Country People," *Munsey's Magazine* LXX, No. 3, 456. *Munsey's Magazine*, credited with being the first mass-marketed American magazine, was published from 1889 to 1929. The article on Squier's club appeared in the August 1920 issue, and highlighted Squier's generosity to his hometown.
38. "Pershing Bust Delivered to U.S.," *New York World-Telegram*, 18 August 1934. Moses Dykaar, the prominent American sculptor who created the bust, was also swindled by Layton and received little of the $5,000 he was promised. Already insolvent because of the Depression, Dykaar committed suicide.

Appendix

1. This could be assumed because ferromagnetic materials were omitted in the transmitter circuit between polarizer and analyzer. Since light was required to travel only a meter's length, a time delay of about 3.33 nanoseconds was introduced, a length of time smaller by a thousand times than the smallest time interval measured with the polarizing photochronograph.
2. Crehore, *Autobiography*, 60–61.
3. J. Mills, *Radio Communication Theory and Methods* (New York: McGraw-Hill, 1917) 63–65.
4. S. Hong, *From Marconi's Black-Box to the Audion* (Cambridge: MIT Press, 2001) 155–56.

Bibliography

Books

Abbe, Truman. *Professor Abbe and the Isobars.* New York: Vantage Press, 1955.

Aitken, H. G. J. *The Continuous Wave.* Princeton, NJ: Princeton University Press, 1985.

_____. *Syntony and Spark: The Origins of Radio.* New York: Wiley, 1970.

Ambrose, Stephen E. *Duty, Honor, Country: A History of West Point.* Baltimore: Johns Hopkins Press, 1966.

Andrews, Lincoln C. *Fundamentals of Military Service.* Philadelphia: J.B. Lippincott, 1916.

Archer, Gleason L. *History of Radio to 1926.* New York: American Historical Society, 1938.

Army Times, ed. *A History of the U.S. Signal Corps.* New York: Putnam, 1961.

Arnold, H.H. *Global Mission.* New York: Harper & Bros., 1949.

Ashburn, P.M. *A History of the Medical Department of the United States Army.* Boston: Houghton-Mifflin, 1929.

Ayres, Leonard P. *The War with Germany: A Statistical Summary,* 2d ed. Washington, DC: Government Printing Office, 1919.

Baker, Ray Standard. *Woodrow Wilson, Life and Letters.* Vol. 5. Garden City, NY: Doubleday, 1935.

Barry-Orth, Cathy D. *Squier-Atwell Family Tree—The Descendants of Samuel Squire,* available at URL http://familytreemaker.genealogy.com/users/b/a/r/Cathy-Diane-Barryorth/BOOK-0001/0000-0001.html.

Beaver, Daniel R. *Newton D. Baker and the American War Effort 1917–1919.* Lincoln: University of Nebraska Press, 1966.

Bemis, Samuel F. *A Diplomatic History of the United States.* New York: Henry Holt, 1953.

Bennett, Geoffrey Martin. *The Battle of Jutland.* Philadelphia: DuFour Editions, 1964.

Blake, G. G. *History of Radio Telegraphy & Telephony.* London: Chapman & Hall, 1928.

Bose, Sir Jagadis Chunder. *Plant Autographs and Their Revelations.* New York: Macmillan, 1927.

Boyd, T. A. *Professional Amateur: The Biography of Charles Franklin Kettering.* New York: E. P. Dutton, 1957.

Bright, Charles. *Imperial Telegraphic Communication.* Westminster: D. S. King & Son, 1911.

_____. *Submarine Telegraphs, Their History, Construction and Working.* London: Crosby Lockwood & Son, 1898.

_____. *Telegraphy Aeronautics and War.* London: Constable & Co., 1918.

Brockett, Paul. *Bibliography of Aeronautics.* Washington, DC: Smithsonian Institution, 1997.

Bruce, Robert V. *Lincoln and the Tools of War.* Indianapolis: Bobbs-Merrill, 1956.

Bullock, Michael J., and Laurence A. Lyons. *Missed Signals on the Western Front—How the Slow Adoption of Wireless Restricted British Strategy and Operations in World War I.* Jefferson, NC: McFarland, 2010.

Burns, R.W. *Communications: An International History of the Formative Years.* London: Institution of Electrical Engineers History of Technology, Series No. 32, 2003.

Caidin, Martin. *Air Force.* New York: Bramwell House, 1957.

Calhoun, Daniel H. *The American Civil Engineer.* Cambridge: Massachusetts Institute of Technology Press, 1960.

Callwell, Charles E. *Field Marshal Sir Henry Wilson: His Life and Diaries.* New York: Scribner's Sons, 1927.

Chandler, Charles, and Frank Lahm. *How Our Army Grew Wings.* New York: Ronald Press, 1943.
"Chronographs." *Encyclopædia Britannica.* 11th ed., Vol. VI.
Clark, George H. *The Life of John Stone.* San Diego: Frye & Smith, 1946.
Clausewitz, Karl von. *Principles of War.* Harrisburg, PA: Military Publishing Service, 1942.
Cochrane, Rexmond C. *Measures for Progress: A History of the National Bureau of Standards.* Washington, DC: U.S. Department of Commerce, 1966.
Crehore, Albert, and Frederick Bedell. *Alternating Currents.* New York: W. J. Johnston, 1895.
Crehore, Albert, and G. O. Squier. *The Polarizing Photo-Chronograph.* New York: John Wiley & Sons, 1897.
Crehore, Albert Cushing. *Autobiography.* Gates Mills, OH: Wm. G. Berner, 1944.
Croon, E. David, ed. *The Cabinet Diaries of Josephus Daniels, 1913–1921.* Lincoln: University of Nebraska Press, 1963.
Crowell, Benedict. *America's Munitions, 1917–1918.* Washington, DC: Government Printing Office, 1919.
Daniels, Josephus. *The Wilson Era, Years of Peace, 1910–1917.* Chapel Hill: University of North Carolina Press, 1946.
Davis, Burke. *The Billy Mitchell Affair.* New York: Random House, 1967.
deForest, Lee. *Father of Radio.* Chicago: Wilcox & Follett, 1950.
Development of Air Doctrine in the Army Air Arm, 1917–1941, The. Washington DC: Extension Course Institute, Air University, 1955.
Dieckmann, Max. *Leitfaden der Drahtlosen Telegraphie für die Luftfahrt.* Munchen u. Berlin: Druck u. Verlag R. Oldenbourg, 1913.
Donaldson, Frances. *The Marconi Scandal.* New York: Harcourt & Brace, 1962.
Doughty, R. A. *Pyrrhic Victory: French Strategy and Operations in the Great War.* Cambridge: Harvard University Press, 2005.
Dunlap, O. E. *Radio's 100 Men of Science.* New York: Harper, 1944.
Dupree, A. Hunter. *Science in the Federal Government.* Cambridge: Harvard University Press, 1957.
Dupuy, R. Ernest. *Men of West Point.* New York: William Sloan Associates, 1951.
Eccles, W. H. *Continuous Wave Wireless Telegraphy.* London: Wireless Press, 1921.

_____. *Electrical Instruments and Telephones of the U.S. Signal Corps* (Revised, 1910). Washington, DC: Government Printing Office, 1911.
_____. *Wireless.* London: T. Butterworth, 1933.
Emme, Eugene M. *A History of Space Flight.* New York: Holt, Rinehart and Winston, 1965.
"Eugene Griffen." *Dictionary of American Biography.* Vol. VII, 1928.
Fahie, J. J. *A History of Wireless Telegraphy (1838–1899).* New York: Dodd, Mead, 1900.
Fletcher, Robert H. *Fort Monmouth History and Place Names.* Fort Monmouth, NJ, 1959.
_____. *The Signal Corps and the Air Service.* Washington, DC: Government Printing Office, 1922.
Foulois, Benjamin D., with C. V. Glines. *From the Wright Brothers to the Astronauts: Memoirs of Major General Benjamin D. Foulois.* New York: McGraw-Hill, 1968.
Freebody, J. W. *Telegraphy.* London: Sir Isaac Pitman & Sons, 1958.
Freudenthal, Elsbeth E. *Flight into History.* Norman: University of Oklahoma Press, 1949.
Ganoe, William Addleman. *The History of the United States Army.* New York: Appleton & Co., 1924.
Gardiner, Martin. *Fads and Fallacies in the Name of Science.* New York: Dover, 1952.
Garratt, Gerald R. M. *One Hundred Years of Submarine Cables.* London: H. M. Stationery Off., 1950.
Genesis of Military Air Power in the United States, The (Serial no. 4096). Alabama: Maxwell Air Force Base, n.d.
"George Owen Squier." *National Cyclopaelia of American Biography.* Vol. XXIV, 1935.
Glines, Carroll V. *The Compact History of the United States Air Force.* New York: Hawthorn Books, 1965.
Goddard, Esther C., and Edward G. Pendray, eds. *The Papers of Robert H. Goddard.* New York: McGraw-Hill, 1961.
Goddard, Robert H. *Autobiography of Robert Hutchings Goddard.* Worcester, MA: Achille J. St. Onge, 1966.
Goetzmann, William H. *Exploration and Empire.* New York: Alfred Knopf, 1966.
Goldberg, Alfred. *A History of the United States Air Force, 1907–1957.* Princeton, NJ: Van Nostrand, 1957.
Goldsmith, Alfred N. *Introduction to Line Radio Communication* (Radio Communica-

Bibliography 263

tion Pamphlet, No. 41). Washington, DC: Government Printing Office, 1923.

———. *Radio Telegraphy*. New York: The Wireless Press, 1918.

Gorrell, E. S. *The Measure of America's World War Aeronautical Effort*. Northfield, VT: Norwich University Press, 1940.

Greely, Adolphus W. *Reminiscences of Adventure and Service*. New York: Scribner's Sons, 1927.

Griffith, P. *British Battle Tactics on the Western Front 1916–1918*. New Haven CT: Yale University Press, 1994.

Hagedorn, Hermann. *Leonard Wood: A Biography*. New York: Harper & Bros., 1931.

Haigh, K. R. *Cableships and Submarine Cables*. London: Adlord Coles, 1968.

Hartcup, G. *The War of Invention: Scientific Developments 1914–1918*. London: Brassey's, 1988.

Hausmann, Erich, ed. *Radio Phone Receiving for Everyone*. New York: Van Nostrand, 1922.

Hausrath, Alfred H. *Simulation and Security in War and Peace*. McLean, VA: Research Analysis Corporation, 1968.

Hendrick, Burton J. *The Life and Letters of Walter H. Page*. Vol. I. Garden City, NY: Doubleday, 1922.

Hennessey, Juliette A. Research Studies Institute, Air University, *The United States Army Air Arm, April 1868 to April 1917*, USAF Historical Studies: No. 98. Mobile, AL: Gunther Air Force Base, 1958.

Hittle, J. D. *The Military Staff: Its History and Development*. Harrisburg, PA: Military Service Publishing, 1949.

Holley, Irving B. *Ideas & Weapons*. New Haven: Yale Press, 1953.

Hong, S. *From Marconi's Black-Box to the Audion*. Cambridge, MA: MIT Press, 2001.

Howeth, L. S. *History of Communications-Electronics in the United States Navy*. Washington, DC: Government Printing Office, 1963.

Hughes, Thomas P. *Elmer Sperry: Inventor and Engineer*. Baltimore: The Johns Hopkins Press, 1971.

Hume, Edgar Erskine. *Victories of Army Medicine: Scientific Accomplishments of the Medical Department of the United States Army*. Philadelphia: J. B. Lippincott, 1943.

Hurley, Alfred F. *Billy Mitchell*. New York: Franklin Watts, 1964.

Huston, James A. *The Sinews of War: Army Logistics, 1775–1953*. Washington, DC: Government Printing Office, 1966.

Jessup, Philip C. *Elihu Root*. New York: Dodd, Mead, 1938.

Josephy, Alvin, Jr., ed. *The American Heritage History of Flight*. New York: American Heritage Publishing, 1962.

Kahn, David. *The Codebreakers*. New York: Macmillan, 1967.

Kelly, Fred C. *The Wright Brothers*. New York: Harcourt, Brace, 1943.

Kennelly, A. E. "George Owen Squier." *National Academy of Sciences Biographical Memoirs*. Vol. XX, 1934.

Knappen, Theodore Macfarlane. *Wings of War*. New York: Knickerbocker Press, Putnam's Sons, 1920.

Landrum, Charles M. *Michigan in the World War*. Ann Arbor: Michigan Historical Commission, 1924.

Lehman, Milton. *This High Man: The Life of Robert H. Goddard*. New York: Farrar, Straus, 1963.

Lessing, Lawrence. *Man of High Fidelity: Edwin Howard Armstrong*. Philadelphia: J. B. Lippincott, 1956.

Loening, Grover. *Our Wings Grew Faster*. New York: Doubleday Doran, 1935.

———. *Takeoff Into Greatness*. New York: G. P. Putnam's Sons, 1968.

Low, A. M. *Modern Armaments*. London: John Gifford, 1939.

MacClosky, Monro. *The United States Air Force*. New York: Praeger, 1967.

MacFarland, Marvin, ed. *The Papers of Wilbur and Orville Wright, Vol. II, 1906–48*. New York: McGraw-Hill, 1953.

MacLaurin, W. Rupert. *Invention & Innovation in the Radio Industry*. New York: Macmillan, 1949.

Magnus, Philip. *Kitchener: Portrait of an Imperialist*. London: Butler & Tanner, 1958.

Manufacturers Aircraft Association. *Aircraft Yearbook, 1919*. New York: Manufacturers Aircraft Association, 1919.

———. *Aircraft Yearbook, 1920*. Garden City, New York: Doubleday, 1920.

Marconi, Degna. *My Father, Marconi*. New York: McGraw-Hill, 1962.

Marcosson, Isaac F. *Colonel Deeds: Industrial Builder*. New York: Dodd, Mead, 1947.

Marshall, Max L. *The Story of the U.S. Army Signal Corps*. New York: Franklin Watts, 1965.

Matloff, Maurice, ed. *American Military History*. Army Historical Series. Washington, DC: Office of the Chief of Military History, United States Army, 1969.

May, Ernest R. *The World War and American*

Isolation, 1914–1917. Cambridge: Harvard University Press, 1959.

Miller, Francis. *World in the Air.* Vols. 1 & 2. New York: Putnam, 1930.

Mills, J. *Radio Communication Theory and Methods.* New York: McGraw-Hill, 1917.

Millikan, Robert A. *Autobiography of Robert A. Millikan.* London: MacDonald, 1951.

Mingos, Howard. *The Birth of an Industry.* New York: W. B. Conkey, 1930.

Mixter, G. W., and H. H. Emmons. *United States Army Aircraft Production Facts.* Washington, DC: Government Printing Office, 1919.

Morison, Elting E., ed. *The Letters of Theodore Roosevelt.* Vols. 1–8. Cambridge: Harvard University Press, 1954.

———. *Men, Machines, and Modern Times.* Cambridge: The MIT Press, 1966.

Morris, Lloyd, and Kendall Smith. *Ceiling Unlimited: The Story of American Aviation from Kitty Hawk to Supersonics.* New York: Macmillan, 1953.

Mott, T. Bentley, trans. *The Personal Memoirs of Joffre, Field Marshal of the French Army.* New York: Harper, 1932.

Murchison, Carl, gen. ed. *A History of Psychology in Autobiography.* Worcester, MA: Clark University Press, 1930.

Palmer, Frederick. *Newton D. Baker: America at War.* Vols. 1–2. New York: Dodd, Mead, 1931.

Pershing, John J. *My Experiences in the World War.* Vols. 1–2. New York: Frederick Stokes, 1931.

Phillips, Thomas R., trans. *The Instructions of Frederick the Great for His Generals, 1747.* Harrisburg, PA: Military Service Publishing, 1944.

———. *Roots of Strategy.* Harrisburg, PA: Military Service Publishing, 1940.

Projet de Notice sur le Materiel et les Liaisons Téléphoniques des Compagnies d'Aerostiers. Report #48/I. 5 March 1917.

Pupin, Michael. *From Immigrant to Inventor.* New York: Scribner's Sons, 1957.

Raines, Rebecca. *Getting the Message Through.* Washington, DC: Office of the Chief of Military History, U.S. Army, 1996.

Rhodes, Frederick L. *John J. Carty: An Appreciation.* New York, 1932.

Risch, Erna. *Quartermaster Support of the Army: A History of the Corps, 1775–1939.* Washington, DC: Office of the Quartermaster General, 1962.

Roseberry, C. R. *Glenn Curtiss: Pioneer of Flight.* Garden City, NY: Doubleday, 1972.

Rowland, Henry. *The Physical Papers of Henry A. Rowland.* Baltimore: The Johns Hopkins Press, 1901.

Sayre, Farrand. *Map Maneuvers and Tactical Rides.* Springfield, MA: Springfield Printing and Binding, 1911.

Scott, J. B. *The Hague Conventions and Declarations of 1899 and 1907.* New York: Oxford University Press, 1915.

Scriven, George P. *The Service of Information.* Circular #8, Office of the Chief Signal Officer. Washington, DC: Government Printing Office, 1915.

Seymour, Charles. *The Intimate Papers of Colonel House.* Boston: Houghton-Mifflin, 1926.

Sheffield, G., and D. Todman. *Command & Control on the Western Front: The British Army's Experience 1914–1918*. Stroud, UK: The History Press, 2007.

Singer, Charles, et al., eds. *A History of Technology.* Vol. V. Oxford: Clarendon Press, 1967.

Squier, George O. *Aeronautics in the United States at the Signing of the Armistice, November 11, 1918.* An address before the American Institute of Electrical Engineers. New York: American Institute of Electrical Engineers, 1919.

———. *Elementary Principles of the Alternating Current.* Compiled as a guide to the lectures upon this subject to the class of student officers, U.S. Artillery School. Fort Monroe, VA: Artillery School Press, 1898.

———. *A Few Facts of My Own History*, by 2nd year Cadet George Owen Squier, West Point, available at the URL http://familytreemaker.genealogy.com/users/b/a/r/Cathy-Diane-Barryorth/BOOK-0001/0099-0001.html.

———. *Field Equipment of Signal Troops.* Fort Leavenworth, KS: Staff College Press, 1907.

———. *Multiplex Telephony and Telegraphy by Means of Electric Waves Guided by Wires.* Professional Paper of the Signal Corps, U.S. Army. Washington, DC: Government Printing Office, 1911.

———. *Telling the World.* Baltimore: Williams & Wilkins, 1933.

———, ed. *The Easy Course in Home Radio.* New York: M. H. Ray & the Review of Reviews, 1922.

Squier, George O., and Albert C. *Waves of Pressure in the Atmosphere Recorded by an Interferometer Barograph.* Washington, DC: Government Printing Office, 1911.

Stockbridge, Frank Tarker. *Yankee Ingenuity in the War.* New York: Harper, 1920.

Sweetser, Arthur. *The American Air Service.* New York: D. Appleton, 1919.
Thomson, Joseph John. *Notes on Recent Researches in Electricity and Magnetism.* Oxford: Clarendon Press, 1893.
Tillman, Stephen. *Man Unafraid.* Washington, DC: Army Times Publishing, 1958.
Toulmin, H. A. *Air Service, American Expeditionary Force, 1918.* New York: Van Nostrand, 1927.
Towers, Walter Kellogg. *From Beacon Fire to Radio.* New York: Harper, 1924.
Tribolet, Leslie B. *International Aspects of Electrical Communications in the Pacific Area.* Johns Hopkins University Studies in Historical and Political Science. Cambridge: Johns Hopkins Press, 1929.
Vagts, Alfred. *The Military Attaché.* Princeton: Princeton University Press, 1967.
Walcott, Charles D., ed. *Above the French Lines: Letters of Stuart Walcott, American Aviator.* Princeton, NJ: Princeton University Press, 1918.
Waterman, Talbot H., and Harold J. Morowitz. *Theoretical and Mathematical Biology.* New York: Blaisdell, 1965.
Weevers, Theodore. *Fifty Years of Plant Physiology.* Amsterdam: N. V. Van de Garde, 1949.
Weigley, Russell F. *History of the United States Army.* New York: Macmillan, 1967.
Whitnah, Donald L. *A History of the United States Weather Bureau.* Urbana: University of Illinois Press, 1961.
Wilson, Andrew. *The Bomb and the Computer: Wargaming from Ancient Chinese Mapboard to Atomic Computer.* New York: Delacourt Press, 1968.
Wood, Leonard. *Facts of Interest Concerning the Military Resources and Policy of the United States.* Washington, DC: Government Printing Office, 1914.
_____. *Leonard Wood on National Issues.* Garden City, NY: Doubleday, 1920.
_____. *Our Military History.* Chicago: Reilly & Britton, 1916.
_____. *Universal Military Training.* Collier Classics, Vol. 12. New York: P. F. Collier & Son, 1917.
Woodhouse, Henry. *Textbook of Military Aeronautics.* New York: Century, 1918.
Woods, David L. *A History of Tactical Communication Techniques.* Orlando, FL: Orlando Division, Martin Co., Martin-Marietta, 1965.
Yerkes, Robert M., ed. *The New World of Science.* New York: Books for Libraries Press, 1969.
Yoakum, Clarence S., and Robert M. Yerkes. *Army Mental Tests.* New York: Henry Holt, 1931.
Young, John P. *A Survey of Historical Developments in War Games.* Bethesda, MD: Operations Research Office, The Johns Hopkins University, 1960.

Periodicals

"Above the Ground and on It: Our American Ace Extraordinary." *Air Travel,* Vol. 1, No. 8 (May 1916): 353, 383.
"An American Pacific Cable (Editorial)." *Electrical Review and Western Engineer* XXXVI (January 1900): 33.
"Automatic Air Bomber." *Army-Navy Journal,* Serial No. 2909 (1919): 1314–1516.
"Brig. Gen. George O. Squier Is Promoted." *Air Service Journal* (October 14, 1917): 438.
Brittain, James E. "The Introduction of the Loading Coil: George A. Campbell and Michael I. Pupin." *Technology and Culture* XI (January 1970): 36–57.
Brown, Ralph. "Wartime Development of Vacuum Tubes." *Electrical World* LXXIII (February 1919): 358–63.
Chanute, Octave. "Recent Aeronautical Progress in the United States." *The Aeronautical Journal* (London) XII (July 1908): 52–55.
Clark, Paul Wilson. "Early Impacts of Communications on Military Doctrine." *Proceedings of the IEEE* (September 1976): 1407–13.
Claudy, C. H. "Telephone to Europe? What General Squier Has Done by Way of Utilizing Bare Wire Even in Water." *Scientific American* (May 8, 1920): 513–15.
"Col. Squier Arrives." *Aerial Age* (May 22, 1916): 297.
Colpitts, E. H., and O. B. Blackwell. "Carrier Current Telephone and Telegraphy." *Electrical Engineering* XX (April 1921): 301–15.
_____. "Telephonie et Telegraphie Multiplex." *Annales des Postes Telegraphes et Telephones* X, No. 3 (1921): 415–54.
"A Country Club for Country People." *Munsey's Magazine,* August 1920.
Coursey, D. R. "Wired Wireless Transmission." *The Wireless World* VIII, 21, 22 (1921): 699–701, 731–35.
Crehore, Albert C. "A Reliable Method of Recording Variable Current Curves." *Transactions, American Institute of Electrical Engineers* XI (October 1894): 507–22.

Crehore, Albert C., and George O. Squier. "An Alternating Current Range and Position Finder." *Journal of the United States Artillery* VII (January-February, 1897): 42–61.

———. "Discussion of the Currents in the Branches of a Wheat-stone's Bridge, Where Each Branch Contains Resistance and Inductance, and There is an Harmonic Impressed Electromotive Force." *The London, Edinburgh and Dublin Philosophical Magazine and Journal of Science* (5th Series) XLIII, No. 262 (1897): 161–72.

———. "Experimental Determination of the Motion of Projectiles Inside the Bore of a Gun with the Polarizing Photo-Chronograph." *Journal of the United States Artillery* V (May-June, 1896): 325–52.

———. "Experiments with a New Polarizing Photo-Chronograph, Applied to the Measurement of the Velocity of Projectiles." *Journal of the United States Artillery* IV (July 1895): 409–52.

———. "The New Polarizing Photo-Chronograph at the U.S. Artillery School, Fort Monroe, Va., and Some Experiments with It." *Journal of the United States Artillery* I (November-December, 1896): 271–316.

———. "Note On a Photographic Method of Determining the Complete Motion of a Gun During Recoil." *Journal of the United States Artillery*, Vol. IV (July 1895): 470–76.

———. "A Practical Transmitter Using the Sine Wave for Cable Telegraphy, and Measurements with Alternating Currents Upon an Atlantic Cable." *Electrical Review* XXXVI (June 1900): 632–36.

———. "A Practical Transmitter Using the Sine Wave for Cable Telegraphy, and Measurements with Alternating Currents Upon an Atlantic Cable." *Transactions of the American Institute of Electrical Engineers* XVII (May 1900): 385–443.

———. "The Synchronograph." *Journal of the United States Artillery* VIII (July–August 1897): 19–50.

———. "The Synchronograph." *Transactions, American Institute of Electrical Engineers* XIV (April 1897): 93–124.

———. "Tests of the Synchronograph on the Telegraph Lines of the British Government—A Report to the Postmaster-General of the United States." *Journal of the Franklin Institute* CXLV, No. 2 (1898): 161–62.

Culver, Charles A. "Guided Wave Telegraphy." *Journal of the Franklin Institute* CXCI (March 1921): 301–28.

Dougherty, Emmet. "Army's Greatest Inventor." *Popular Mechanics*, September 1927.

Duncan, R. D., Jr. "Broadcasting by Wired Radio." *Electrical World* LXXXIV (July 1924): 157–60.

Dunlap, Knight. "Psychological Observations and Methods." *Journal of the American Medical Association* LXXI (October 26, 1918): 1392–93.

———. "Psychological Research in Aviation." *Science*, New Series XLIX, No. 1256 (1919), 94–97.

"Duplex Telephony Announced as Practical." *Telephony* (January 7, 1911): 21.

"Editorial on Squier's Promotion of Aeronautics." *Engineering News* XXXVI (July 1908): 126.

"Ernest R. Cram, Obituary." *Proceedings, Institute of Radio Engineers* XXXIX (November 1951): 1469.

Espenschied, Lloyd. "The Origin and Development of Radio Telephony." *Proceedings, Institute of Radio Engineers* XXV (September 1937): 1101–23.

"Experiences con el neuve fotochronografo polarizador de los Drs. Crehore y Squier." *Memorial de Artilleria*, 4th Series, VII (February 1897): 165.

Fuller, Lt. A.C. "The Ordinary Versus the Quenched Spark in Wireless Telegraphy." *The Royal Engineers Journal* XVII (January 1913): 7–10.

———. "Wireless Telegraphy." *The Royal Engineers Journal* XI (February 1910): 95–106.

———. "Wireless Telegraphy. Some Up-to-Date American Methods of Reception." *The Royal Engineers Journal* XII (August 1910): 145–48.

"Gen. Squier Talks of Aerial Program." *Air Service Journal* I, No. 5 (1917): 146.

"Gen. Squier Tells of Air Achievements." *Air Service Journal* IV, No. 3 (1919): 3, 12.

"Great Names in the Signal Corps: George Owen Squier." *Technical and Tactical Training Aid of the U.S. Signal Corps* (September 1959): 33–39.

Gross, Charles J. "George Owen Squier and the Origins of American Military Aviation," *The Journal of Military History* (July 1990): 281–305.

Harris, William, Jr. "Giving the Public a Light Socket Broadcasting Service." *Radio Broadcast* (October 1923): 465–70.

"An Invaluable Contribution to Popular Knowledge." *Flight* I (February 1909): 112.

"Is Wireless the Future of Broadcasting?" *Radio Broadcast* (October 1923): 457.

Kennelly, A. E. "George O. Squier." (Obituary). *Science*, New Series, LXXIX (May 1934): 470-71.
Kevles, Daniel J. "Testing the Army's Intelligence: Psychologists and the Military in World War I." *Journal of American History* LV (December 1968): 565-81.
"Kleine Mittheilungun: Nordamerika." *Militär-Wochenblatt* 29 (April 1, 1899): 793-94.
Lahm, Frank P. "The Wright Brothers as I Knew Them." *Sperry Scope* (April 1939): 1-5.
LeMay, Curtis E. "U.S. Air Force: Power for Peace." *National Geographic*, September, 1965.
"Lieut. Col. George O. Squier." *Aerial Age Weekly*, January 29, 1917.
"Lieutenant Colonel George O. Squier." *Aviation and Aeronautical Engineering* I (September 1916): 114.
"Lieut. Col. George O. Squier: The Man in Charge of the Aeronautical Branch of Our Land Defenses." *Flying*, September 5, 1916.
MacFarlane, Peter Clark. "Putting a Fleet in the Air; Our Biggest Task and the Man on the Job." *Collier's Weekly*, June 2, 1917.
Mallet, E. "Wired Wireless." *Wireless World* X (May 1922): 169-73.
Mauborgne, J. O. "High Frequency Current on Wires." *Physical Review*, 2nd Series, XIV, No. 5 (1919): 452-56.
_____. "High Frequency Signals Over Telegraph Wires." *Telegraph and Telephone Age* (August 1, 1919): 364-65.
"More Than a Billion for Aviation; Chief Signal Officer Squier Asks for $1,032,260 for Aviation Program." *Automotive Industries* (December 6, 1917): 1018, 1026.
"Multiplex Telegraphy and Telephony with High Frequency Currents Along Wires." *Revue Generale de l'Electricite* IX (February 19, 1921): 590.
"Multiplex Telephony in Canada." *Telegraph and Telephone Age*, August 1, 1919.
Nyquist, H. "Certain Factors Affecting Telegraph Speed." *Bell System Technical Journal* 3 (April 1924): 324-46.
O'Meara, Major W. A. "The Practical Problem of Telegraph Transmission from a New Angle of View." *The Royal Engineers Journal* XXII, No. 4 (1915): 183-92.
_____. "Recent Developments in Telegraphy and Telephony." *The Royal Engineers Journal* XIII (May 1911): 341-58.
_____. "Various Systems of Multiplex Telegraphy." *The Royal Engineers Journal* XIV (December 1911): 353-70.

Page, Frank C. "Lt. Col. G. O. Squier, USA, Inventor." *World's Work*, June 1916.
"Pershing Bust Delivered to U.S.," *New York World-Telegram*, August 18, 1934.
"Personalities in the Wireless World: Lt. Col. George Owen Squier." *The Wireless World* (London), New Series, II (December 1914): 545.
"Plans for Aviation Personnel." *Aerial Age Weekly*, May 14, 1917.
Pomey, J. B. "Submarine, Telegraphic Transmissions." *L'Electricien* LII (January 15, and May 1, 1921): 31-33, 202-05.
Ruhmer, Ernst. "Multiplex Telephony." *Electrical Review and Western Electrician* LIX (July 1, 1911): 28-29.
Schwartz, Mischa. "Origins of Carrier Multiplexing: Major George Owen Squier and AT&T," History of Communications Column, *IEEE Communications Magazine* 46, no. 5 (May 2008): 20-24.
"Signal Corps in Relation to the Communications Industry." *Army and Navy Journal* LXXII (June 29, 1935): 948.
Slaughter, Nugent H. "The Production of Vacuum Tubes for Military Purposes." *The American Physical Society* XIV, No. 5 (1919): 453-56.
_____. "Radio Development during the World War." *Electrical World* LXXIII (February 1919): 322-23.
Slaughter, Nugent H., Francis G. Gray, and John W. Stokes. "Radio-Telephone Development in World War I." *Electrical World* LXXIV (August 16, 1919): 340-43.
Squier, G. O. "Advantages of Aerial Craft in Military Warfare," *Aeronautics* II, No. 1 (1908): 17-18.
_____. "Aerial Locomotion in Warfare." *Scientific American*, January 2, 1909.
_____. "Aeronautics." *Engineering World* XIV (January 15, 1919): 33-35.
_____. "Aeronautics in the United States." *Electrical Review* LXXIV (January 25, 1919): 131-33.
_____. "The Air Deficiency Appropriation." *Air Service Journal* II (March 21, 1918): 389-90.
_____. "Airplane Direction Finder: A Development of the War." *Electrical World* LXXIII (February 1, 1919): 222.
_____. "Airplane Radio-Telephone Set." *Electrical World* LXXIII (January 18, and February 1, 1919): 130, 222.
_____. "Airplane Radio-Telephone Set." *Electrical Review* LXXIV (January 25, 1919): 132-33.

———. "Airplane Radio-Telephone Set." *Transactions, American Institute of Electrical Engineers* XXXVIII (February 1919): 49–51.
———. "Alternating Current Submarine Telegraphy." *Electrical Review* LXVIII (April 1, 1916): 586–88.
———. "An American Pacific Cable." *Electrical Review and Western Engineer* XXXVI, No. 2 (1900): 35–37.
———. "American Pacific Cable." *Scientific American*, February 17, and 24, 1900.
———. "Army Aeronautics." *American Aeronaut*, October 1909.
———. "Comparative Illustrations of French, German, English, and American Dirigibles." *Fly*, February 1909.
———. "Details of Liberty Motor Development." *American Machinist*, January 16, 1919.
———. "Developments in Radio Apparatus." *Electrical World* LXXIII, No. 3 (1919): 129–30.
———. "Discussion on 'Multiplex Telephony and Telegraphy by Means of Electric Waves Guided by Wires.'" *Transactions, American Institute of Electrical Engineers* XXX (August 1911): 1666–81.
———. "Electrical Methods of Intercommunication for Military Purposes." *Journal of the Franklin Institute* CL (II), No. 1037 (1911): 545–57.
———. "Electric Waves Directed by Wires." *Scientific American*, December 16, 1911.
———. "Electric Waves Directed by Wires for Intercommunication Purposes." *Electrical Review* LIX, No. 21 (1911): 1045–46.
———. "Electricity and the Art of War." *Journal of the U.S. Artillery* II (January 1893): 99–101.
———. "Electro-Chemical Effects Due to Magnetization." *American Journal of Science*, 3rd Series XLV, No. 270 (1893): 443–58.
———. "Electrochemical Effects Due to Magnetism." *The London, Edinburgh, and Dublin Philosophical Magazine and Journal of Science* (5th Series) XXXV, No. 217 (1893): 473–89.
———. "High Frequency Signals Over Telegraph Wires." *Telephone and Telegraph Age*, August 1, 1919.
———. "Influence of Submarine Cables Upon Military and Naval Supremacy." *National Geographic Magazine*, January, 1901.
———. "Influence of Submarine Cables Upon Military Naval Supremacy." *Scientific American*, April 20, 1899.
———. "The Influence of Submarine Cables Upon Military and Naval Supremacy—Lecture at Naval War College, New Haven." *Proceedings of the U.S. Naval Institute, Annapolis* XXVI, No. 4 (1900): 231–36.
———. "The International Electrical Congress of 1893, and Its Artillery Lessons." *Journal of the U.S. Artillery* III (January 1894): 1–13.
———. "Liberty Aircraft Engine." *Transactions, American Institute of Electrical Engineers* XXXVIII (February 1919): 104, Ill.
———. "Liberty Engine: One of America's Main Wartime Achievements." *U.S. Air Services*, March 16, 1931.
———. "Meteorological Service of the Army." *Monthly Weather Review*, February 1919.
———. "Multiplex Telephony." *Outlook*, February 11, 1911.
———. "Multiplex Telephony and Telegraphy." *Scientific American*, July 1, 8, and 22, 1911.
———. "Multiplex Telephony and Telegraphy by Means of Electric Waves Guided by Wires." *Electrical Review* LIX, No. 2 (1911): 59.
———. "Multiplex Telephony and Telegraphy by Means of Electric Waves Guided by Wires." *Transactions, American Institute of Electrical Engineers* XXX (May 1911): 857–905.
———. "On an Unbroken Alternating Current for Cable Telegraphy." *Proceedings, Physical Society of London*, XXVII, Part V (August 15, 1915).
———. "On Coast Artillery Fire Instructions." *Journal of the United States Artillery* III (April 1894): 243–49.
———. "On the Electrical Congress of 1893." *Journal of the United States Artillery* IV (January 1890): 154–56.
———. "Open-air Route to Germany." *Flying*, August, 1917.
———. "Physiological Study of the Flier." *Transactions, American Institute of Electrical Engineers* XXXVIII (February 1919): 79–85.
———. "The Present Status of Military Aeronautics." *Engineering News* LX (December 1908): 632–46.
———. "Present Status of Military Aeronautics." *Flight* (London) I, Numbers 9–13 (1909): 121–23, 137–38, 149–50, 166–67.
———. "Present Status of Military Aeronautics." *Fly*, June, 1909.
———. "Present Status of Military Aeronautics." *Journal of the American Society of Mechanical Engineers* XXX, No. 12 (1908): 1571–1642.

———. "The Problems of Aeronautics." *Aerial Age Weekly*, January 22, 1917.
———. "Production of Helium on a Commercial Scale." *Electrical Review* LXXIV (January 25, 1919): 131.
———. "Production of Helium on a Commercial Scale." *Engineering and Mining Journal* CVII (February 8, 1919), 273.
———. "Production of Helium on a Commercial Scale." *Proceedings, American Institute of Electrical Engineers* XXXVIII (February 1919): 75.
———. "A Question of Nomenclature—Wire Radio." *The Electrician*, December 17, 1920.
———. "Recent Progress in Aeronautics." *Science* (New Series) XXIX (February 1909): 281–89.
———. "Scientific Research for National Defense as Illustrated by the Problems of Aeronautics." *Journal of the Franklin Institute*, CLXXXIII (January 1917): 35–40.
———. "Scientific Research for National Defense as Illustrated by the Problems of Aeronautics." *Proceedings, National Academy of Sciences* II (December 1916): 740–42.
———. "Scientific Research for National Defense as Illustrated by the Problems of Aeronautics, Abstract." *Nature* XCVIII (February 1917): 440–41.
———. "Scientific Research Needed for Solution of Aeronautical Problems." *Aviation and Aeronautical Engineering* I (December 1916): 287–88.
———. "Ships—Air and Water." *Scientific American*, February 6, 1909.
———. "Some Experiments in 'Wired-Wireless' Telegraphy for Field Lines of Information for Military Purposes." *Journal of the Franklin Institute* CLXXIII (April 1912): 333–39.
———. "Some Tests of the Magnetic Qualities of Gun Steel." *Journal of the Military Service Institution of the United States* XXIII (July 1898): 35–52.
———. "Some Tests of the Magnetic Qualities of Gun Steel." *Journal of the United States Artillery* III (October 1894): 559–75.
———. "A System of Multiplex Telephony." *Electrical Review and Western Electrician* L (January 14, 1911): 97.
———. "System of Multiplex Telephony and Telegraphy; Squier's Gift to the Public." *Scientific American*, January 21, 1911.
———. "'Teamwork' in War." *U.S. Cavalry Journal* XVI (July 1906): 5–10.
———. "Tests Between Signal Corps War Laboratory and Bureau of Standards." *Electrical World* VII (February 1921): 291.
———. "Tree Telephony and Telegraphy." *Electrician* LXXXIV (January 30, and February 6, 1920): 111–12, 147–49.
———. "Unification of Communication Engineering." *Scientific American*, August 1922.
———. "A United States Government Pacific Cable." *Journal of the United States Artillery* XIII (March-April 1900): 152–76.
———. "The United States Pacific Cable." *The Independent* LII (February 1900): 359–62.
———. "What Mechanical Flight Means to an Army." *Harpers Weekly*, November 21, 1908.
———. "What Will Take the Place of Today's Broadcasting?" *Popular Radio*, October 1922.
———. "Wired Wireless for Military Purposes." *Electrical Review* LVIII (January 14, 1911).
———. "Wired Wireless for Military Purposes." *Engineering Magazine*, July, 1912.
———. "The Wright Brothers—A Bit of History." *Flight* (London) V (June 14, 1913): 651–52.
Squier, G. O., and Albert C. Crehore, Jr. "A Horizontal-Base Rang and Position Finder for Coast Artillery." *Journal of the United States Artillery* X (November-December, 1898): 1–8.
———. "Note on Oscillatory Interference Bands and Some Practical Applications." *Bulletin of the Bureau of Standards* VII, No. 1 (1911): 131.
Stone, John Stone. "The Practical Aspects of the Propagation of High Frequency Electric Waves Along Wires." *Journal of the Franklin Institute* CLXXIV, No. 4 (1912): 353–84.
"A System of Multiplex Telephony." *Telephony, Telegraphy, and Wireless* XVIII (January 1911): 97.
Todd, D. W. "The International Radio Telegraphic Conference of London." *Journal of the American Society of Naval Engineers* XXVI (August 1912): 1330–35.
Wagner, K. W. "Multiplex Telegraphy by Means of High Frequency Alternating Current." *Telegraphen und Fernsprech Technik* VIII, No. 3 (1919): 29.
"Wired Wireless Telephony." *La Nature* XLVIII (August 7, 1920): 93–94.
"The Wireless Equipped Airplane." *Wireless Age*, July 1917.
"Wireless Telegraphy and Telephony." *Journal of the Royal United Service Institution* LVI (July-December 1912): 997–1011.

Wright, Allen Henry. "Aero Radio Telegraphy." *Aerial Age Weekly*, December 11, 1916.
Zabel, Max W. "From the Patent Office." *Telephony*, January 14, 1911.

Government Publications

Historical Section, The Army War College. *The Signal Corps and Air Service: A Study of Their Expansion in the United States, 1917–1918, Monograph No. 16*. Washington, DC: Government Printing Office, 1922.
Office of the Chief Signal Officer, War Department, *Equipment for Aero Units of the Aviation Section (Signal Corps)*. Washington, DC: Government Printing Office, 1916.
Office of the Chief Signal Officer, War Department. *Reports of the Chief Signal Officer* [for the years 1893–1924]. Washington, DC: Government Printing Office, n. d.
U.S. Bureau of Aircraft Production. *History of the Bureau of Aircraft Production* (Vols. 1–8). Dayton, OH: Wright Patterson Air Force Base, 1951.
U.S. Congress, House, Committee on Interstate and Foreign Commerce. *Cables Between the United States and Hawaii, Guam, and Philippine Islands, Hearings* (H. Doc. 568, Serial no. 4401, 57th Cong., 1st Sess.), 1902.
U.S. Congress, House, Military Affairs Committee. *Pioneer Aviators, Hearings* (before a subcommittee of the Military Affairs Committee on N. R. 11273, 70th Cong., 1st Sess.), 1928.
U.S. Congress, House Select Committee. *Expenditures in the War Department, Hearings* (before House Subcommittee No. 1 [Aviation], House of Representatives, on H.R. 637, Serial 7652; 66th Cong., 2d Sess.), 1920.
U.S. Congress, Senate, Committee on Interstate Commerce. *Cable Landing Licenses, Hearings* [before a subcommittee of the Committee on Interstate Commerce, Senate, on S. Bill 4301, a bill to prevent the unauthorized landing of submarine cables in the United States, 66th Cong., 3rd. Sess.], 1921.
U.S. Congress, Senate, Committee on Naval Affairs. *Relating to Construction of Telegraphic Cables Between the United States, Hawaii, Guam, Philippine Islands, and Other Countries, Hearings* (S. Doc. 141, Serial no. 4231, 57th Cong., 1st Sess.), 1902.
U.S. Congress, Senate. *Multiplex Telegraphy and Telephony* [S. Doc. 172, Serial 8712, 69th Cong., 2d Sess.], December 9, 1926.
U.S. Navy Department. *Annual Report, 1899* [Report of the Chief of the Bureau of Equipment]. Washington, DC: Government Printing Office, 1899.
U.S. President of the United States. *In Matter of Application of Commercial Pacific Cable Company for Permission to Land on Shores of United States, Hawaiian Islands, Midway Islands, Guam and Philippine Islands, Telegraph Cable to Be Laid Between United States and Philippine Islands, and to China.* [S. Doc. 24, Serial 4417, 57th Cong., 2nd Sess.], 1902.

Manuscript Collections

Case Western Reserve University. Archives, Benedict Crowell Papers.
Library of Congress, Washington, DC:
  Newton D. Baker Papers.
  Octave Chanute Papers.
  Adolphus Greely Papers.
National Academy of Sciences and National Research Council, Washington, DC, Archives, George Owen Squier File.
National Archives and Records Service, Washington, DC:
  Adjutant General Officer's File, Record Group 94.
  Chief of Staff File, Record Group 165.
  National Academy of Sciences File, Record Group 189.
  National Advisory Committee for Aeronautics File, Record Group 255.
  National Bureau of Standards File, Record Group 167.
  Office of the Chief Signal Officer, Record Group 111.
Saint Louis Records Repository, Saint Louis, Missouri, George Owen Squier File.
Smithsonian Institution, Washington, DC, Archives:
  Secretary of Smithsonian Files, 1905–1919.
  Division of Electricity and Nuclear Energy.
United States Air Force Academy, Special Collections:
  Mason Patrick Papers.
  George Owen Squier Collection.
United States Army Historical Research Collection, U.S. Army War College, Carlisle Barracks, PA, Archives:
  George Owen Squier File.
  Military Journals.

Bibliography   271

United States Army Signal Corps Museum, Fort Monmouth, New Jersey: George Owen Squier Collection (Now at U.S. Army Communications-Electronics Command (CECOM) Aberdeen, MD).
University of Michigan, Historical Society of Michigan, Rackham Hall, Archives: George Owen Squier Collection.

Reports

Arnold, Bion J. "Secret Report on Automatic Carrier; Flying Bombs (F.B.), Aerial Torpedos (A.T.)." Directed to Secretary of War, January 31, 1919. Office of the Chief Signal Officer Files, Record Group 111, National Archives, Washington, DC.
Association of Graduates of the United States Military Academy. "George Owen Squier." *Sixty-fifth Annual Report.* New York: Moore Printing Company, June 11, 1934.
Association of Graduates of the United States Military Academy. "Henry Clarence Davis." *Sixty-third Annual Report.* Newburgh, NY: Moore Printing Company, 1932.
Association of Graduates of the United States Military Academy. "John Wilson Ruckman." *Fifty-third Annual Report.* Newburgh, NY: Moore Printing Company, 1922.
National Academy of Sciences. *Annual Report, 1917.* Washington, DC: Government Printing Office, 1917.
National Advisory Committee for Aeronautics. *Annual Report, 1917.* Washington, DC: Government Printing Office, 1917.
Squier, G. O. "On the Absorption of Electromagnetic Waves by Living Vegetable Organisms." *Major General Arthur MacArthur's Report to the War Department on the Military Maneuvers in the Pacific Division, 1904.* Private Reprint, 1904.
_____. "The Present Status of Military Aeronautics. *Annual Report, Smithsonian Institution, 1908.* Washington, DC: Government Printing Office, 1908.
_____. "Report of Lieut. Col. George O. Squier and Prof. Albert C. Crehore, on Experiments and Discoveries in Sine Wave Telegraphy." *Report of the Chief Signal Officer.* Washington, DC: Government Printing Office, 1899.
_____. "Reports of Military Attaché on British Army in the Field." War College Division, Chief of Staff Files, Record Group 165, National Archives, Washington, DC.

Unpublished Materials

Arnold, H. H. "Outline of History, Aviation Section, Signal Corps & Division of Military Aeronautics." Typescript, April 1917–October 1918.
Burge, V. L. "Early History of Army Aviation." Manuscript, undated, USAF Academy Library.
Camfield, Thomas Marley. "Psychologists at War: History of American Psychology in the First World War." Unpublished Ph.D. dissertation, the University of Texas at Austin, August 1969. University Microfilms, No. 70-10, 766.
Clark, C. H. *Radio in War & Peace.* Unpublished Manuscript, Clark Radio Collection, Class 100, Smithsonian Institution, MHT, Division of Electricity and Nuclear Energy.
Clark, Paul W. "Millikan and the Rice-Webster Gun Controversy." This treatise was presented before the annual meeting of the Society for the History of Technology, December 1971.
Frey, Royal D. "Evolution of Maintenance Engineering, 1907–1920." Unpublished study, Serial #5-1877-1A, undated, Maxwell AFB, Alabama.
Futrell, Robert Frank. "Ideas, Concepts, Doctrine: A History of Basic Thinking in the United States Air Force, 1907–1964." USAF Historical Study No. 139, unpublished manuscript, USAF Library.
Milling, T.D., Major. *A Short History of the United States Army Air Service.* Typescript, Serial #167.401-11A, undated, Maxwell AFB, Alabama.
Myer, Samuel C. *The Impact of Signal Communications Upon the Art of War, 1860–1942.* Unpublished thesis submitted to Princeton University, January 15, 1943.
Purtee, Edward O. *History of the Army Air Service 1907–1926.* Unpublished study prepared by Historical Office, Executive Secretariat Air Materiel Command, Wright-Patterson Air Force Base, May 1948.
Reingold, Nathan. *Science and the United States Army.* Unpublished manuscript, held in the Office of the Chief of Military History, Washington, DC.
Thompson, George Raynor. "How Development of Radiotelephone (Voice Radio) in World War I Promoted the Radio Broadcast Industry in the 1920's." Unpublished, undated manuscript, Fort Huachuca, AZ.
Traditions of the Signal Corps. Unpublished

draft manuscript, Fort Monmouth, NJ: Signal Corps Museum, April 1959.

Newspapers

London Times
The Monmouth [NJ] Message
New York Times
Washington Post
Washington Star

Interviews

David P. Gibbs, Major General, U.S. Army (Retired). Private interview at his home in Colorado Springs, CO, February 1971.
Oliver Wendell Holmes, M.D., Private interview held in Washington, DC, February 1970.
Joseph O. Mauborgne, Jr., Major General, U.S. Army (Retired), former Chief Signal Officer of the Army. Private interview in Atlanta, GA, February 1971.
I. I. Rabi, Interviewed by Thomas S. Kuhn at his New York home, December 8, 1963, described at http://www.aip.org/history/ohilist/4836.html.
George Raynor Thompson, Ph.D., former Chief Historian of the Signal Corps. Private interview at Fort Huachuca, Arizona, December 1970.
John F. Victory, first civilian employee of NACA. Private interview at his home in Tucson, AZ, December 1970.

Film

Chief Signal Officers of the Army, Signal Corps. Revised Version, 1936, 7 min., silent, black and white, 35mm (Historical Film, no. 1216). Script available.

Index

Aberdeen, MD 1, 37, 41, 48, 170, 179–180, 184, 252, 271
acoustics 76, 221
aerodynamics 83, 98–99, 190
aeronautics 4, 72, 74–75, 77–78, 80–83, 86, 89, 91–99, 115, 117, 128, 135, 138–139, 141–146, 153, 156–157, 164, 166–167, 176, 178, 194, 208, 234–239, 243, 247–248, 251, 253–255, 257, 261, 264–265, 267–269, 271
aeroplane 86, 90, 117, 133–135, 147, 149, 235, 238, 246, 248–249
ailerons 164
Ainsworth, Adj. Gen. 131
Air Service, American 3, 156, 158, 189, 194, 204–206, 234–235, 245, 248, 250–254, 262, 265–267, 270–271
airborne radio 179, 185–188, 190, 192, 194–195, 203, 255
aircraft 4, 79, 83–84, 86, 92–94, 117, 127–129, 134–135, 140, 143, 146–153, 156–160, 163–165, 167, 178–179, 181, 184, 186–188, 190, 192–193, 204, 238, 247–251, 254–256, 263–264, 268, 270
airmen 127, 186
airplane 60, 77, 79, 81–86, 89–91, 93, 97–98, 116–117, 126–127, 134, 147–151, 155, 157, 159, 162, 164–165, 167, 170, 178, 187–189, 192–195, 203–204, 234–236, 238, 243, 246–249, 253–255, 269; manufacturers 149, 238; radio 184, 189, 192, 254–255
airpower 91–94, 163
airship 77, 84, 86
Alexanderson, E.F.W. 5, 103–105, 240–241
Allegheny, PA 40
Allen, Brig. Gen. James 50, 53–55, 61–62, 64, 72, 74–75, 78–79, 81–85, 88–91, 94, 96–99, 101–107, 109, 113–114, 196, 200–201, 233–240, 242, 270
allies 3, 125–126, 131, 136, 151, 156–158, 184, 192

alternating current (AC) 50, 57–58, 119–120, 236, 259
alternator 5, 46, 102–105, 121, 146, 219, 240–241
Ambrose, Stephen E. 223, 226
American Expeditionary Force (AEF) 126
American Telephone and Telegraph Company (AT&T) 110–111, 118–119, 205, 209–210, 212–213, 240–241, 267
Americans 3–6, 20, 24, 28, 30, 35, 43, 46–49, 55, 61–62, 65–70, 85–86, 89–91, 97, 99, 106, 109–114, 117–118, 123–124, 126–127, 129–131, 135–136, 138, 144, 146–148, 150–151, 153, 155–159, 163, 169–171, 173–175, 178, 181, 184, 189, 191–192, 194–195, 198–200, 206, 208–211, 214, 216, 226–231, 234, 240–242, 244–246, 248–249, 251–252, 255, 257–269, 271
Ames, Joseph S. 20, 24, 38, 142, 146, 171, 228
Amityville, NY 162–163, 166
amplitude modulation (AM) 58, 126, 166, 211, 222, 242, 245
antenna 118, 188–189, 221, 223, 256
anti-aircraft 128–129
Antietam, Battle of 49
apparatus 29, 37, 43, 46, 78–79, 102, 105, 109, 121, 147, 168, 189–193, 255, 268
appropriation, Congressional 3, 6, 38, 83, 89, 97–99, 101–103, 107, 144–145, 152–153, 157, 162, 168, 178, 238, 240, 247, 252–253, 267
Archer, Gleason L. 261
archipelago 52, 64, 200
Arlington, VA 188, 256
armistice 158, 166–167, 205, 209, 251, 264
armor 28, 30, 32, 93, 127, 129, 255
Armstrong, Edwin Howard 191, 223, 256, 263
Army, U.S. 1–6, 9–10, 17–19, 21–29, 31–35, 37, 40–41, 43, 45, 47–52, 54–55, 58–59,

62–64, 66–67, 69, 71, 73–87, 89–93, 95, 97–99, 102, 106–107, 112–113, 116–118, 123–125, 127–131, 133, 135–146, 149, 152–159, 163–167, 171, 173–175, 177–191, 194–195, 197–212, 215–216, 225–228, 230, 233–239, 241, 243–247, 249–250, 253–254, 256–259, 261–265, 267–272
Arnold, Bion J. 164–167, 243, 251, 254, 261, 271
artillery 3–5, 19–20, 22–23, 25–35, 37–38, 42–43, 49, 53, 92–93, 102, 126–128, 134–135, 152, 157, 163–164, 186–188, 192, 199, 226–230, 236, 239, 254, 264, 266, 268–269
astronomy 17, 20, 252
autobiographies 225, 228, 231, 249–250, 252, 260, 262, 264
automobile 84, 132, 146, 149–152, 159–160, 221, 237, 246, 248
aviation 1, 3–6, 73–78, 80–81, 83, 85, 87, 89–99, 112, 116–117, 124, 133–135, 137–144, 146–148, 151–159, 162, 164, 174–181, 187–189, 195, 205, 208, 216, 233, 237, 246–251, 253, 264, 266–267, 269, 271
aviator 76, 89, 147, 247, 250, 265

Babcock (née Parker), Lavinia 7
bacteriological 129
Baekeland, Leo H. 145
Baker, Newton T. 140, 152, 155–157, 163, 178–180, 191, 208, 237, 245–251, 253–254, 259, 261, 264, 270
Baldwin, Capt. Thomas Scott 82, 86, 91
Balkan Wars 5, 117
ballistics 4, 19, 34–38, 40, 45, 164, 228
balloon 73–75, 82–84, 90, 97, 143, 257
Baltimore, MD 3, 226, 251–252, 261, 263–264
barograph 264
barometer 164
barracks 72, 75, 270
battalions 129, 173, 184
battery, artillery 19, 21, 22, 24–26, 29, 31, 33–35, 42, 53, 56, 106, 108, 121, 188, 192–193, 226–227, 231
battle 10, 27, 42, 49, 74, 82, 92, 127, 138, 170, 184, 197–198, 201–202, 245, 257–259, 261, 263
battleship 88, 90
Baxendale, A. 120–122, 244
bearing 90, 249
Beck, Lt. Paul 97, 237
Bedell, Frederick 46, 56, 230–231, 262
Belgium 38
Bell, Alexander Graham 2, 58, 77, 79, 85, 89, 94–96, 98–99, 105, 109, 143, 173, 181, 198, 209, 212, 231, 233, 235–239, 267

Bemis, Samuel 244, 261
Bennet, James Gordon 66
Bennett, Geoffrey 259, 261
Berlin, Germany 28, 32–33, 82, 171, 262
Bermann, Rufus R. 106
Bethesda, MD 258, 265
biplane 116, 164
Bishop's Palace, Rouen, France 137, 235
bisulphide, carbon 217
Blake, G.G. 241, 261
Blenheim, W.J. 241
Bliss, Gen. Tasker 24, 33, 181, 226, 253
blockade of Germany 131
Bohr, Nils 59
bomber 251, 265
Bonaparte, Napoleon 197–198
Boston, MA 147, 239, 246, 258, 261, 264
Boyd, Lt. Carl 126, 251, 261
Bristol, Capt. Marl L. 142, 248, 253
Britain 28, 38, 41, 47, 131, 135, 146, 153, 158, 189, 200, 210, 223, 236
Brittain, James B. 241–242, 265
broadcasting 70, 266
Brockett, Paul 94, 96, 116, 181, 236, 261
Bruce, Robert V. 228, 230, 261
Buckley, O.E. 192
Bullock, Mike 225, 254, 261
Burleson, Albert S. 209, 259
USS *Burnside* 4, 63–65, 215
buzzerphone 50

cablegram 233
cables, undersea 2, 4–5, 45, 48–52, 54–58, 61–70, 75, 78, 105, 112, 118–121, 125, 199, 230, 232–233, 241–244, 257–259, 265–266, 268–270
California 4, 19, 55, 70, 72, 75, 167, 169, 249–250
Calliaferro, Lt. 246
Callwell, Charles E. 244, 261
Cambridge, MA 100, 226, 232, 244, 249, 257, 260–265
Camp Vail, NJ 173, 184, 209, 254, 259
Campbell, George A. 106, 209, 241–242, 265
Canada 9, 38, 210, 267
cannon 4, 32, 44
Carlisle, PA 244, 270
carrier frequency 4–5, 58, 100, 110–111, 119, 166, 169, 213, 222, 240–241, 267, 271
Carty, John J. 119, 156, 173, 187, 191–192, 210, 243, 252, 264
cavalry 5, 20, 27–28, 72, 74–75, 126, 152, 204, 233, 269
Cervera, Adm. 196, 200
Ceylon 64
Chambers, Maj. L.B. 184

Index

Chandler, Capt. Charles 83, 90, 233–234, 238, 262
Chanute, Octave 78–79, 85, 89, 94–95, 98, 234–235, 237–238, 265, 270
chemistry 19–20, 176, 252
chemists 202
Chicago, IL 25, 29, 32, 46, 109, 113, 145, 176, 192, 234, 240, 252, 262, 265
Chief Signal Officer (CSO), U.S. Army 230–234, 236–240, 242–244, 246, 249–258
China 65, 67, 70, 233, 258, 270
chronograph 35–39, 41, 46, 229, 266
chronoscope 229
Cincinnati, OH 180
Clark, George H. 100–101, 239–240, 251–255, 262, 264–265
Clark, Paul Wilson 2, 179, 225, 267, 271
Clark, Capt. Virginius, USN 179
Clark University, Worcester, MA 167–168
Clark's *Radioana* 240, 254
Clausewitz, Karl von 197–198, 257, 262
Cleveland, OH 40, 52–54, 211, 214, 250
Cochrane, Rexmond C. 239–240, 248, 262
code, telegraph 50, 56, 219, 221
codebreakers 257, 263
Cohen, Dr. Louis C. 183
coherer 221
Colorado 21, 230, 233
Colpitts, E.H. 240–241, 265
Columbia University, New York 46, 85, 111, 143, 174, 252, 256
Columbian Exposition 25, 29
Columbus, OH 141
Commercial Cable Company 48–49, 51, 54, 56, 58, 66, 69, 199
communications 1–2, 4–6, 51–52, 60–62, 67, 70, 73–74, 78, 83, 100–102, 105, 107, 112, 119, 129, 132, 136, 138, 186–190, 192, 194, 196–205, 210, 214, 216, 225, 232, 240–241, 246, 254–259, 261, 265, 267
Communications-Electronics Command (CECOM), U.S. Army 13, 18, 23, 27, 61, 63, 72, 87–88, 108, 113, 124, 133, 139, 154, 172, 183, 191, 206–208, 213, 254, 271
commutator 193
computer 258, 265
Conant, Jennet 229
Connecticut 9
Constantinople, Turkey 117
continuous wave (CW) 39, 121, 164, 188, 202, 220, 222–223, 241, 254, 261–262
Corliss, John B. 65–69, 232–233
Cornell University, Ithaca, NY 45–46, 58, 192
Cornwall, VT 9
Corregidor, PI 54, 62

Cox, Charles 235, 239, 262
Craig, Professor 23, 32–33, 35
Cram, Ernest R. 100, 106, 239–240, 255, 266
Crampton Range-Finder 229
Crawder, Gen. 249
Crehore Albert C. 4, 35–42, 45, 47–48, 50–56, 58–59, 62, 73, 121, 146, 217–219, 228–231, 233, 258, 260, 262, 265–266, 269, 271
Crowell, Benedict 250, 254–257, 262, 270
Crozier, Gen. 85, 132, 234–235
cryptography 138, 257
crystal 100, 221, 223
Cuba 195–196, 198, 200
Culver Col. Charles C. 111, 189, 191, 194, 204, 240, 242, 255, 266
Curtiss, Glenn H. 83, 95–98, 145, 148–150, 160, 234, 237–238, 248–249, 264

damped sine wave 222
Daniels, Josephus 155, 162, 180, 248–249, 262
Dargue, Lt. Herbert 139–140, 188, 246
Dartmouth University, Hanover, NH 35, 37
Davis, Lt. Henry C. 26–27, 34–35
Dayton, OH 77, 79, 164–167, 192, 249, 252–253, 270
decipherment 257
Deeds, E.A. 153, 156, 174, 250, 253, 263
deForest, Lee 100, 104, 112, 119–122, 209, 223, 240, 243–244, 256, 262
Delaney, P.B. 56, 231
Delaware 111
Delco corporation 153
Dellinger, J. Howard 104
Detroit, MI 10, 13, 164, 174, 182, 253
Dewey, Admiral George 20, 84, 234
Dillon, John F. 106
Dilno, Jemima 9
direct current (DC) 58, 225–227, 229–232, 234–236, 238–241, 243, 246–250, 252–255, 261–265, 270–272
dirigible 79, 82–84, 86, 88–92, 236, 239
Donaldson, Frances 262
Douhet, Gen. Giulio 92, 163
Drachmann, Richard C. 237
Dreyfus, Capt. Alfred 257
Dryden, MI 3, 9–12, 14–15, 17, 215, 218, 225
Dublin, Ireland 226, 266
Duncan, Louis 21–22, 24, 46, 226, 266
Dunne biplane 116
Dupree, A. Hunter 142, 226, 247, 252, 262
Dupuy, Ernest R. 17, 226, 262
Dykaar, Moses 213, 216, 260
dynamo 26, 193

Index

Eccles, Dr. William 115, 242, 246, 262
Eckert, Gen. Thomas E. 230
Edgar, C.G. 50, 174, 232–233, 263
Edgerton, Harold E. 36, 228
Edinburgh, Scotland 266, 268
Edison, Thomas A. 21, 145
electrode 39, 241
electrolysis 227
electromagnetic waves 17, 35, 40, 56–59, 101–102, 106, 115, 119, 136, 178, 188, 199, 206, 209, 218–222, 231, 240–241, 243–244, 254, 258, 261–262, 266, 271
electromotive force 30, 56, 229, 266
electron 223
engines 3, 28, 76, 83–84, 117, 146, 150, 158, 160, 248, 255
England 18, 43, 47, 49, 66–67, 113, 118–119, 123–124, 132, 136, 150, 209–210, 243
Espenscheid, Lloyd 194, 254, 257

Fessenden, Reginald 102–103, 240–241
Fleming, Dr. Ambrose 223, 242
Flexner, Dr. Simon 20
Florida 167
Foch, Marshal Ferdinand 198, 255
Fokker synchronized machine gun 134
Fort Hamilton, NY 54
Fort Huachuca, AZ 271–272
Fort Leavenworth, KS 4, 71–77, 81–82, 97, 143, 187, 233–234, 237, 258, 264
Fort McHenry, Baltimore, MD 3, 19, 21, 23–26, 31, 33, 228
Fort McPherson, GA 29, 31
Fort Meyer, VA 257–258
Fort Monmouth, NJ 1, 172, 182–184, 186, 204, 225, 245, 254, 257, 259, 262, 271–272
Fort Monroe, VA 4, 23, 26–27, 31, 33–35, 37, 42–43, 45–46, 49, 60, 64, 128, 146, 192, 203, 226, 228–229, 231, 264, 266
Fort Riley, KS 50, 187
fortress 25, 31, 226
Foulois, Maj. Gen. Benjamin D. 76, 151–152, 235, 247–249, 262
France 3, 5, 18, 38, 43, 82, 123, 125–126, 128, 130, 136–137, 150, 153–154, 158, 166–167, 170, 174, 176, 184, 191, 207, 237, 248, 253, 255, 257
Franklin Institute 3, 97, 142, 147, 207, 230, 233, 240, 242, 248, 251–253, 259, 261, 263, 266, 268–269
frequencies 5, 100, 102, 104–107, 110, 114, 118, 136, 206, 213, 221–222, 240–241, 256
frequency 36–37, 57–58, 102–106, 110–111, 188, 221–222, 240, 255–256, 267–269
frequency modulation (FM) 222

Galilei, Galileo 36
galvanometer 42, 50, 229–230
Galveston, TX 246
Gardiner, Martin 231, 262
Gardner, Emily 12, 14
Gatling gun 43
genealogy 225, 261, 264
General Electric 21, 30, 56, 109, 156, 176, 193, 210, 240, 253, 256–258
George, David Lloyd 123
Germany 18, 38, 43, 119, 131, 155, 157, 161–162, 170, 210, 236, 241, 261, 268
Gerstner Field, Lake Charles, LA 194
Gibbs, Lt. Gen. David 228, 233, 272
Gilman, Daniel Coit 20, 24, 226
Glasgow, Scotland 48
Goddard, Esther 251, 262
Goddard, Robert 161, 167–170, 251, 262–263
Goodier, Lt. Col. Lewis E. 140, 246
Gorgas, Gen. William C. 129–130
Grant, Gen. Ulysses Simpson 10, 66, 144, 167–168, 198, 209
Greely, Gen. Adolphus W. 4, 49–55, 60–65, 68, 70, 72, 74, 77, 78, 80, 83, 96, 138, 195, 196, 199–201, 225, 230–234, 238, 257, 258, 270
Griffen, Eugene 21, 226, 262
Guam 62, 70, 230, 232–233, 270
gunnery 30, 228
gunpowder 163
Guynemer, Georges 134
gyroscope 162, 164

Hagedorn, Hermann 245, 263
Hague Convention 93, 236, 245, 264
Haig, Gen. Sir Douglas 207
Hampton, VA 179–180, 252
Harris, Col. F.E. 164–166, 251, 266
Harrisburg, PA 257, 263–264
Hartcup, Guy 263
Hartman, Lt. Col. Carl F. 183
Harvard University, Cambridge, MA 109, 226, 244, 262, 264
Hawaii 62, 66, 115, 188, 228, 230, 232, 270
Hayford, John F. 143
Hazeltine Corporation 194, 256–257
Helmholtz, Dr. Hermann von 30, 32–33, 37
Henderson, Sir David 117, 126, 128, 135
Hendrick, Burton 137, 242, 244–246, 263
heterodyne 256
Hewitt, Peter Cooper 84–85, 162–163, 235
Holmes, Dr. Oliver Wendell 235, 272
Honeycut, 1st Lt. F.W. 126, 245
Hong Kong 64, 67, 260, 263
Honolulu, HI 62, 70
USS *Hooker* 4, 54, 61–63, 200
Hooper, Lt. S.C., USN 136

Index

Hoover, Herbert 206
Houston, Edwin J. 30
Houston, TX 167
Howeth, L.S. 196, 241, 244, 254, 257, 263
Huertley, E. 119, 243
Hughes, Thomas P. 250–251, 263
Hungary 38
Hunsaker, Jerome C. 116, 144, 253
hydroplane 162

Indianapolis, IN 230, 261
infantry 20, 74–75, 97, 152, 233
Ingram, William M. 180
interphone 255, 256
Irwin, Col. LeR. 26
Italians 171, 246, 258

Jackson, Adm. Sir Henry B. 120–121, 244
Jacksonville, FL 52
Jamestown, VA 82
Japan 62, 67, 125, 259
Jenkins, Francis C. 206
Jessup, Philip C. 233, 242, 263
Jewett, Frank B. 109–111, 173, 187, 189, 241
Joffre, Marshal Joseph 125, 197, 257, 264
Johns Hopkins University, Baltimore, MD 3–4, 19–26, 29, 31–33, 35, 38, 46, 76, 103, 142, 171, 212, 226, 242, 251–252, 258, 261, 263–265

Kahn, Albert 254, 257, 263
Kansas 4, 50, 71, 79, 187
Keever, A.M. 23
Kelly, Fred C. 234, 263
Kelvin, Lord 47, 50, 56, 230
Kenly, Brig. Gen. William 164, 166
Kennelly, Arthur C. 57, 226, 242, 257, 259, 263, 267
Kentucky 51
Kettering, Charles Franklin 161, 163–165, 167, 251, 253, 261
Kiel, Germany 157
Kimball, W.R. 23, 84–85, 235
Kitchener of Khartoum, Lord Horatio H. 5, 123–127, 136, 244, 263
Knappen, Theodore MacFarlane 250, 263
Kodak 214
Kolster, Frederick A. 100
Koutchnif, Russia 99
kriegsspiel (wargame) 258
Kuhn, Thomas S. 231, 253, 272
Kuroki, Gen. 201

Lady Franklin Bay Expedition 49
Lafayette Escadrille 167
Lahm, Lt. (later Gen.) Frank P. 75, 82, 86, 89, 91, 233–236, 238, 262

Lamb, John M. 11–12, 15
Lanchester, F.W. 135
Langley, Samuel 1, 60, 76, 78, 94–97, 112, 116, 143, 146–147, 159, 172, 177, 181–182, 184–185, 190–191, 193, 235–237, 239, 247, 253
Lapeer County, MI 11, 14–16, 215
Layton, Alfred 216, 260
Le Havre, France 128, 136
Lebaudy Dirigible 236
Leipzig, Germany 258
LeMay, Gen. Curtis E. 76, 267
Lexington, KY 51
Liberty Eagle 161–167, 170, 251
Lincoln, NB 181, 230, 261–262
Livingston, Col. LaRhett L. 31
Lodge, Sir Oliver 95, 115, 237
Loening, Grover 152, 233, 249, 263
London, England 4–5, 9, 47–49, 54, 62, 67, 82, 112–122, 124–126, 128–130, 133, 135–137, 140, 144, 154, 157, 207, 226, 228–229, 231, 236, 239, 241–244, 249, 259, 261–269, 272
Loomis, Alfred 37, 41
Lovelace, Capt. T.T. 85
Ludlow, Israel 82, 85
Luzon, PI 62, 70
Lyons, Laurence A. 225, 254, 261
Macarthur, Gen. Arthur 72, 198–199, 201, 230, 232–233, 271

Macfarland, Marvin 263
Mackay, John, "Bonanza King" 65–70, 232
Macomb, Gen. M.M. 10–11, 132, 137, 141, 245–246
Malcolm, H.W. 120, 243
Manchester, England 113
Manila, PI 27, 54, 62, 64–65, 67–69
Marconi, Guglielmo 4–5, 43, 54, 112, 114–115, 118, 136, 199, 221, 223, 242, 260, 262–263
Martin, Glenn 229, 231, 249, 255, 261–262, 265
Marvin, Charles 95, 143, 208, 233, 253, 259, 263
Mauborgne, Maj. Gen. Joseph O. 141, 187–188, 254, 267, 272
Mauwaring, Joshua 16
Maxwell AFB, Montgomery, AL 234, 255, 262, 271
McConnellisville, OH 22
McCook Field, Dayton, OH 159, 252
McIntyre, Gen. Frank 177, 253
McKinley, William 61, 65
McMeen, S. G . 109
Memphis, TN 180
Mendenhall, C.E. 29, 175–176, 227

Index

Menoher, Maj. Gen. Charles T. 253
Messer, Mr. 15
Messinger, Sarah 9
Meuse-Argonne 257
Mexico 21, 141
Michie, Gen. Peter S. 17–18, 20
Michigan 1, 3, 9–11, 18, 65–66, 89, 215, 225–228, 230, 232–233, 238, 240–244, 246–247, 259–260, 263, 271
Millikan, Robert A. 152–153, 155–156, 175–177, 249–250, 252–253, 264
Milling, Capt. Thomas DeW. 179, 250, 253–254, 271
Mindanao, PI 64
Mineola Field, NY 235, 252
Missouri 270
Mitchell, Col. William, "Billy" 75, 92, 140, 163, 195, 233, 238, 246, 248, 250, 253, 257, 262–263
Moltke, Marshal Helmuth von 197–198, 257
Morison, Elting E. 196, 249, 257, 264
Morse Code 50, 101, 121, 174, 219, 221
Muirhead and Company, UK 112, 118–120, 243–244
multiplex telephony 100, 105–109, 205, 207, 209, 241–242, 264, 269
multiplexing 5, 100, 110–111, 213
Muzak 2, 213–214

Nebraska 261–262
New Zealand 67
Newcomb, Professor Simon 20, 95, 173
Nicol prism 217
Norman, Henry 114, 234, 242, 262
Nova Scotia 51, 199, 236
Nyquist, Harry 58, 119, 231, 243, 267

Oklahoma 234, 262
Omaha, NB 74–75
O'Meara, W.A.J. 119, 243, 267
Ontario, Canada 9
Oxford, England 9, 129, 229, 232, 264

Page, Walter Hines 5, 29, 46, 113, 124–126, 130–131, 136–137, 226, 239, 242, 244–246, 263, 267
Panama 89
Panizzardi telegram 257
Paris, France 30, 55, 59, 64, 75, 82, 89–90, 116, 125–126, 129–130, 188, 191–192, 194, 206, 245, 256
Parkhurst, Lt. Charles D. 26–27, 32, 34–35
Pasadena, CA 169, 251
patents 5, 38, 41, 52, 77, 83, 95, 98, 101, 107–110, 112, 120–121, 148–151, 209–213, 215, 221, 230–231, 234, 238, 243–244, 249, 259, 270

Pennsylvania 40, 97, 104, 106, 183
Pershing, Gen. John J. 109, 166, 181, 185, 216, 264
Philadelphia, PA 30, 55, 97, 109, 147, 227, 256, 258–259, 261, 263
Philippines 4, 54–55, 59–70, 115, 124, 141, 198, 200
Piatt, Thomas C. 54–55, 231
Pickering, Charles 95, 237
Pittsburgh, PA 78, 252
polarizing photochronograph 4, 35, 37–40, 45–46, 50, 217–218, 228–229, 260
Powell, John Wesley 142, 236
Preece, William H. 30, 47
Presidio, CA 97
Princeton, NJ 233, 242, 251, 257, 261–262, 265, 271
Proctor, Redfield 21, 25
Prussia 258
Pupin, Michael 46, 56, 143, 146, 148, 219, 242, 248, 264–265

Rabi, I.I. 231, 272
radiotelegraphy 68, 101–102, 104, 112, 127, 188, 194, 206, 223, 239
radiotelephony 101–102, 186, 187–188, 190, 192–195, 203–204, 206, 223, 239, 254, 255–256, 271
Rayleigh, Lord 47, 99, 115, 239, 242–243
Reber, Capt. Samuel 50–51, 54, 116, 125, 135, 137, 140, 143, 145, 243, 246–247
regenerative receiver 223
Remsen, Ira 20, 22–23, 226
Rhode Island 59
Ribot, Alexander 150, 152
Rice, E.L. 177, 253
Richardson, Holden C. 143
Robins, Benjamin 36, 228
Rochester, NY 228
Rockwood Sprinkler Company 168, 251
Rodman rifle 22
Roosevelt, Franklin Delano 65–67, 70, 80–81, 131, 138, 142, 153, 230, 248–249, 264
Roosevelt, Theodore 65, 138, 153, 173, 209, 249–250, 263–265
Root, Elihu 55, 68, 71, 74, 76–77, 107, 111, 138, 200–201, 203, 231–233, 242, 263
Rouen, France 137
Rowland, Henry A. 20, 23–25, 30, 32, 227, 264
Rowland, Dr. Sydney 129–130, 191, 245
Royce, Josiah 20
Ruckmann, John W. 26–27, 35, 227
Rumford, Count 36
Russell, Col. Edgar 113, 227, 232–235, 243, 255, 258, 265
Russia 33, 38, 67, 99, 119, 125, 137

Index 279

Saltzman, Brig. Gen. Charles 75–76, 168, 236, 251
San Diego, CA 140, 178, 181, 188–189, 194, 239, 247, 255, 257, 262
Santiago, Cuba 196, 198
Sarnoff, David 206
Satterlee, Herbert 54–55, 58, 209, 231
Saunders, W.L. 162
Sautter, Harle & Company, Paris 30–31
Schultz Chronoscope 229
Schwartz, Mischa 111, 213, 241, 260, 267
Scotland 126
Scott, Gen. Hugh 82, 141, 143, 153–154, 191, 236, 246–247, 253, 264
Scriven, Gen. George 84, 124–125, 135, 140–145, 171, 178, 181–182, 242, 246–247, 251–253, 264
Selden Automobile Patent case 150–151
Selfridge, Lt. Thomas 89, 235–236
Selfridge Field 89, 235–236
Shafter, Gen. William R. 195, 198
Sheridan, Gen. Philip Henry 18
Sherman, Gen. William Tecumseh 17–18, 28, 76, 245
Shinkle, Col. E.M. 168–169, 251
Siemens, Alexander 30
Sierra Madre 141
Signal Corps, U.S. Army 4, 6, 49–56, 58, 60–62, 64, 68, 70–85, 88, 91, 96–97, 99, 101–102, 104–107, 109, 112–113, 115–118, 124, 127, 132, 135, 138–143, 145–147, 152–153, 155–158, 163–164, 168–169, 171–176, 178, 181–184, 187–196, 198–201, 203, 207, 210, 213–214, 222–223, 225, 228, 230–241, 246–247, 249–250, 252, 254–259, 261–264, 266, 269–272
Singapore 67
Slaughter, Nugent H. 131, 187, 189, 194, 255, 257, 267
Slocum, Major Stephen 113
Smithsonian Institution 77, 81, 94, 97–99, 112, 115–117, 141–143, 146, 167–169, 177–178, 180, 216, 236–239, 242, 244, 246–251, 254, 261, 270–271
Sorbonne University, Paris, France Selfridge Field 89, 235–236, 256
Spanish-American War 44
Sperry, Elmer A. 145, 162, 251, 263, 267
Sprague, Frank J. 145
Squier, Almon 11–15
Squier, Emily 12–14
Squier, Ethan 9–15
Squier, Maj. Gen. George Owen 1–7, 9–66, 68–100, 102–196, 198–219, 225–256, 258–260, 262–267, 269–271
Squier, Hiram 9
Squier, Jemima 9

Squier, Lodama (also Lodemia) 9
Squier, Lovinia 11–12
Squier, Luman 9–10
Squier, Mary (married name Parker) 7, 14–16, 207, 225–226
Squier, Nathaniel 9–10
Steinmetz, Charles 30, 32, 56
Stettinius, Edward R. 156
Stockton, Capt. C.A., USN 55, 231
Stone, John S. 100–101, 109, 209, 239–241, 262, 269
Stratton, Samuel W. 101, 103, 143, 145, 149, 156, 168, 176, 208, 248, 253
stroboscopic 36
Sweden 38
Switzerland 18–19, 38
Sykes, Col. F.H., RFC 116–117, 128
synchronograph 45–50, 219, 229, 266
syntony 221–222, 261

Tampa, FL 195, 198, 201
tank 189, 204, 255–256
Taylor, Maj. John R.M. 117, 155, 207–208, 259
telegram 150, 152, 166, 226, 233, 236, 242–246, 249, 251, 253, 257
telegraphy 4, 5, 27, 45, 47, 48, 50–53, 56, 57, 58, 60, 62, 64, 66–67, 69, 72, 73, 75, 78, 99, 100–101, 106, 107, 109–112, 117, 118–122, 138, 139, 145, 173, 184, 184, 188, 191–192, 195, 198, 200–202, 205–207, 209, 219, 221–222, 225, 227, 230–233, 240–244, 242–244, 252, 254, 255, 258–260, 261–270, 266–268
telephony 2, 5, 30, 46, 50, 69, 100–112, 117, 118–119, 122, 156, 173, 184, 191–193, 198, 203, 205, 207, 209, 211, 213, 221–222, 227, 239–242, 252–253, 255–257, 260, 264, 265, 266–270
telescope 41–42, 103
television 206, 239
Texas 21, 167, 252, 271
thermionic valve (vacuum tube) 193
thermodynamics 23
Thompson, George Raynor 256–257, 271–272
Thompson, Sylvanus P. 30, 47, 242
Thurston, Capt. G.C. 22, 25–26, 29, 226
Le Tonnant, French Ironclad 227
torpedo 90, 163–167, 170
transmitter 5, 44–46, 48, 50, 52–53, 55–56, 64, 102, 106, 119, 136, 188–189, 199, 217–219, 221–223, 240, 256, 258, 260, 266
transmitting 43, 57, 102, 110, 190–191, 211, 213, 221, 239, 243, 255–256
triode (vacuum tube) 104–105, 108, 223, 240–241
Tucson, AZ 234, 272

University of South Dakota, Vermillion 24

vacuum tube (called valve by British) 100, 104–105, 108, 164–165, 186, 188, 190, 192–194, 204, 209, 222–223, 239–241, 255–256, 267
Venezuela 47
Virginia 4, 22, 37, 82, 86, 179–180, 182, 188, 190, 253
voice 74, 100–102, 105, 142, 185–188, 190, 192, 194, 221, 239, 241, 254, 256–257, 271
voice-commanded squadrons 186–187, 189, 191, 193–195, 197, 199, 201, 203–204, 257

Walcott, Charles D. 94–96, 98–99, 116, 120, 134, 142–145, 147–150, 154–156, 167–169, 178–181, 236–239, 244, 246–251, 253, 265
Walcott, Lt. Stuart 167, 251, 265
Waldon, Sidney D. 156, 174
warfare 3, 5, 26, 80, 90, 92–94, 127–129, 135, 152, 156, 161, 186, 189, 267
wargaming 258, 265
Warner and Swasey 40, 52–54
warship 239
Washington, D.C. 4–6, 21, 47, 51–53, 55, 62, 65, 70, 79–81, 83, 85–86, 89, 96, 101, 103–105, 109, 111, 116, 118, 127, 131, 136–137, 139, 144–145, 150, 152–154, 159, 167, 169, 182, 211, 225–227, 229–232, 234–236, 238–241, 243, 246–252, 254–255, 261–265, 270–272
Waterbury, CT 118–119, 243
Waterloo, Belgium 157
Waterville, Ireland 199
Watson, John B. 175
weapons 34, 80, 134, 161–163, 166–167, 187, 195, 203
Wellington, Duke of 157
Wesleyan University 103
Western Electric 113, 118–119, 174, 189, 192–193, 210, 253, 256, 267, 270

Western Union 52, 173, 231, 252, 258
Westinghouse Corporation 210
Wheeling, WV 180
White House 207, 235, 254, 256, 272
Whitney, Willis R. 145, 156
Wieczorek, Lt. 78–79
Wiedman, Maj. 118, 243
Wilkins, Mr. 120–121, 238, 244, 252, 264
Wilkinson, Lt. Comm. T.S. 162
Willcox, Gen. Sir James 132
Wilson, T. Woodrow 20, 116, 131, 142, 145, 225, 227, 242, 244–245, 256, 258, 261–262, 265, 271
wireless 5, 99–102, 106, 109, 122, 206, 211, 213, 221–223, 225, 238, 240–243, 254–255, 261–263, 265–267, 269
Wisconsin 104, 211, 238
Woodhouse, Henry 148, 248, 265
Woolwich Arsenal 128
Worcester, MA 168, 262, 264
Wright, Orville 4, 85–86, 88–89, 98, 145, 161, 163, 208, 233, 235–236, 253, 259, 263
Wright, Wilbur 85, 87, 90, 98, 233–237, 252, 263
Wright Brothers 2, 4, 77–80, 83, 85–91, 95–98, 145, 148, 150, 161, 163–164, 167, 208, 233–239, 242, 246–247, 249, 252–253, 255, 259, 262–263, 267, 269–271

Xenia, OH 165

Yerkes, Robert M. 252, 257, 265
Yermoloff, Lt. Gen. 125
Yokohama, Japan 62
Yorke, G.M. 173

Zahm, Dr. Alfred 94, 96–99, 116, 143, 145, 181, 236, 238–239, 242–243, 247
zeppelin 135

www.ingramcontent.com/pod-product-compliance
Lightning Source LLC
Chambersburg PA
CBHW051211300426
44116CB00006B/525